ENGINEERING SYSTEMS INTEGRATION

Theory, Metrics, and Methods

ENGINEERING SYSTEMS INTEGRATION

Theory, Metrics, and Methods

Gary O. Langford

CRC Press
Taylor & Francis Group
Boca Raton   London   New York

CRC Press is an imprint of the
Taylor & Francis Group, an **informa** business

CRC Press
Taylor & Francis Group
6000 Broken Sound Parkway NW, Suite 300
Boca Raton, FL 33487-2742

© 2012 by Taylor & Francis Group, LLC
CRC Press is an imprint of Taylor & Francis Group, an Informa business

No claim to original U.S. Government works

Version Date: 20120314

International Standard Book Number: 978-1-4398-5288-0 (Hardback)

Library of Congress Cataloging-in-Publication Data

Langford, Gary O.
　Engineering systems integration : theory, metrics, and methods / Gary O. Langford.
　　p. cm.
　"A CRC title."
　Includes bibliographical references and index.
　ISBN 978-1-4398-5288-0 (alk. paper)
　1. Systems engineering. 2. Systems integration. 3. System analysis. 4. Product life cycle. I. Title.

TA168.L315 2012
620'.0042--dc23   2011043570

Visit the Taylor & Francis Web site at
http://www.taylorandfrancis.com

and the CRC Press Web site at
http://www.crcpress.com

I dedicate this book to Teresa, Terry Todd, and Whitney.

Contents

Disclaimer

Any views, opinions, findings, conclusions, or recommendations expressed or implied in this book are those of the author and do not reflect or represent the official policy or position of the United States Government, the United States Department of Defense, the United States Navy, the Naval Postgraduate School, and the National Aeronautics and Space Administration, and nor do they reflect or represent the official policy or position of the University of South Australia, its School of Electrical and Information Engineering, or its Defence and Systems Institute. Correspondence concerning this book should be addressed to Taylor & Francis Group LLC/CRC Press, 6000 Broken Sound Parkway Northwest, Suite 300, Boca Raton, Florida 33487.

Foreword

Systems engineering is the glue that holds together complex creations and enables them to perform beneficial functions. The current demand for systems engineers outpaces all but a very few other fields, and Gary Langford is eminently qualified to inform those who seek to work or want to work effectively in this crucial endeavor.

Engineering Systems Integration: Theory, Metrics, and Methods fills a glaring gap faced when we consider and then carry out systems integration. This book offers a sound approach to planning for systems integration. It treats integration as a fundamental approach based on common-sense rationale. The activities of integration are discussed in a clear and straightforward manner.

Gary Langford brings his practical experience in integrating large and small systems to a wide audience in different fields and disciplines. He has woven considerable technical skills honed in industry with the right blend of academics to deliver a much needed and most readable textbook. There is a great need for this book.

Norm Augustine
Chairman and Chief Executive Officer (Retired)
Lockheed Martin Corporation

Preface

Our past and our plight are in the hands of dreamers and pragmatists. Dreamers rule the gateways to our future, concerned with a world that could be. Pragmatists build our next reality, driven by the success or failure of their previous work. *If* they solve the wrong problem, or *if* in the wrong way they solve the right problem, neither is a credit to mankind. Their solutions may be clever, apt, and ingenious—exemplars of accomplishment—but they may have failed a crucial test: failure to appreciate systems integration. The lessons are stern. Systems integration, misunderstood and ineptly executed, wreaks havoc with other systems. We live with those other systems.

This book concerns the principles and practices of integrating parts to form a system. This book is not a rehash of integration platitudes or without mettle. The aim is to discuss the fundamental nature of integration, expose the subtle premises to achieve integration, posit a substantial theory framework that is both simple and clear, and elaborate on the discussion of integrating in ways in which you may not be accustomed. The practices of integration are substantially more than a narrative of fiddle-faddle banter casually dispensed during planning sessions or abused when directing other people's work. Managers share the greater responsibility to not just say, "just go integrate it." The practicalities of integrating parts when we build or analyze systems mandate an analysis and evaluation of existing integrative frameworks of causality and knowledge. Integrating is not just a word that describes a best practice, an art, or a single discipline. The act of integrating is an approach, operative in all disciplines, in all we see, in all we do. Integration is not found everywhere, but when it exists, we find systems.

Philosophy and reason, curiosity and questioning, and mystery and contemplation are not enough to disentangle the concepts of integration. Nor will they ever be enough. An idea begins with a single notion—a notion that something is either right or wrong. That notion might be an extension to an established natural or social "law," a nuance imposed on theory or knowledge structures of our thinking, or just a plain intuition, however inspired. No discernable method need be followed. Whether right or wrong, the notion persists, perhaps long after a spell of acquiescence or self-doubt. Yet the notion yearns for completeness. The unrequited notion has no self-determination, no humbleness; it just eats at you. Right requires justification, wrong requires conviction. If you stay on the path of right, the guideposts are many, and you never stray. But there is no one path for wrong, no lit way, no signposts, no guide. To be wrong, you must glean from whatever right you allow as your encumbrances. The greater the right, the greater the burden you carry, and the less you can stray. The burden you carry is all that has come before you.

Forever question what is right, find it where you may. Right can offer a hint at perspective, a dash of method, a snippet of theory, a glint of an idea. But the right you need is never enough to grasp, it is ever fleeting, never entombed or made comfortable. Were right to be right, there would be no problems to solve, no discord between practice and theory, no contemplation of experiments, no compassion for research, no "what if." Order and resolution would prevail. As right cannot be right and never wrong, neither can right be *all* wrong. There must be some right in right as there is some wrong in right. Ask, "What is wrong with right?" But beware, the way of right is numbing, but the way of wrong is treacherous. The fact that you have strayed off the path of right is taken as irreverent and is offensive. You will sense ire from the gentlest soul. Persist and you may or may not discover, you may or may not be better for your journey. There may not be an end, you may never find completeness. It is the journey that pits you against yourself.

I view everything as if it is flawed, whether by error in assumption, perception, logic, or judgment. Although the natural consequence of my thinking might be perceived as mistrusting of all that is said and written, that contrivance is untrue. Assumptions are tested, perceptions changed, logic analyzed, and judgment is made malleable. The result is a staunch commitment to intense study and reflection, development of countertheories, conjuring novel ideas, exploring new approaches to solving problems, and forging new ways to pose and answer questions with context more proper and fit.

I am deeply indebted to my wife Teresa whose endearing patience made this book a reality. No other person could have such tolerance and indulgence, for I am a person oft referred to as "stubborn" and "difficult." I think I heard "and" and not an "or." And to the few who tolerate my ideas, thank you for your labor of patience: John Osmundson, James Lake, Tim Ferris, Eduardo Kujawski.

Author

Gary O. Langford teaches systems engineering and system integration, and is a practicing systems engineer; a NASA Ames fellow; visiting lecturer in the Aeronautics and Astronautics Department, Stanford University; founder and president of four U.S. corporations (one publicly traded on Nasdaq); owner of an international consulting firm; earlier member of the boards of directors of seven corporations (holding three positions as chairman of the board); executive vice president of a merchant bank; manager of an aerospace systems engineering department; contract research scientist; foreman of a cannery nightshift operation; and lifelong learner. His work of thinking in systems integration spans his education in physics, astrophysics, geophysics, electrical engineering, sociology, business management, and systems engineering. He has served as the principal investigator for contracts and grants from the U.S. Navy, the U.S. Army, the U.S. Air Force, the U.S. National Aeronautics and Space Administration (NASA), U.S. Customs and Border Protection Agency, and the Temasek Defence Systems Institute, Singapore. In addition to extensive work with corporations and universities in the United States, he has engaged in collaborative research with researchers from Australia, Singapore, South Korea, Japan, United Kingdom, Canada, Turkey, and France.

Gary Langford is a senior lecturer in the Systems Engineering Department at the United States Naval Postgraduate School and a doctoral candidate at the Defence and Systems Institute, University of South Australia, Australia. He has an AB astronomy degree from the University of California, Berkeley and an MS physics degree from California State University, Hayward. Since 1976, Gary Langford has worked in all facets of systems engineering and systems integration on projects ranging from $200 thousand to $1 billion. His research interests include the creation and sustainment of systems.

1

Importance of Integration

Introduction

Society has the need to integrate. People build and integrate products and services. Specialists engineer complicated products and services, while systems engineers work with many domain specialists (e.g., physicists, biologists, chemists, sociologists, economists, psychiatrists, information specialists, corporate managers, workers, and decision makers) to tackle complex systems and system of systems. In essence, building systems is nothing more than integrating parts into a whole. Objects interact with other objects through energy, matter, material wealth, and information (EMMI). EMMI expresses the interactions between objects. To span the types of interactions and integrations that are observed in engineering, biology, sociology, and economics (essentially all things natural and human-built), the forms of interaction need to cover all that we do and all that we see. Different forms of enablers are or can be similar, can embody one another, and can exist in various combinations simultaneously (Burgin 2003). To the most common sets of the three forms of EMMI (energy, matter, and information) we add material wealth as a further initiator of an object's mechanism (broadly speaking). The actions of interaction and objects are covered in more detail in Chapter 2. Many authors* discuss interactions based on various inputs: energy and matter (Sage and Armstrong 2000); energy and information (Morris and Pinto 2004); energy, material, and information (Wieringa 1996; Oliver 1997; Kossiakoff 2003); and energy, matter, and information (Miller 1978; Bornemann and Wenzel 2006; White 2007; Edwards 2009; Tan et al. 2009; Wells and Sage 2009). Our reference to EMMI will be used throughout this book to represent the inputs that drive mechanisms to transform EMMI into outputs as well as into the losses of EMMI needed to achieve those outputs. Outputs are the performances exhibited by the object that transforms the input EMMI.

Society has a need for knowledge about the fundamentals of integration and the use of integrated physical and intellectual entities (referred to as "objects"). Generally speaking, physical objects are those entities that have

---

* Only a few of the more recent references are indicated.

causal relations to other objects through the interactions of EMMI. This notion suggests that no physical object is isolated from other physical objects. Intellectual objects are entities by reason or principle. Physical objects have boundaries, and intellectual objects have justification, motive, impetus, and explanation. Humans, buildings, ships, factories, chairs, trees, and molecules are physical objects. A software code or algorithm that drives a scanner or is the command to print is an intellectual object embodied in physical objects. Objects are covered in more detail in Chapters 2 through 4.

In its own right, integration stands head-to-head with discovery, application, and teaching (Green 2008). Whether for academics or for anyone in society (regardless of role), integration (along with discovery, application, and teaching) is the key determinate of success in everyday life. Here we distinguish between integration and interaction, with integration providing adoption of ideas and causal changes in being in contrast to interaction which offers only the potential for integration. Emphasis may change with role, but both interaction and integration are essential. Everyone discovers and learns; everyone applies what they learn and some teach others; and everyone integrates data and information to discover and apply what they know.

Fundamentally, integration is a method* that facilitates outcomes that are beyond what an individual object can do either individually or by a number of objects acting independently, that is, makes things happen that would otherwise not happen. The whole is crucially greater than the sum of its parts. Integration makes things happen faster than with individual, interacting objects. Individual objects are presumably optimized for their particular needs. Transferring EMMI between individual objects that are not optimized for such transfer will have nonoptimum transfer compared to a system that is structured for passing EMMI. This is not to say that all paths in the integrated whole are optimized compared to that of individual objects. Rather, those paths that are optimized will outperform objects that are independent, but merely interacting. An example of this independence is a team working to complete a task versus individuals competing with a team. A few individuals may outperform a team in many respects, but a team's performance will in general exceed that of any one individual. Integration provides an efficiency of operation that reduces the overall EMMI required to perform various tasks (assuming the task were even achievable by the individual objects). A team's effort takes less energy than individuals working as individuals. Consider a sports team playing against a group of individuals. The sports team will have practiced together and have developed techniques of play that combine to produce an economy of actions versus that of an individual who must exert

---

* The logic of integration embodies reasoning greater than acts (the elemental structures of activities). The enactments of integration reflect more judgment than processes (the aggregative effects of activities). The rationale and coherence of integration is systematic as well as step-wise logical. As such, integration is a method. However, we refer to integration as a process throughout the text. The reason reflects the emphasis on the details of implementing integration (the method) through its various processes.

individual performances above that of the opposing team. Building further on the example of teams versus individuals, integration links related (and often integral) objects into the same context to provide an overall management of effort that benefits the team through efficiencies of communication, planning, organization, directing actions, controlling their positions, and exhibiting structured teamwork. Integration supports seamless action between objects that promotes scale-wise fast and effective action for a range of EMMI. However, the process of integration is neither inexpensive nor instantaneous. And the result of integration may have unintended consequences such as operations that are unsafe, change in failure rates due to usage, and an overemphasis on one aspect of a decision or trade space to the detriment of another aspect. A benefit of systems engineering integration is to invest upfront to derive efficiency, economy of scale, or capability that would otherwise be unobtainable. In essence, integration shows the manner and means of putting objects to work in an efficient, collaborative environment. Often, integration is the only means of doing what is otherwise not possible by individual objects. As such, integration refers to the outcome—an outcome that is observable by both demonstration and measurement. The simple notion of being able to accomplish a task by the use of one object as opposed to accomplishing that same task for less EMMI with an integration system advocates for and stimulates an interest in integration. From the user's perspective, integrated systems offer higher cost savings and efficiencies than the outcome of what can be extracted from a set of unrelated products or services. Integration makes EMMI available to objects throughout the system, as is appropriate to the movement of EMMI from object to object. Just the distribution process of information alone changes the meaning and context of the information as EMMI moves from one object to another. Each object transforms the input EMMI into an output EMMI by the process of changing the input through the actions of the object's mechanism(s). Information integrity is essential and integration can provide the requisite integrity by its structures and processes of building the whole by the synergistic combination of its parts (i.e., the objects that comprise the system). Sometimes, distribution of EMMI needs to occur quickly and integration can provide it. The transformation of EMMI from the initial set of input EMMI through a structured form of distribution throughout the structure of objects is an economy of action fostered by the process of integration. In general, that economy of action is unequaled by the actions of individual interacting objects.

Case Study Introduction

Oftentimes, lessons from previous projects (referred to as cases) can be assembled and reviewed to glean lessons. In developing an appreciation for how certain aspects of these projects seemed to affect or be affected, the power of hindsight is often too critical of the progress from one stage to the next. By the knowledge of the results of the project work or by ignorance of what actually transpired, lessons taken from these cases can be extracted and applied to similar, representative examples of current work studied. After a bit of review and introspection, patterns of behavior or events may develop that suggest a commonly occurring set of variables and outcomes. At some point, a behavioral model might be constructed that represents a more detailed examination of a portion of the lessons, grounded in a set of perspectives, measurement theory, and the objective actions. We refer to such a set as a case study. A more formal discussion of case studies, from the point of view of what is misunderstood about case study research, is written by Flyvbjerg (2006) wherein he points out, "Forget the conventional wisdom, go ahead and do a case study." There is valuable information to be gleaned from case studies: they can be useful for formulating hypotheses, hypotheses testing, theory construction, and developing general theories (Eisenhardt 1989). Yet, the ultimate learning comes from practice. Systems engineering is a discipline of employing practices that have proven useful in various situations. Learning to integrate, to prepare the engineered objects for integration (systems engineering), and to manage integration (systems integration management) are steeped in practice without much theory to guide improvements. Knowing the limits of what one can do is just as important as knowing what to do. Stated bluntly, following a set of practices (or best practices) neither guarantees nor implies your project will be better or as good as those from the retinue of projects from which the practices were derived. Rather, it is knowing how to satisfice the perfect product or service *and* know how to deal with the constraints of development time and budget *as well as* meeting the lifecycle costs that defines what the best practice should be. Perfect products or services are difficult to come by—often unachievable due to negotiations or compromises to cope with key stakeholders, sundry problems caused by applying inappropriate or inadequate skills to the engineering activities, and ineffectual management discipline or knowledge to do or know what needs to be done. Regardless of the historical precedence, the application of best practices, or the systems engineering and management skills, products, and services (perfect or not so perfect) embody the key principles of systems integration. Examining these principles (that are evident in one or more case studies) exposes the actions and circumstances that have major influence on the outcomes of system integration. Perhaps the most one should expect from a case study is to observe the aftermath of principles being followed. A method founded on an appropriate set of principles provides managers, systems engineers, and engineers with a practical guide for action and a set of heuristics that should be a stalwart guide.

When attempting to integrate two objects where one or a combination of both objects requires an amount of rework that is more constrained by cost or time than starting anew, the result is a failure to integrate. Failure to integrate objects may have several root causes. But integration failures can be classified into two categories: those failures that are a result of all that transpired before integration and those failures that are a direct consequence of the integration activities. Integration failures that are inherent in the development work prior to integration can be a symptom of a poorly defined problem (i.e., redefined or corrected before integration is completed); a missed stakeholder or stakeholder requirements and therefore unknown, incorrectly interpreted, or ignored needs; a system design or architecture that inadequately responds to the stakeholder requirements or needs; or mismanagement of the efforts leading up to and including the integration activities (specifically any subprocess of planning, directing, controlling, communicating (Kasser and Shoshany 2000), organizing, and team-building). Integration failures that occur during the process of integration have several root causes, including mismanagement (focused on poor alignment of resources with required tasks, poor communication, and unreliable functionality which is attributable to one or another object) and poor integration skills, tools, or test equipment. Within the integration stage of work, communication between team members needs to be called out specifically as a key issue to succeed with integration. Beyond simply communicating what members of the team are doing and why, systems engineers with broader knowledge of the stakeholders, the implications of the design and the architecture, acceptance criteria, manufacturing, user intentions, and key sensitivities of stakeholders will improve the effectiveness of the integration team in achieving functional interoperability. And, engineers with specialization and expertise will improve the efficiency of integration. Both background specialties are important to successful integration (Nissen et al. 2006).

Identifying the category of integration failure from an analysis of case studies helps (1) to recognize the relation between formative work in systems engineering and system integration and the integration activities, (2) to provide insight into identifying what to look for and improve in the formative work, and (3) to develop a better sensitivity to the need for recursive thinking rather than iterative thinking.

Hubble Space Telescope Systems Engineering Case Study

Introduction

Even with adaptive optics,* viewing stellar objects from Earth-based telescopes is limited by the Earth's atmosphere and environmental effects.

---

* Adaptive optics compensate for much of the Earth's atmospheric turbulence that reduces the "seeing" to less than the diffraction-limited performance of a telescope. The mechanism of adaptive optics is to compensate for errors in a spherical wavefront by changing the longitudinal location of segments of the optical system. The effect is to correct for deviations in the incoming radiation by distorting one or more of the optical elements in the optical train of the telescope (Tyson 1991).

On April 24, 1990, the Hubble Space Telescope* was launched from Kennedy Space Center, Florida and carried in the Shuttle Orbiter payload bay aboard the STS-31 mission of the Space Shuttle Discovery. Some three-and-a-half years later in December 1993, the crew on STS-61 from the Space Shuttle Endeavor installed subsystems of corrective optics and replaced solar arrays and gyroscopes, added new instruments and computer equipment, and extended the space telescope's lifecycle. This second mission corrected the problems with initial implementation of the primary mirror (housed within the Ritchey–Chretien Cassegrain telescope). The optical flaw had reduced the performance of the telescope's resolving power. The result had been a blurring of the images which used the primary mirror (Mattice 2005). The development and operational teams overcame the engineering and technical obstacles, the external influences, and the political guidance to provide a magnificent system that was part of a spectacular system of systems (independent researchers from around the world, astronauts, and space shuttle with its infrastructure support, space tracking network, and the Hubble Space Telescope (to highlight but a few components)). With new equipment and improvements brought and implemented by the space shuttle mission STS-125 in April 2009, the Hubble Space Telescope is expected to continue operations into 2014. The Hubble Space Telescope was described (Mattice 2005) in terms of its technical characteristics, its mission, and development and integration issues.

Hubble Space Telescope Description

The primary mirror is a first-surface optical slab 2.4 m in diameter, providing 10 times better resolution than Earth-based telescopes, nominally 0.1 arc-seconds. Several other major telescopes are integrated into the Hubble Space Telescope's physical, electrical, thermal, and data electronics structure. Hubble's dimensions are: length of 13.3 m (43.6 ft) and diameter of 4.3 m (14 ft); its weight (measured at Earth-sea level) is 24,973 lb; its solar arrays deliver 4400 W; and its data rate to support science is 1 megabits per second.

Hubble's mission is to serve as a permanent space-based observatory at 600 km (320 ± 5 nm) altitude (low-Earth orbit) with an orbital inclination of 28.4°. One Earth orbit takes 95 min to complete, and while the Hubble Space Telescope operates 24 h, seven days a week, not all of the time is spent observing sectors of the sky to support the schedules of observers. Pointing to new sectors, avoiding the Sun or Moon, switching electronic modes and communicating data (from and to Earth), and calibrating instruments constrain both the ground support operations and the available "seeing" time.

---

* In the early days of work on the Hubble Space Telescope, the project was called the Large Space Telescope (LST) (1973/4).

Integration Issues

Optical systems designed to operate on ground-based mounts endure as exemplars of stability and precision. Building a very large, meticulously crafted optical instrument into a package (that has passed thousands of tests simulating launch and deep space temperatures and pressures) that will then be accelerated to three times the acceleration of gravity in the Shuttle Orbiter payload bay and ferried to orbit is a time-consuming, exacting systems engineering, engineering, and systems integration feat.

The types of subsystems on the Hubble Space Telescope include physical, electrical, optical, electronic, thermal control, power (generation, distribution, and management), communications, computer processing and storage, pointing, orbital stability, and operational/housekeeping configurations. Representative performance issues focus on optical reflectivity and light baffling, pointing accuracy, weight management, power consumption, and adhering to proper temperature design parameters. Representative quality issues focus on the size of the optical airy disk, pointing variances, power fluctuations, and temperature excursions beyond design variances from specifications.

Integration Problems

There is a veritable rule for systems that are planned to launch into space and expected to operate unattended for years—if they do not work on the ground, they will not work in space. Unfortunately, these systems can work perfectly on the ground, and still not work in space. The space environment is one that is difficult to emulate on Earth. Specifically, large vacuum chambers that have full instrumentation to test the completed system are extremely rare. Large-sized vacuum chambers with thermal and vibration controls to mimic the intense heat of the Sun and the deep cold of dark space are even rarer. The Hubble Space Telescope was larger than existing chambers and thus not subjected to the full range of tests that would cover launch conditions and the gravitational anomalies that would occur over the course of its 83,000 orbits around Earth. Smaller, unit-sized elements up to subsystems were tested instead. Additionally, the practice had been to build two "payloads," one a qualification unit for launch and one a backup unit, both built to near the same specifications. In case of a component or subsystem failure of one, the backup was a ready solution to stay on schedule. However, for the Hubble Space Telescope, the systems integration approach was to "design–build–test–fix" in an iterative fashion that conformed to the prime contractor's previous work experience.

Compounding the systems integration problems, the project was over budget* and severely behind schedule, and subsystems were failing their acceptance tests. Everything from the scientific instruments to the ground and control system needed considerable work to improve reliability to sustain

---

* Original budget estimates were $200 million, compared to a final cost near $2 billion (Mattice 2005).

the 24-h, 7-day on-orbit operations. Lifecycle issues including costs, operational tempo, refurbishment strategy (transport the Hubble Space Telescope back to Earth for upgrades and routine 3-year maintenance versus maintaining the system on-orbit from the space shuttle), and retention of key personnel needed to be integrated into all planning, including the development planning, integration planning, and operational planning.

Additionally, the space shuttle crews needed to equip and train for space walks of long duration, operate the remote arm to load and unload the Hubble Space Telescope, and manage the controls that linked the Hubble Space Telescope to the space shuttle.

Integration Management

The overall responsibility of management of systems engineering and systems integration was given to the prime contractor, Lockheed Missiles and Space Company (Lockheed), and Perkin Elmer (contractor for the primary mirror). Lockheed management needed to align the interests of a broad-base of key stakeholders, including various NASA (National Aeronautics and Space Administration) centers, the European Space Agency, representatives from the U.S. Air Force, astronomers and scientists, contractors, and employees. One of the more innovative actions by Lockheed was to search through the skills of their existing engineering staff to determine who could help uncover problems with the design, architecture, and implementation of the subsystems. Sometimes, small teams of Lockheed scientists and engineers were formed with a mix of senior systems engineers and their recent hires just out of college. Other times, lone systems engineers set out to consider the consequences of various designs and specifications. The mix of skills and enthusiasm on these teams helped inspire a quest for excellence that left nothing unturned within Lockheed's domain as the prime contractor. Even when the results of a mathematical model indicated that the specification for the cleaning of the primary mirror was insufficient to meet the design requirement, Lockheed confirmed the model predictions and informed relevant parties about potential problems with the cleaning specification for Hubble Space Telescope's primary mirror. In a strongly collaborative environment, the integration of the Hubble Space Telescope moved forward. But in spite of resolute persistence and scrutiny, a defect in the curvature (a spherical aberration flaw) of the primary mirror went undetected at Perkin Elmer until the Hubble Telescope was placed in operation after its 1990 launch into orbit. The users were the first to point out the defect. The test configuration that was designed to detect such an error had been set up incorrectly which resulted in the primary mirror being polished to the wrong shape. The integration and test procedures at Perkin Elmer were inadequate and documentation necessary to reproduce the test configuration could not be found (Allen et al. 1990).

Some rules of thumb regarding integration activities for software provide insight into the integration activities of the Hubble Space Telescope.

Approximately half of the first demonstrable software objects pass their unit tests, while the failed units undergo rework. However, nearly half of all software units that eventually pass this early testing fail their integration and system test (Mishler et al. 2007). Given the "design–build–test–fix" approach, integration for the Hubble Space Telescope probably began as early as possible to use the success and failures from testing as valuable feedback to help guide planning and to indicate progress. As such, integration and testing was used as part of developing the subsystem, blurring the distinction between development, integration, and testing.

Principles

Beginning the discussion on systems integration with principles may seem a bit stilted, as it seems to weigh toward an academic discussion away from any practical value. However, when the range of phenomena is very broad, the subject complicated, the practice encumbered with lore, legends, and superstitions, principles help defeat the viciousness of myths. The "bird's-eye view" enables us to distill the exactitudes without "losing our bearings" (Ashby 1962).

A principle is a means of organizing thoughts to articulate a pattern of behavior that frames or structures action. In essence, a principle represents both the context and concepts* that enable us to classify and interpret a situation in terms of previous situations. A situation is a sequence of events where an event describes an activity that relates an input EMMI to an output EMMI through a causal mechanism. An event is something that happens, an action that takes place, or an occurrence. An event is a change in an object due to the transformation of input EMMI into output EMMI. Changes in an object may be causal to changes in other objects. In this manner, events can be identified as precipitating or triggering other events. Events can result in interactions or integrations. Events have structure, can be sequenced through mechanistic transformations of EMMI, and are distinctively recognizable as objects (i.e., physical or intellectual).

The framework in which a principle is described has both observation and measurement, that is, both subjectivity and objectivity. The scope (or in this case, the applicability) of a framework determines the completeness for a given event or sequence of events. For systems integration, the events are grouped or evaluated by principle, each underpinned with a fundamental or undisputed law, doctrine, or assumption that is justifiable on the basis of another principle (e.g., treat others as you would like to be treated); a rule;

---

* Including interactions and integrations of the concepts.

a heuristic process*; a factual basis or underlying set of data or circumstances; or a commonly accepted truth that has validity in certain circumstances. Therefore, a principle is a description of that which has to be, or that which is, or guidance that is arguably sound, or inevitably consequential of something else. Principles are verifiably correct—as proven by both experience and experiment. Principles can be combined, worked into a coherent body of knowledge, and accepted as a basis for reasoning, logic, and action.

Ultimately, the test of a principle is through its applicability for a given circumstance and context. The validity of a theory and its framework for evaluating theoretical efficacies determines the utility of the principle for a particular fitness of use. The fitness of a principle to a circumstance is a measure of its completeness.

Principles of integration extract insular thinking from the quagmire of confusion, help guide decisions based on the sound and the good of a particular situation, and present a logic that creates an air of confidence. In as much as the principles help others justify their actions, the mere fact of discussing and employing principles facilitates decision fitness[†] and decision making throughout an organization (Howard and Matheson 1984), reinforces the teamness that inspires success, and provides a top-level perspective that retains purpose, objectivity, and planning prowess.

Principles of Integration

From a variety of case studies and reports on integration failures (Allen and Cohen 1967, 1977; Cooper 1979; Hoopes and Postrel 1999; Zaitun et al. 2000; Richey 2004; Burkatzky 2007; Leblond 2007), a few guiding principles can be derived. For our purposes, principles are defined as the root causes of events. We hold such root causes to be valid indicators of related, subsequent events.

Knowledge is valuable. This is a *principle* that relates knowledge and value, where knowledge is the object and value is the modifier of that object. A change in the value of knowledge indicates a change in knowledge; however, a change in knowledge may neither be an indicator of a change in the value of the knowledge nor the value of the change in the knowledge. Therefore, a principle is a widely held "truth" based on inviolability.

---

* Heuristics can be thought of generally as the steps of solving a problem by trial and error rather than by the distinctness of rules or factors thought through (Wu, citing and elaborating on the definition of heuristics from the Encarta dictionary, Wu and Adams 2006).
† Decision fitness (as described by Strategic Decisions Group) is illustrated as a chain of steps that describe the appropriate frame for the decision, the creative and doable alternatives that are possible, the meaningfulness and reliability of the information used, the clear values and tradeoffs, the logically correct reasoning, and the commitment to action (Howard 1983).

The empirical setting for integration for human-built systems is product or service development; for nature, it is chemistry, physics, and biology. Both human-built and naturally occurring systems are either involved with life-forms* or not. Empirical sociology offers well-developed tools and method-ologies to help explore the subjective nature of systems engineering, systems integration, and systems engineering integration management to uncover the key principles (Lazarsfeld 1993).

Principle 1: The Principle of Alignment

Alignment of strategies for the business enterprise, the key stakeholders, and the project results in better outcomes for product or service development.
The integration plan should align with the strategies of the project, of the business enterprise in which the project is supported, and of the dealings with key stakeholders. Knowing the needs of the project and how those needs are supported by the business enterprise and the key stakeholders is important in keeping high-level visibility with the decision makers. The business enterprise has two requirements that the project must satisfy: (1) providing revenue and profits consistent with the enterprise policies, and (2) operating within the limited and constrained environment imposed by the realities of the project. Fundamentally, the enterprise is in business to make money with the projects being one of the means of fulfilling the needs for revenue and profitability. Stakeholders have their own perspectives about the results of the project, ranging from providing the deliverable product or service within the limitations of budgets and schedules; satisfying some political or social need; or perhaps responding to an issue of safety or sur-vival. Stakeholder and enterprise needs are laid out in a contractual form before the project is begun. Laying out the requirements for the product or service is the objective basis on which the contract is formulated and agreed. Systems engineers place great importance on working with stakeholders to determine requirements, verify that the work accomplished satisfies the requirements, and deliver to satisfy the user requirements. Presumably, the requirements satisfy a need and solve a problem for the stakeholder. Often, one of the steps in systems engineering is to assure the customer, the user, and other key stakeholders that the requirements have been evaluated and matured to cover all aspects of the problem. In this basic manner, systems engineering is fundamentally iterative. Reanalyzing, reevaluating, and relooking the requirements help develop a comprehensive set of parameters that can be modeled by decomposing seemingly complex objects (physical and intellectual) into simpler objects with greater detail. These objects are comprehensible by their nature, more easily dealt with because of their limited scope, and readily parsed into tasks due to their definability and tractability

---

* No definition attempted. Life-forms are presumed to have a modicum of cognitive structure(s) and consciousness.

by common procedures and tools. The subject of objects has been a topic of great concern with contributions from before Plato to Einstein (Einstein 1950), Gödel (Gödel 1951 [1995]), and after (Korman 2011). For this presentation, objects can be either physical or intellectual. All that is not physical is intellectual. A person is a physical object, but the thoughts of a person are intellectual objects. The paper on which numbers can be written is a physical object, but the idea or concept of numbers is intellectual. Physical objects always appear to have a location in space (Everett III 1957) and are only measurable in that position space (Lyre 1995), whereas intellectual objects exist mentally. Physical objects can move or be moved from one location to another, whereas intellectual objects move only when represented as physical property. Software listings (in some readable form) are intellectual property that is represented in physical media. And, physical property can be represented as intellectual property. The digital media which contains an algorithm is the physical expression of the algorithm before it is placed into operation within the executable environment of a computer microprocessor. Physical objects are made up of component objects that may be themselves divisible. A unit of material wealth may be physical in its representation as money made of paper or coin and the intellectual aspect is what decisions can be made after spending the money on a physical object (i.e., product or service). A whole can be represented as an object, either physical or intellectual. Changing the physical location of a physical object may concomitantly change the location of its parts. If the physical object is connected, tightly coupled, and with high cohesion, then the constituent parts change location as a whole. In this regard, consider as one the physical constituent objects of the solar system as being comprised of Sun, planets and their moons, comets, asteroids, and space dust. A gravitational disturbance of the trajectory of the Sun and the orbiting bodies around it caused by a passing star might break the connection between the Sun and its orbiting bodies by reducing the coupling and cohesion of some or all of the physical objects orbiting the Sun. Similarly, intellectual property (as an object) may have constituent parts that are separable by EMMI.

Physical and intellectual objects comprise products and services. Physical and intellectual objects facilitate inputs and outputs of EMMI. As such, integration is accomplished by the EMMI that is received by an object.

In spite of the quite dissimilar needs of the key stakeholders (e.g., business enterprise, project, customer, user, and funding sources), general agreement is reached through a negotiated contract. Alignment of strategies of key stakeholders with the goals of the project and the delivery of the agreed product or service is paramount for success.* Integration of the various stakeholder strategies into an integration plan (as well as into the project plan and the systems engineering plan) for objects follows from this alignment.

---

* Success is defined as delivering the functional requirements and the performance requirements, within the budget and time constraints.

Principle 2: The Principle of Partitioning

Partitioning of objects can create tractable problems to solve if and only if boundary contiguity is achieved.

Integration success thrives on simplicity. Simplicity is often achieved by decomposing a high-level concept that embodies a few high-level objects into multiple low-level objects in a hierarchical fashion. The high-level objects (i.e., intellectual or physical) set the limits for all of the object boundaries, that is, no object within the hierarchy of objects will have a boundary limit that exceeds that of its logical high-level object. An additional constraint for these multiple lower-level objects is their individual subboundaries do not overlap or underlap each other's boundaries, *if the objects are on the same level*. Further, an individual object (regardless of its level in the hierarchy) is distinguishable by its mechanism, that set of actions that converts an input into an output. The results of an object's transformation of EMMI through the actions of its mechanism(s) are the object's contribution to the performance of itself as well as the larger aggregation of objects. Partitioning an object or a set of objects into more objects can create more manageable work packages to build and integrate the objects. The ease of integration is facilitated, if and only if the object's boundaries are contiguous in terms of adjoining physical structures, enabled functions that do not overlap or underlap with other functions, and with whom user behaviors are uniquely identifiable. The object or objects that are partitioned at the top level must be uniquely distinguishable and must cover the complete domain of the higher level partition(s). There should be nothing left out of this top-level partitioning that does not extend to the boundaries of the system. In other words, every object needs to fall into a partition and stay within the boundaries of the top-level object. The three boundaries (i.e., physical, functional, and behavioral) of each object are by themselves the maximum extent of the object's presence. In aggregation, the objects form the system and its boundaries. When decomposing an object to its component objects, the combined boundaries of the components must extend to the top-level boundaries of the parent object. This condition of contiguity ensures that all that was conceptualized at the top level was indeed included, nothing more (i.e., overlap condition) and nothing less (i.e., underlap condition). Figure 1.1 illustrates the overlap and underlap conditions.

Integration success requires that partitioning be carried out according to this principle. Overlapping or underlapping boundaries between objects creates shared control over object mechanisms (overlapping condition) or lack of control over a portion of an object (underlapping condition) which is identified as causing problems during development and integration. These conditions that portend integration problems are not normally found in interfaces or interface specifications as both of these conditions are symptomatic rather than causal for such problems.

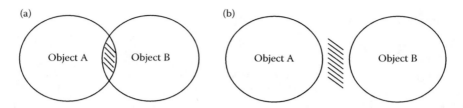

FIGURE 1.1
(a) Overlapping and (b) underlapping boundaries.

Principle 3: The Principle of Induction

Inductive reasoning should guide integration management and recursive thinking.

Using a process-driven approach to the development of a product or service has been shown to be effective assuming that setting objectives and orchestrating work toward accomplishing those objectives is causal to achieving these objectives. If one assumes this objective-driven paradigm to be an acceptable practice, then it is important to recognize that managing integration and application of systems engineering processes to build products or services involves several distinctly different types of thinking. While systems engineering thinking is primarily iterative in nature (as the principal way of getting consensus on a reasonable set of requirements), the thinking to accomplish integration is principally recursive, that is, enabling a forecast of events based on one or more of the preceding events or emerging patterns of behaviors associated with individual or sets of objects ("one set means more"). For systems engineering, iterative means getting to consensus, or in essence one set of requirements fits most stakeholders ("one set fits most"). Management of both the systems engineering activities and the systems integration activities requires inductive reasoning. Induction deals with the inferential processes that increase knowledge given the uncertainty (Holland 1986).

A process can be conceived as capturing a level of abstraction that carries with it the conceptualization of all its included activities. The result(s) achieved when these activities are accomplished are not only expected to be greater than any one activity but they are also presumed to be greater than the simple sum total of the tasked roles and assignments. For activities, a meaningful (yet indirect) measure of one activity versus another activity is an evaluation of the relative differences between results for the two activities. In essence, the "plan ahead" series of activities taken to achieve an outcome distinct from that of an unorganized approach is termed as a process.

Inductive thinking is often mischaracterized as rule based or rule driven. There are essentially two types of rules that drive the management of systems engineering and systems integration: rules of thumb and rules of dumb.

Rules of thumb (i.e., "know what is best") span a shared view of the nexus of goal-driven technology developments (Burns and Machado 2007) and social interactions. These rules of thumb include: small groups can be managed easier than large groups; single points of contact prevent miscommunications (e.g., is easier to deal with one supervisor rather than two supervisors); communications is vitally important (both as an indicator of collaboration and the structure in which the communication is provided); and do not over- or underdelegate authority and responsibility. Rules of dumb present a different image of the social organization as these rules sometimes reflect or embody "project legend"—the organizational culture, social interactions, and interpretations of policies and rules of behavior. Rules of dumb (i.e., "hope for the best") include patterns of behaviors that are readily detectable with minimal effort on the part of management. These rules of dumb include not informing management when falling behind in progress toward an objective (attempting to hide a missed deadline), and delivering partially completed or inadequately completed work (expecting to continue development during rework cycles). In essence, rule-driven execution of work requires inductive reasoning to keep abreast of both systems engineering work and systems integration work. In general, rules are helpful in keeping track of models that represent reality. But in keeping with those models, their accuracy is questionable in any specific instance, for example, "how close the budget or schedule is to the actuals." Therefore, inductive reasoning more accurately captures the tenor of work by generalizing approaches, posing and investigating ideas, and collecting evidence that suggests the dynamics of interplay between social and physical processes, rather than mere guesswork based on urban or office rules (Haas 1992).

Principle 4: The Principle of Limitation

Integration is only as good as architecture captures stakeholder requirements.

Not all architects have the presumed advantage of preparing an optimum architecture based on the inputs of key stakeholder and an expert system designer (or team* of designers). Some architects must work with very little information or sift through conflicting information. And not all system integration efforts benefit from the work that goes into the system design or the system architecture. However, in both instances, the architecture remains a key ingredient for successful integration. The other key ingredient for integration is a concept of operations. The concept of operations and the system design imbue the architecture with its primary emphasis as represented by

---

* A team is defined as members of an organization who have only common interests, where "an organization is a group of people whose actions (decisions) agree with certain rules that further their common interests" (Marschak and Radner 1972).

the stakeholder needs for solving their problem.* The conceptual architecture (high-level aspects) and the concept of operations are hand-in-hand related to the system design. The conceptual architecture, the concept of operations, and the system design are highly influenced by the budget limitation. Consequently, the lower-level architecture is commensurably cost-constrained. Once the ramifications of the budget have limited the architecture (at the top level), each architectural component at the lower level(s) are constraint driven by allocation of resources. It is the duty of the architect to capture the full measure of implications from the requirements and inculcate the requisite functions into the object descriptions within the system architecture. The purpose of the architecture is to combine the requisite functions at the highest level (built on the subfunctions that have been partitioned within the lower levels) to satisfy the needs of the key stakeholders.

The essential steps in using the architecture and the concept of operations for integration planning begin first by developing a prioritized listing of system-level functions that are key to satisfying the stakeholders. These functions should reflect the customer and user requirements as well as the derived requirements that have come about from the system design work and elaboration of the concept of operations. The aim is to identify the objective(s) that must be achieved in the delivered product or service.

Second, the relations between objects in the architecture are identified and summarized through their connectivity, coupling, and cohesion. Connectivity is the physical connection between objects. Connection is established by an interaction of one object with another through EMMI. In the case of a physical connection (one where there is a physical touching of one object with another object), there might be additional connectivity through other EMMI (e.g., with a computer circuit) that passes electrical energy that carries information regarding financial information. Coupling is the characterization of the strength of interaction between two objects. With high coupling, the individual depositing money into their bank's automated teller machine expects a high degree of coupling between the transaction and the crediting of the deposit to their bank account. The display on the automated teller machine may indicate the deposit, displaying

---

* Depending upon the expectations from the acquirer of systems engineering skills at the beginning of the systems engineering development work, the system design and the concept of operations may be either peripherally related to each other or, in certain instances, strictly dependent. The product or service needed by the acquirer can be specified in advance of beginning the systems engineering development which orchestrates the building, integration, and delivery. The starting point for systems engineering development may be stipulated in terms of top-level requirements and a concept of operations. However, the system design may or may not be described in much detail. If the system design is described in more than simple and general terms, the system design may reflect only the top-level requirements without detailing specific details that define instances of implementation. The alternative is to have a reasonably detailed first draft of the system design. In either case, the operational concept may reasonably represent the stakeholder needs with or without specifying too many details in the system design. (See Endnote 1 at the end of this chapter.)

the new balance in the bank accounting records. However, if there is a problem with the printing of a receipt (i.e., the written record of the account number, date, time, and location of the deposit), then there may be low coupling between the event of the deposition transaction and the written confirmation. Cohesion is the characterization of the measure of binding between two objects through their interaction(s). The strength of interaction means that two objects can be coupled under various conditions in which their interactions can change each other, whereas tight coupling presents as the observed causality between two objects. Loose coupling implies either that many variables are at work and therefore observations show weak causality between actions of the two objects, or that the observations do not reveal the linkage(s) between the two objects (although the connectivity and the relation between the two objects are represented or known). Coupling is a measure of the relation between objects. Unlike coupling which is determinable by a relation-in-fact through connection(s) that are direct and causal, cohesion is the consequence of a relation-by-degree. The relation-by-degree is observable, identifiable, and referenced to a particular object. Tight coupling can spawn dependencies that are distinguishable as causal. Tight coupling presumes a high degree of "trust" between objects. In contrast, cohesion is the manner in which one object relates to another object. Cohesion is formed by interactions of EMMI across boundaries of two or more objects. Some of these interactions are by process, some by functions, some by behaviors, some coincidental, and some temporal. All these types of interactions can be summarized as cohesion by structure and coupling by circumstance. Any manner and means of interaction may result in a form of cohesion. An operative definition of cohesion is the minimum number of objects which, if removed from a group of objects, would disconnect the interactions of the group. This definition is a modification of that presented by Moody and White (2003). The many pertinent definitions of coupling and cohesion can be applied to systems integration as the overall point and expectation of integration are to bring together objects to produce cooperative, unified system-level functionality (Pikula and Siemion 2007).

In essence, the measurable concepts of connectivity, coupling, and cohesion are other ways of expressing interactions between objects. Coupling and cohesion are defined as measurable concepts rather than specific measures (Darcy et al. 2005). As such, coupling and cohesion are direct indicators of interaction. If either coupling or cohesion is of small consequence, then two objects are interacting and have met the threshold for interaction.

From an integration perspective, connectivity, coupling, and cohesion are designed into the structure and action(s) of each object and exhibit the characteristics expected by the design, concept of operations, and the architecture. Integration testing confirms that the objects behave in the manner expected by demonstration of their functionalities with various performances.

Third, the physical, functional, and semantic structures of the architecture need to be identified and included as an integral part of systems integration. The semantic structures (i.e., data model (Feng and Yang 1994)) should reveal the perspective of the user within the system (or system of systems) architecture to recognize and facilitate an appreciation for the utility of the data that crosses the interfaces between system objects (Fritz 2006). The meaning of the data (distinct from the information that is transmitted) is an important aspect of providing the user with a mapping of knowledge to system physical and system functional domains.[*][†]

Fourth, all the mappings of functions to physical entities, semantics to functions, and user behaviors to functions support the integration activities through various views of the architecture. Architecture is the venue for connectivity, coupling, and cohesion to be explained in terms of the requirements that the product or service must be operative. Architecture is the venue for representing the key stakeholder needs and requirements as modified by their values, preferences, and desires. System integration depends on architecture, without which system integration cannot succeed.

Fifth, integration can be seen as a rationalization of the functions of the product or service and its operational use. Once the product or service is put into operation, the issues relating to integration change to process integration. Since the use of a product carries with it its own history of events, processes will evolve to accommodate those events. Existing operations are changed by the architecture of new products or services. When architecting a system, the difficulties in creating a common set of parameters and relations to describe the architecture are compounded by interpreting the divergence of the operational situation and the objectives when the product or service is put into use. Combining this divergence with organizational changes, new objectives that may result in new products and services, and the missing common working processes, transferring architectural knowledge from an old or legacy system to a new system may be problematic for the user and the associated interactions with the product or service in the user's environment and enterprise. Specifically, the integration of knowledge from the legacy operations to the new operations are skewed by the new product's or service's architecture. The ownership and interpretation of information from the newly operational system become an issue for the design and integration level of the enterprise data information system. Integration must take place in multiple domains at multiple levels to complete the integration of a new product or service into the user's enterprise. At

---

[*] The term *information* is used in the sense of data with a context. When information is combined with a model for relating the implications of scaling to a set of defined metrics and interpreting the data within a context that is representative of logic and reason, we have knowledge. The word *information* is used widely in this book to refer to data, information, and knowledge.

[†] The physical and functional domains are also an integral part of the product or service architecture.

this juncture, the integration may take on the proportions of a system of systems integration. The more ambitious the integration, and the more out of control are the interfaces (i.e., not under change control or management), the more difficult the integration of the new product or service into the existing user's environment and enterprise.

In addition to the product's or service's functional aspects of integration, the overall determination of the process view of the integration level needs to be determined explicitly.

Principle 5: The Principle of Forethought

Integration is a primary, key activity, not an afterthought considered as the result of development.

Integration must neither be considered nor treated as an afterthought or as a consequence of development. The key determinants of integration must be considered during the planning stages for integration: the defining of requirements; the considerations of the problem solution incorporated in the system design; the achievement and satisfaction of key stakeholder needs carried out by architecture; and the building of physical entities that embody the expected functionalities (and their performances), and engender the desired behaviors from the users. Integration must not only be planned upfront but also used to guide scheduling of development tasks so the culture, skills, and style of the development team are incorporated into building and testing of the product or service. In this manner, the organizational processes carried out by the team and the politics embedded in the tasking are not just assumed but considered important for integration work which is exactly where the results of every factor that can impact on the work is finally revealed. In essence, the system-level perspective is embedded in these formative tasks to reflect not only the specific instances for object development, test, and integration, but further to incorporate the system-level view in the integration activities (Ring et al. 2007). To avoid unpleasant surprises, the satisfactory technical solution is modified by the social and political environment (Brooks 1972).

Fundamental to integration planning and execution is the notion of forethought and measurement. Integration requires the structures of knowledge, the benefit of information, and meaningful data to determine the alternative ways in which to integrate a product or service. Thinking in systems to make the key decisions before planning and development begin results in more meaningful integration with complex products (Dirk 1994) and services. Measures and measurement are the only substantive means of testing and verifying the results of building and assembling objects. Ineffective measures confound the integration efforts (Bullock 2006).

Integration is a daily focus, with periodic updates to the integration plan. Integration is an explicit goal of acquisition, with key stakeholders who represent the buyer, the seller, and the user. Often, integration requirements are left unstated, as they are sometimes assumed. Systems engineers are trained to focus on the task at hand (as reinforced by the systems engineering process models) and most find it expedient to not think too far ahead as the current work is most demanding. However, planning for integration alleviates problems that surface during integration that are caused by ineffectual measurements, trade-offs that show preference for one design or a particular decision versus another, and ill-conceived schedule or allocation of resources to overcome technical problems. Moreover, if the allocated baseline changes (occasionally, frequently, or continually), the integration plan will no longer hold the advantage of being the plan, but will be subjugated to a piece of historical rhetoric of no utility or value (Collens and Krause 2004).

Reliability of the product and service is based on the aggregation of the reliability of the product or service components in addition to the interactions of the components. Developing the aggregate reliability of a group of objects begins with forethought about the system design. That forethought includes exploratory thinking about how to produce a system design that carries with it a set of meaningful alternatives (meaningful from the perspectives of the different stakeholders, such that each alternative emphasizes a major component of a stakeholder's needs and position on requirements). When selecting a system design, these meaningful alternatives help determine the context for further exploration of the design space. Often, these discussions surface additional requirements, modify the concept of operations, and sometimes suggest not so subtle changes in the systems architecture. Integration planning carries those system design parameters through architecting and development with the aim of preparing the objects for integration. Incorporating object reliability into the integrated structures is not an afterthought (Ferris 2007).

In those instances of product or service development of a large effort in which integration includes a great number of transfers of data across subsystem interfaces or in support of the interfaces with users, integration planning is substantially more than merely identifying, designing, and managing interfaces. The semantic architecture (the structures, interactions, and preferences that are made meaningful by data that is made interoperable by the design and implementation of the system objects) exposes information relevant to the user, a portion of which is presented to the user either in summary fashion or in the form of an analytical depiction of key results. Greater than half of the software in a project can be integral to the user interface (Oliver et al. 1997 citing Brown 1988).

At best, this technique of building systems is a guess at building objects so that performances are met. Support for this practice comes from technology that has been shown to move data at sufficient rates and quantities that many such transfers are done within reasonable periods. In other words, the users have not complained too much. Modeling and simulation certainly help in the determination of meeting performances but the guess-and-try-again approach to integration is time-consuming and dollar-expensive. Consequently, integration as an afterthought is time-consuming and dollar-expensive (McKenna et al. 2006). The issue of forethought is completely masked by the predominant influence of interfaces between objects as the central focus for integration. Much of the planning that is done for integration is concerned with the interfaces between objects. Early in the development stage, much attention (and therefore tasks and documentation) is placed on linking the subsystems with various connections across an interface. Interfaces are managed to be consistent with the planned interactions between objects, each iteration adding more detail and refining the actions of the transfers of data. The expectation of the engineers is that by moving data through objects, subfunctions and therefore system-level functions are enabled to satisfy the performance requirements. While planning for interfaces is indeed a necessary and essential process, it should neither mask nor undermine the importance and enactments of forethought.

Principle 6: The Principle of Planning

Integration planning is predicated on pattern scheduling (lowest impact on budget), network scheduling (determinable impact on budget), and ad hoc scheduling (undeterminable impact on budget).

Integration planning requires knowledge of the completion dates for developing objects. But knowing which tasks are to be completed is not the key for planning. Rather, the problem for planning for integration is in identifying how long the tasks will take to complete. Systems engineering management planning works within the systems engineering process models to lay out the tasks (break down the tasks) into a structure (referred to as the work breakdown structure or WBS). The key issue is in determining the type and amount of resources, the delays due to factors internal and external to completing the task, and the overall impact(s) of missing a scheduled milestone or delivery.

Task durations should be assessed only based on inputs from engineers, systems engineers, and management specialists who have direct and applicable experience with integration planning. Still, the theory of planning for integration rests principally on one of two fundamental premises: integration planning predicated on the scheduled needs for the optimum sequencing of objects or ad hoc by the dictates of what is completed. Planning for integration involves knowing what technological problems need to be resolved, recognizing the skills and resources that are available to prepare objects for integration, and protecting the project team from disruptions that have significant impact(s) on the progression toward milestones. In this regard, integration can be thought of as having explanatory variables for propitious aggregation, starting with the systems design and architecture baselines.

For deterministic scheduling (or pattern scheduling), that is, all that is known regarding the task durations, the planning can be based on a sequence vector in a time-domain set, which is optimized for costs, or alternatively a sequence vector in a cost-domain set, which is optimized for task times. Deterministic scheduling has the lowest impact on budgets. Therefore, the sequencing can be defined for every task duration and need not be dependent on a preference under either the budget or temporal constraints. The sequencing can be modeled as tasks, where every task has the same duration but with different resource requirements and risks. The results of such modeling in this simple case are either unrealistically simple or very dependent on the demands for resources (and as reflected in higher risks). In this manner, integration can be modeled after aggregation theory (Hildenbrand 2008).

A common variation of the simplest case is to develop a network model, such as that used to plan projects. Network scheduling techniques that are based on the theory of constraints and the concept of the critical chain* (Goldratt and Fox 1984) and are extended by adding additional tasks to absorb risk have the highest chance of predetermining the integration sequence and schedule. Network scheduling predicts the impacts on budgets and schedule for various sequences and durations of objects that are planned to be integrated. Adding additional tasks to accommodate risk is an alternative to simply padding individual task and project estimates with what is often referred to as "management reserve." Management reserve can be swept up by the business enterprise, squandered on "essential" but noncritical issues, and seen as a psychological cushion to deal with the uncertainties associated with problematic or inept management. The management reserve is often thought of as "good enough," as "insurance sufficient," or as "the lifeboat to save the project." However, if 8% of the management reserve is allocated at the beginning of the project to the early identified risk areas, with the remaining 20% retained for a second look at risk before the start of development, then the network schedule can incorporate risk in terms of additional tasks that are allocated according to the areas of risks, each task having budgets and schedules. For integration planning, the sequencing of the objects is derived from the network schedule in the same manner as with the simplistic deterministic scheduling.

Given the uncertainties associated with object development, many large systems engineering projects are resigned to on-demand (ad hoc) scheduling with whatever objects become available. In this ad hoc fashion, integration efforts are saddled with emulators that are built to represent what an in-progress object should be like when completed. Integration to emulators allows a first look at the issues of integration to build subfunctions. However,

---

* A critical chain is "the longest chain of dependent events and takes into account both task dependencies and resource conflicts" (Goldratt and Fox 1984).

by its nature these early first-look integrations are designed to be an iterative means of filling in until the test-ready object can be made available. Ad hoc scheduling reduces the orderliness of integration to a somewhat chaotic clustering of objects that seem ready but lack their reciprocal objects from which to build functions. There are two strategies from which to proceed. First, the objects that seem ready for integration may be shelved until their counterpart objects also seem to be ready to be integrated, at which time the two objects may begin the integration process. Alternatively, an emulator may be constructed that presumes to match the as-yet-completed object and integration may proceed with the seemingly finished object with that of the simulator. When the as-yet-completed object seems ready, the two objects can be integrated together or, should there be other circumstances that prohibit that type of integration, the second object can also be integrated to a like-kind emulator. The procedures of using emulators to perform integration is time-consuming, problematic, and never quite the same as integrating the intended objects. The most significant difference between an emulator and the "real" object is noticed in the performance of the emulator. Typically, the performance of an emulator is significantly slower than the "real" object. For example, during the development of a new microprocessor chip, emulators are used by software developers to test their code before the microprocessor chips have been returned from the foundry, packaged, and tested.

Principle 7: The Principle of Loss

When two objects are integrated, both objects give up some measure of autonomous behavior.

For every action there is a loss (the law of action*). That loss is quantifiable as EMMI. When two objects interact, there is a loss. When two objects are integrated, energy is expended (there is a loss). When EMMI is transferred from one object to another across an interface, EMMI is expended. Interaction is different from integration and integration is different from interaction. There must be interaction to accomplish integration; however, an interaction does not portend the integration of two objects. In either case, EMMI is expended. Interaction and integration are covered in more detail in Chapter 2.

Whether integration for a system or integration for a system of systems, there are fundamental questions regarding the losses incurred to sustain operations of the system. An analysis of losses is a means of discussing what it takes at the system level to operate. What is the amount of EMMI expended to achieve various levels of performance?

Further, there are fundamental losses that arise in a system of systems integration that are not apparent in integrating a system. In a system of systems, integration is different than that of a system. Integrating a system into

---

* This could be considered Newton's fourth law of motion.

a system of systems results in a set of systems that are both integrated and interoperable to achieve a set of metasystem functions in which all the component systems participate (to varying degrees). Bringing people together with a system can be thought of in terms of system of systems integration. Typical of system of systems integration is a preponderance of highly complex exchanges of EMMI between systems sometimes coupled with a loss of individual system capability. For example, what is the effective mix of integration and autonomy? Giving up autonomy is another form of loss. In systems integration, the relation between objects is expressed as connectivity, coupling, and cohesion. However, for system of systems integration, the notion of autonomy is more germane at the system level. For the system of systems, what flexibility is sacrificed at the individual system level when an interface is integrated? The integration of one system with another system (i.e., a system(s) of systems) brings about challenging issues, such as joint operations, joint interoperability, and the dimensions of distributed command ('to direct'*) and control. The systems engineer considers the political, social, and technical integration of system(s) of systems. The secondary domain of systems engineering is that of developing the structures and overseeing the engineering tasks to build objects, and integrating object by object to achieve functionalities and performances. Beyond functions and performance, integration requires adaptability and flexibility to reach the required degree of system stability. The intended result of integration is to facilitate the exchange of EMMI that is required to achieve the system(s) objectives. For every interaction, there results a loss of EMMI. For every integration into a system, there results a loss of EMMI. For every integration of a system into a system of systems, there results a loss of EMMI.

Endnote

1. The type of acquisition, the amount of detail in the system design, and the degree of specificity for the concept of operations are typical considerations for starting the systems engineering development work. The U.S. government acquisitions for the Department of Defense often presents a set of general system requirements (with some degree of specificity for core aspects) and a concept of operations as the starting point for systems engineering development work. As a consequence of the acquisition process, which at some point asks potential bidders to propose their ideas and means for building and integrating a product or service to a set of requirements, reasonably explicit and detailed descriptions of a system design and a concept of operations are included in the proposal submission. The systems engineering work for the U.S. Department of Defense is required by regulation

---

* Function are designed by single quote marks; whereas processes are designed by double quote marks.

DoD 5000.2-R to transform the required operational need into an integrated system design, but does not specify whether the degree of linkage between the system design and the concept of operations. As such, the acquirer presents the designated systems engineering and integration contractor with toplevel requirements and a concept of operations that reasonably reflects the consensus view of what is to be built. The specifics of the system design are left to the systems engineering development work to determine, explicate, or refine.

However, contrast this style of acquisition with one in which the top-level requirements and a concept of operations are the starting points for systems engineering development. The range of outcomes for the product or service may vary considerably from that resulting from an upfront limitation on the resultant system design. The trade spaces for the constraints of cost, schedule, and functional performances, the interaction to further define the end product or service, and the degree to which more creativity can cast new thoughts to solve the problem of the customer(s) and uses is broader reaching. The overall difference between greater or few limitations is perhaps a bit more time spent on upfront design and architecture, but the development and integration may be significantly less problematic.

The basis for this optimism for reducing costs for systems engineering development and integration work is that by adding more thinkers upfront deals directly with one of the primary reasons (if not *the* primary reason) why systems engineering was initially conceived as the means to solving increasingly complex problems. There are two issues that premise this discussion: First, the system design represents the consensus view of what the stakeholders need to solve their problem. Getting to the system design takes considerable analysis and thought. Without due consideration by many, varied thinkers, the *right* objects that make-up the system (system components) will not be identified and the architecture will not be optimized to provide what is needed. Second, "The first need for systems engineering was felt when it was discovered that satisfactory components do not necessarily combine to produce a satisfactory system" (Schlager 1956).

The solution to the customer and user needs requires integrating the right components in the right way. Integration requires a sound system design and a sound architecture as a minimum to provide the requisite product or service. Allowing systems engineers to assist in the upfront work of assessing the current and future trends in capabilities (explicitly determining what the problems are that need to be solved), determining the need for the system, analyzing the gaps in future systems or system of systems, identifying the top-level requirements, specifying the initial capabilities, developing the key performance parameters, and posing a system of systems architecture that shows the fit of a future capability into the trend line of current, soon to be concurrently interoperable, systems.

References

Allen, L. Commission. 1990. *The Hubble Space Telescope Optical Systems Failure Report.* National Aeronautics and Space Administration.

Allen, T. J. 1977. *Managing the Flow of Information Technology*. Cambridge, MA: MIT Press.

Allen, T. J. and Cohen, S. 1967. Information flow in R&D labs. *Administrative Science Quarterly* **14**: 12–19.

Ashby, F. G. 1962. Principles of the self-organizing system. *Principles of Self-Organization: Transactions of the University of Illinois Symposium*. London: Pergamon Press.

Bornemann, F. and Wenzel, S. 2006. Managing compatibility throughout the product life cycle of embedded systems—Definition and application of an effective process to control compatibility. *INCOSE 2006—16th Annual International Symposium Proceedings: Systems Engineering: Shining Light on the Tough Issues*. Toulouse: International Council on Systems Engineering (INCOSE).

Brooks, H. 1972. A framework for science and technology policy. *IEEE Workshop on National Goals, Science Policy, and Technology Assessment*. Warrenton, VA.

Brown, C. M. L. 1988. *Human-Computer Interface Design Guidelines*. Norwood: Ablex Publishing.

Bullock, R. K. 2006. *Theory of Effectiveness Measurement*. PhD thesis, Wright-Patterson Air Force Base, Air Force Institute of Technology, 188pp.

Burgin, M. 2003. Information: Problems, paradoxes, and solutions. *Triple C: Cognition, Communication, Co-operation—Vienna University of Technology (http://tripleC.uti. at)* **1**(1): 53–70.

Burkatzky, F. H.-H. 2007. *Development of Measurement Scales for Project Complexity and Systems Integration Performance*. PhD thesis, School of Management, Walden University, 161pp.

Burns, T. R. and Machado, N. 2007. *Technology and Complexity: The Perspective of ASD on Comlex Sociotechnical Systems, Uncertainty, and Risk*. Stanford: Stanford University, 48pp.

Collens, J. R. and Krause, B. 2004. *Theater Battle Management Core System: Systems Engineering Case Study*. Center for Systems Engineering, Wright-Patterson Air Force Base, Air Force Institute of Technology, 74pp.

Cooper, R. G. 1979. The dimensions of industrial new product success and failure. *Journal of Marketing* **43**: 93–103.

Darcy, D. P., Kremerer, C. F., Slaughter, S. A., and Tomayko, J. E. 2005. The structural complexity of software: An experimental test. *IEEE Transaction of Software Engineering* **31**(11): 982–996.

Dirk, M. J. 1994. Systems thinking for product development. *NCOSE Symposium*, San Jose, pp. 231–237.

Edwards, M. 2009. The impact of technical regulation on the technical integrity of complex engineered systems. *INCOSE International Symposium*. Singapore: International Council on Systems Engineering.

Einstein, A. 1950. *The Meaning of Relativity*. Princeton: Princeton University Press.

Eisenhardt, K. M. 1989. Building theories from case study research. *The Academy of Management Review* **14**(4): 532–550.

Everett III, H. 1957. "Relative state" formulation of quantum mechanics. *Review of Modern Physics* **29**(3): 454–462.

Feng, S. C. and Yang, Y. 1994. A dimension and tolerance data model for concurrent design and systems integration. *Factory Automation Systems Division*. Gaithersburg: National Institute of Standards and Technology, 38pp.

Ferris, T. L. J. 2007. Some early history of systems engineering—1950's in IRE publications (Part 1): The problem. *INCOSE 2007: System Engineering: Key to Intelligence Enterprises*. International Council on Systems Engineering.

Flyvbjerg, B. 2006. Five misunderstandings about case-study research. *Qualitative Inquiry* **12**(2): 219–245.

Fritz, D. 2006. The semantic model: A basis for understanding and implementing data warehouse requirements. www.TDAN.com.

Gödel, K. 1951 [1995]. *Collected Works, Volume III*. Oxford: Oxford University Press.

Goldratt, E. and Fox, J. 1984. *The Goal*. Croton-On-Hudson: North River Press.

Green, R. G. 2008. Tenure and promotion decisions: The relative importance of teaching, scholarship, and service. *Journal of Social Work Education* **44**(2): 117–127.

Haas, P. M. 1992. Introduction: Epistemic communities and international policy coordination. *International Organization* **46**(1): 1–35.

Hildenbrand, W. 2008. Aggregation theory. *The New Palgrave Dictionary of Economics*, 2nd edition, 26pp.

Holland, J. H., Holyoak, K. J., Nisbett, R. E., and Thagard, P. R. 1986. *Induction: Processes of Inference, Learning, and Discovery*. Cambridge, MA: The MIT Press.

Hoopes, D. G. and Postrel, S. 1999. Shared knowledge, "glitches", and product development performance. *Strategic Management Journal* **20**: 837–865.

Howard, R. A. 1983. The evolution of decision analysis. In: R. A. Howard and J. E. Matheson (Eds), *The Principles and Applications of Decision Analysis*, Vol. 1. Menlo Park, CA: Strategic Decisions Group, pp. 6–13.

Howard, R. A. and Matheson, J. E. 1984. Influence diagrams 1981. In: R. A. Howard and J. E. Matheson (Eds), *Readings on the Principles and Applications of Decision Analysis*, Vol. 2. Menlo Park, CA: Strategic Decisions Group, pp. 719–762.

Kasser, J. E. and Shoshany, S. 2000. Systems engineers are from Mars, software engineers are from Venus. *Thirteenth International Conference "Software & Systems Engineering & Their Applications"*. Paris.

Korman, D. Z. 2011. Ordinary objects. *Stanford Encyclopedia of Philosophy*. published online by Stanford University, http://plato.stanford.edu/archives/spr2007/entries/ordinary-objects/. Retrieved August 13, 2011.

Kossiakoff, A. and Sweet, W. N. 2003. *Systems Engineering Principles and Practice*. Hoboken: John Wiley and Sons, Inc.

Lazarsfeld, P. F. 1993. *On Social Research and Its Language*. Chicago: University of Chicago Press.

Leblond, P. 2007. The fog of integration: Reassessing the role of economic interests in European integration. *Economic Interests and European Integration*. Edinburg: University of Edinburg, April 8, 2006, 32pp.

Lyre, H. 1995. The quantum theory of ur-objects as a theory of information. *International Journal of Theoretical Physics* **34**(8): 1541–1552.

Marschak, J. and Radner, R. 1972. *Economic Theory of Teams*. New Haven: Yale University Press.

Mattice, J. J. 2005. *Hubble Space Telescope: Systems Engineering Case Study*. Center for Systems Engineering, Wright-Patterson Air Force Base, Air Force Institute of Technology, 90pp.

McKenna, B., Gualtieri, J., and Elm, W. 2006. Joint cognitive systems: Considering the user and technology as one system. *INCOSE 2006—16th Annual International Symposium Proceedings*, International Council on Systems Engineering.

Miller, J. G. 1978. *Living Systems*, New York: McGraw-Hill Inc.

Mishler, J., Carleton, A., and Nichols, B. 2007. How to talk to a program manager. *10th Annual Systems Engineering Conference*. Software Engineering Institute, Carnegie Mellon University.

Moody, J. and White, D. R. 2003. Structural cohesion and embeddedness: A hierarchical concept of social groups. *American Sociological Review* **68**(1): 103–127.

Morris, P. W. G. and Pinto, J. 2004. *The Wiley Guide to Managing Projects.* Hoboken: John Wiley & Sons, Inc.

Nissen, M. E., Orr, R. J., and Levitt, R. E. 2006. Streams of shared knowledge: Computational expansion of organization theory. Monterey: Naval Postgraduate School, 40pp.

Oliver, D. W., Kelliher, T. P., and Keegan, J. G. 1997. *Engineering Complex Systems with Models and Objects.* New York: McGraw-Hill.

Pikula, M. and Siemion, A. 2007. Design patterns in application integration based on messages. *Software Engineering.* Warsaw: Polish-Japanese Institute of Information Technology, Master: 101.

Richey, G. K. 2004. *F-111: Systems Engineering Case Study.* Center for Systems Engineering, Wright-Patterson Air Force Base, Air Force Institute of Technology, 90pp.

Ring, J., Bahill, T., Blackmon, T., Cloutier, R., Clymer, J., Hodgson, R., Jacoby, C., Krane, S, Lloyd, K., Newman, R. A., Orr, J., Rose, C., Skipper, J., Sorenson, R., and Steiner, R. 2007. Discovering a strategy for whole systems modeling—Panel 4.3.0. *INCOSE 2007: Systems Engineering: Key to Intelligence Enterprises; 17th Annual International Symposium, International Council on Systems Engineering, 17th Annual International Symposium Proceedings,* San Diego.

Sage, A. P. and Armstrong, J. E. 2000. *Introduction to Systems Engineering.* New York: John Wiley & Sons, Inc.

Schlager, K. J. 1956. Systems engineering—Key to modern development. *Institute of Radio Engineering* **EM-3**: 64–66.

Tan, Y. H., Teo, S. H., Lee, K. S., and Lim, H. S. 2009. DSTA's journey in systems architecting. *3rd Asia-Pacific Conference on Systems Engineering (APCOSE).* Singapore: Asia-Pacific Conference on Systems Engineering.

Tyson, R. K. 1991. *Principles of Adaptive Optics.* San Diego: Academic Press.

Wells, G. D. and Sage, A. P. 2009. *Engineering of a System of Systems.* Hoboken: John Wiley & Sons, Inc.

White, D. R. 2007. *Innovation in the Context of Networks, Hierarchies, and Cohesion.* Dordrecht: Springer Methods Series.

Wieringa, R. J. 1996. *Requirements Engineering: Frameworks for Understanding.* Amsterdam: John Wiley & Sons, Inc.

Wu, M. and Adams, R. 2006. Modelling mathematics problem solving item reponses using a multidimensional IRT model. *Mathematics Education Research Journal* **18**(2): 93–113.

Zaitun, A. G., Mashkuri, Y., and Wood-Harper, A. T. 2000. Systems integration for a developing country: Failure or success? A Malaysian case study. *The Electronic Journal on Information Systems in Developing Countries* **3**(5): 1–10.

2

Essences of Interaction

Without Boundaries: Oneness

The first one to have advocated that nature had oneness of character and a reality as only a whole was Parmenides* (Kirk et al. 2009). The view ascribed to Parmenides for philosophical inquiry is captured in three questions: What is it that is? What is it that is not? What is it that cannot be? This reinforces unity as the object of knowledge—the universal element of nature. That unity as an object is what we term as integration. The process of integration is to unify.

Since 500 BC Parmenides' tenor of logic has inspired thinking about time, motion, change, and unity. But not everyone agreed. In a lecture on the philosophy of science in 1967, Professor Paul Feyerabend (University of California, Berkeley) showed respect for Parmenides' logic of enquiry, but condemned his Monist logic. Later, Feyerabend wrote, "This theory illustrates a desire that has propelled the Western sciences from their inception up to the present time—the desire to find unity behind the many events that surround us" (Feyerabend 1993). However contorted or prescient the Western philosophical lineage is from Parmenides to the present day, all disciplines and methods of research are affected. But to suggest that the provenance is faulty is counter to the splendid logic that should guide inquiry:

> Knowledge is not a series of self-consistent theories that converges towards an ideal view; it is not a gradual approach to the truth. It is rather an ever increasing ocean of mutually incompatible (and perhaps even incommensurable) alternatives, each single theory, each fairy tale, each myth that is part of the collection forcing the others into greater articulation, and all of them contributing, via this process of competition, to the development of our consciousness. Nothing is ever settled ... Feyerabend (1993)

The profound influence of Parmenides on Socrates and Aristotle speaks to the inherent attractiveness and objective importance of this general notion

---

* Parmenides of Elea (se. 500 BC).

of unity to modern-day thinkers. Parmenides, the first protagonist of unity as the object of knowledge, and Feyerabend, the harbinger of an anarchistic view of method (Lakatos and Feyerabend 1978), sparked the foundations of my reasoning. The result was a realization that thinkers of systems (e.g., systems theorists, systems biologists, and systems engineers) may have tried to answer a question that is unanswerable, and further, they may have struggled against an objective that is unachievable. The way of Parmenides has been debated, contracted, and supported. Yet the fundamental notion of unity, that of integration, has persisted in its historical form. The result is a rather commonplace view of integration that the whole is made up of the parts. At the most general level, a central question posed by the thinkers of systems directed interest to principles that could be gleaned from widely observed patterns, behaviors, and properties. The goal was to interpret said patterns, behaviors, and properties in a prophetic manner. This presentation endeavors to build a consistent, congruous, and symbiotic set of concepts that reveal the robustness of integration as well as its subtle, yet important, qualities.

Boundaries

All objects have boundaries that exhibit piece-wise continuous attributes in the spatial,* functional, and behavioral domains. Some objects have continuous physical boundaries, while a system is comprised of objects that are interconnected and exchange EMMI. Some objects have continuous functional boundaries, while a system is comprised of pair-wise objects that have continuity of functionality. However, from a neophyte sociologist's perspective, the behaviors of objects may have deep-seated relations but may appear as quite discontinuous. The physical, functional, and behavioral boundaries can be characterized by different traits, different properties, and different attributes. But the differences do not negate the requirements for object-to-object connectivity—the matter-to-matter joining that forms the boundary. Connectivity is the condition that is determinable when two objects have had an interaction. The send object is aware of the interaction only if the receive object in some manner communicates actively or passively with the send object. An object that changes its EMMI because it receives a *communication* from another object is said to be aware of an interaction with that object. Awareness may be due to an acknowledgment from the receiving object (a change observed in the receive object is an acknowledgment).

---

* The term "spatial" is used to indicate the physical world; "spatial domain" is used to indicate the physical world of objects that they may be interpreted differently if the context suggests a difference.

Awareness may also be due to a communication from a third-party object who had not participated in the interaction but received a communication from the receive object that indicated that receive object had indeed received EMMI from the send object. Two objects that have interacted and who have communicated with each other are said to be aware of each other. Otherwise the condition of awareness is unsatisfied. Objects may be connected with awareness condition being satisfied. Connection without the awareness condition* satisfied means there is no communication between the receive object and the send object.

Physical objects can be thought of as a solid object, a network of physical objects, and structures with holes, voids, and gaps. The structures may be rigid or flexible or small or large.

Other objects are disconnected, physically, but have a functional or behavioral relation that connects various parts. Objects join to form functions or functional boundaries are identifiable as an extension of the effects of physical boundaries. A functional boundary results from the uses of an object as manipulated by another object via the connection between the two objects. Unlike a physical boundary that follows the materials of one object, all boundaries are porous—they can be crossed, changed, or broken. Physical boundaries can pass EMMI by mechanisms. Functional boundaries (where the action exists due to an interface between two objects) can be changed by changing one or both of the objects. Changing an object's mechanism(s) or physical structure may change the function that results from the juxtaposition with another object. Behavioral boundaries can be changed by changing the interface between objects, or the objects themselves. Remove or break a toaster and change behaviors of individuals who previously used the toaster.

If two objects are in space held in close proximity by mutual gravitational attraction, the physical boundary of each object is localized to the continuous surfaces of each object, respectively. The functional boundary is identifiable (and in this case testable) by removing one of the objects (logical equivalent of falsification). Should the remaining object show (or reveal) no change in its behaviors, no function was established due to the other (now removed) object. In other words, the EMMI (interaction) between the two objects was insufficient to result in a functional dependency. If there is no functional dependency then there are no behavioral changes. If one does not "see" an

---

* For example, the gravitational "tug" by binary planet, for example, the Earth–Moon system is defined as two objects having a reciprocal gravitational connection, where the center of mass of neither object is at the logical center of either's physical structure. The closer the logical center of mass approaches the average distances of their separation, the greater the influence that is exerted by both objects. In other words, there is no observable wobble in the center of mass position. Since the center of mass of the Earth–Moon system is approximately near the surface of the Earth, the wobble should be observable by our technology from outside our solar system. This technique of searching for wobble is one of the approaches for searching for "blue dots"—Earth-like planets orbiting other stars. From a systems integration perspective, oftentimes the observables indicate the dynamics of operations, and therefore illustrate the benefit of inductive logic, Principle 3: The Principle of Induction.

object, there can still be changes in behavior. A person can anticipate objects and possible functions and act differently because of those anticipations. While there may be no commonly accepted EMMI supporting a feeling or attitude, a person may make plans, change direction, set new rules, or do the unexpected "just because," these actions may suggest other mechanisms and other EMMI at work, or below threshold EMMI and mechanisms operative that are suggestive of various events.

Functional boundaries are formed at the interface of objects. If there is no interface, there is no function and vice versa. With an interface arises behavior. Similarly, if there is no function, there is no behavior; but if there are behaviors, there are interface and function as well. A functional boundary exposes the interaction(s) between objects. A physical boundary shows physical matter in space. Objects can come nearby and have distant physical boundaries and not have functions created (as there is no interface). Two airplanes passing at 2000 km distance may have no interface in contrast to two airplanes too close to each other either observed visually or by radar/laser radar (EMMI). Functional boundaries can extend beyond the physical boundaries of objects. Behavioral boundaries can extend well beyond functional boundaries. An event sequence that happens in one location can be deemed a pattern that when recognized or anticipated can be predicative of behaviors at another location. Behaviors of animals (including humans) seem bounded by cognitive recognition of patterns as stimulated by physical objects and functions, all enabled by EMMI.

To differentiate one entity from another or one system from another, we speak of boundaries. A boundary demarks the limit or extent of a defined domain; divides the essential nature of something from that of something else; or restricts properties and traits to one or another entity in some notional or corporal sense. Boundaries are either notably distinct or vague by their lack of uniqueness or reference to something else. A boundary that is arbitrarily set at 500 m from an object (object A) may be designated as a range of interest from that object. That range may be specified by a distance that delineates a domain of concern, but not necessarily to anything that is pertinent at a distance of 500 m. In the region of the boundary at 500 m from object A, the interests at that distance are referenced from object A not to any other object that is at 500 m from object A. In this case, the domain is a radius (of a circle) or ray that is 500 m in length (in a particular direction). The domain is defined by the 500 m distance from object A in some manner that has blithe ignorance and unconcern about objects elsewhere. However, while the 500 m boundary limits the purview of object A, the so-described boundary does not imply a complete disregard for objects outside the boundary, only that the considerations due to these outside-the-boundary objects are dealt with in a different manner. In this example, the boundary changes how object A behaves or is supposed to behave. Boundaries help differentiate the characteristics of one entity from that of another entity.

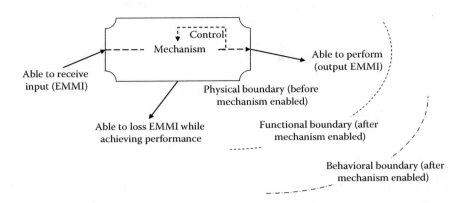

FIGURE 2.1
Object.

One could think of boundaries as ontological entities in themselves. In this view, the object is distinct from a boundary, as a boundary is distinct from an object; and boundaries are infinitely small, yet distinct in their characteristics of indicating a difference between an object and something else (or nothing else). The boundedness of an object encompasses all that is this object and none of which is not this object. The boundary in this view is not object, but can be all else as it is ontologically not object. This view recognizes that boundaries can be different from that of an object or that of not object. An object is depicted in Figure 2.1.

For this presentation, we distinguish between interaction and integration. Objects that interact are in a nonbinding relation with other objects. Objects that interact and create a binding relation with other objects are referred to as integrated. As such, we refer to boundaries as the limit of an object's integration with that of other objects. Therefore, an object includes its boundary (i.e., the boundary is ontologically part of the object). The maximum extent of the boundary of an object is that object. Referring to a whole and its parts (all of which we refer to as objects, without distinction), the whole is made up of parts—all parts are integrated so that the parts act in concert to imbue the whole with characteristics that are unlike its individual parts. A part that is on the boundary of the whole acts like the whole and not like the part, as if the part were separate from the whole and acting on its own. By this construct, we refer to the parts on the boundary as boundary objects. Boundary objects interact with the whole, as well as objects that are outside the whole. Interactions with the whole reflect the properties of integration, whereas interactions with objects outside the whole act as the whole, rather than as the boundary object would have acted were it independent of the whole. Boundaries are not necessarily restricted to just a physical sense of boundedness and further reflect a broader notion of interactions that includes functional and behavioral. Figure 2.2 depicts three objects $e_{1.1.1}$,

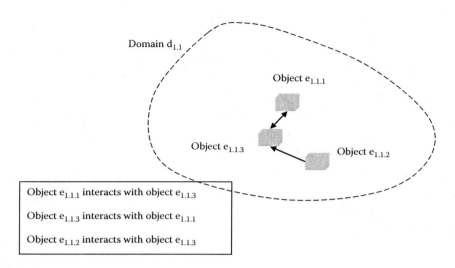

FIGURE 2.2
Interacting objects.

$e_{1.1.2}$, and $e_{1.1.3}$ interacting with each other in the following manner: object $e_{1.1.1}$ interacts with object $e_{1.1.3}$. Object $e_{1.1.3}$ interacts with object $e_{1.1.1}$. Object $e_{1.1.2}$ interacts with object $e_{1.1.3}$.

The interactions are such that the three objects do not change their properties, traits, or attributes.* In other words, the interactions do not result in a causal change that has some degree of permanence and stability in any of the objects. The physical boundaries of these three objects are said to be at the furthest physical extent of the physical object. For the interacting pair of objects $e_{1.1.1}$ and $e_{1.1.3}$, the functional boundary of object $e_{1.1.1}$ extends to $e_{1.1.3}$, and the functional boundary of object $e_{1.1.3}$ extends to $e_{1.1.1}$. The functional boundary of object $e_{1.1.2}$ extends to $e_{1.1.3}$; however, the functional boundary of

---

* A property is embodied in an object that is physical *or* represents something that is physical. A property can be real (physical or material) or intellectual (conceptual, nonphysical, or intangible). A physical property of matter is mass. Intellectual property is a representation of real, physical property, such as software (which represents a process that is enacted through physical objects). A trait is a property within its context. For example, the context may be that the object is moving in which case the activity of moving mass must be responsive to boundary conditions for the moving mass. A trait is the nexus of the property along with its conditions. While an object has a physical boundary, the conditions in which that boundary is effective or by which that boundary signifies the capabilities and capacities of the object, is a trait. Both objects and traits have mechanisms due to physical matter. Attributes are measures and measurements, configuration and structure, and constraints (e.g., time, cost, and scope), performances and losses due to achieving the performances of functions. Systems are made up of objects, the association of properties of objects and their relations (combined properties and contexts are referred to as traits), and attributes. These three terms are used to describe the composition of a system. The definitions of property, trait, and attributes vary significantly by discipline and by author (Blanchard and Fabrycky 2011).

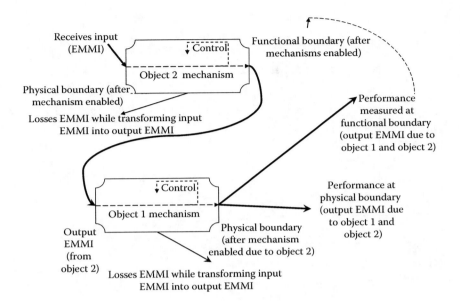

FIGURE 2.3
Functional boundary of two objects.

object $e_{1.1.3}$ does not extend to $e_{1.1.2}$. A view of a functional boundary is depicted in Figure 2.3.

The behavioral boundaries are different than either the physical or functional boundaries, as they are affected not only by the physical and functional boundaries, but also by the anticipation of the physical and functional boundaries. For animate objects such as a person, the anticipation of an object may change the behavior of that person even though neither of the object's physical or functional boundaries extend to the person. For example, a person planning to buy a new bicycle may walk several bike paths that are potential routes for commuting to work to ascertain the scenery and safety issues. The anticipation of purchasing a bicycle has induced certain behaviors. For behavioral boundaries, see Figure 2.4.

Boundaries mark the end of one factor, but not necessarily the beginning of something else. The limit of something is not the same as the beginning of nothing. And, the beginning of something is not the end of something else. Boundaries are predicated on a perspective—delineated to stipulate the outermost domain of interest. Boundaries signify the importance, the maximum extent, essentially, the interest of the one who draws the boundaries. A boundary is about your limits, not those of someone else. It is about your action, not someone else's action. It is about what you believe, not what someone else believes. Consider the metaphor of chess-play as a project. The chess board is a geospatial construct to portray moves, limit the play, and constrain the opponents' strategies. The game-play focuses on the strategy

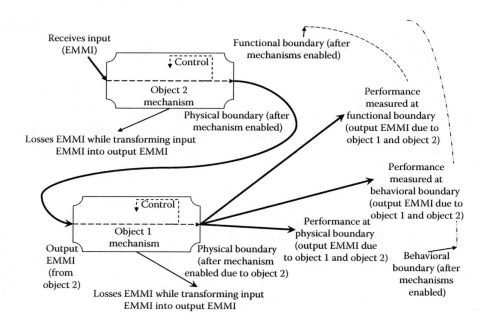

FIGURE 2.4
Physical, functional, and behavioral boundaries.

of two players each moving "pieces" according to a set of rules.* Each piece takes up one square of the board exerting its "influence" over other squares according to rules of movement for its particular type. These rules govern the capability of a player to attack and take the opponent's pieces. The physical boundary of the chess piece is limited to one square. By the nature of the rules governing play for the pieces, a player may use the functions of the pieces (e.g., 1.0 'project power,' 1.1 'protect another piece,' 1.2 'attack an opponent's piece,' or 1.3 'exert influence over a vacant square'). The functional boundaries of the pieces are shown as distinct squares over which the opponent may risk either losing a piece or by disputing its control. The behavior of the players is affected by the physical location and the functional capability of the pieces. From each player's perspective, the game progresses by moving pieces to 1.0 'project power,' 2.0 'force the opponent to change strategy,' 3.0 'construct traps,' or 4.0 'retreat' (for example).

---

* Both players start with the same number and type of pieces, but oppose each other by blocking or attacking and removing "taking" each other's pieces. Winning is defined by either capturing the opponent's piece referred to as the "king" or by forcing the opponent to resign—in either case winning the game. Pieces are differentiated by color and type—one color representing each player, a physical design signifying type of piece. Each player has 16 pieces occupying a single square on a board of 64 equal area squares. Play begins with a particular layout which is identical for both players positioned on opposite sides of the square of squares.

The boundaries drawn by each player are based on the physical location of their pieces, by their functional capability (and uses), and by the behavior of the opponent either due to the physical and functional factors or in anticipation thereof. The functional boundaries of each piece extend beyond the physical boundary of the one square in which the piece is located, while the behavioral boundary due to the physical location and the functional capability are reflected in the immediacy that confronts the position through the progression of pieces or in anticipation of what may occur as a consequence of the moves seen as well as what could be forthcoming. The players can control their own pieces and influence the opponent to move in certain ways. In other words, those who draw boundaries can control only themselves regardless of their intentions of exerting influence on the actions of others. To better appreciate the difference in perspective when a boundary is defined (or moved and redefined as in chess), view the game-play from the opponent's side of the board. The game may appear remarkably different from another perspective. The patterns that are clear from your initial viewpoint may be discernable after considerable study. The strategy of moving pieces to trap an opponent's piece may appear to be problematic from that new perspective. The different vantage point may reveal influences and strategies that went unseen previously.

Drawing or declaring boundaries is sometimes deemed necessary, but at the same time immobilizing. Scoping must be necessary and sufficient. It is necessary to place limits on (i.e., bound) the concerns that in a practical manner cannot go infinitely in all directions for all factors. It is immobilizing to be faced with a poor design and architecture that does not encompass all that is necessary for effective operations once the product or service is used. And it is wasteful not to scope the work effort (e.g., through the WBS) to provide focus and intensity for the project team.

The planners, designers, makers, builders, and users of products or services must consider boundaries—all three boundaries: the physical, the functional, and the behavioral aspects that define an object. Boundaries can represent the primary domain of interest in which a product or service is expected to affect or be affected by objects. The intent of a boundary is to indicate the expected interactions that will occur with a product or service once placed into operations.

Sometimes, boundaries are simply stated or declared by fiat. This being a somewhat arbitrary designation, boundaries may not be perceived as affording the most propitious view of phenomena within the bounded domain. Indeed, there may be altogether unnecessary and insufficient justification for boundedness. That a boundary could be physical (e.g., road barrier that is positioned physically between oppositely moving traffic), functional (e.g., prevent a vehicle from crashing head-on into oppositely moving traffic), or behavioral (e.g., promoting closer scrutiny by the drivers of vehicles that travel in the same direction of traffic, without having to be concerned about appositively moving head-on vehicles)—attempts a sort of isolation by the preponderance of the effects contained within. Alternatively, one could think of the boundedness as those characteristics as not contained in the bounded domain. Yet this boundary presumes or ignores the significance of (the circumstances that arise when boundaries are permeable to energy, matter, material wealth, or information that enters a bounded domain and interacts with object(s) within that bounded domain) that which "bleeds" through (or enacts across) the boundary. The intended and often times prevailing assumption is that the bleed-through influences or phenomena are minor or insignificant. Therefore, bleed-throughs are thought to be ignorable for the "practical" purposes of inquiry or engineering of products. It is exactly this somewhat ill-advised disregard for the consequences of boundaries that confounds the development and operations of systems. Problems arise when developing products and services when the key stakeholders realize that the requirements need revision, additions, or deletions. Such requirement changes are most disruptive for schedules and can add significantly to the costs of development. The greatest impact on cost and schedule is realized, the further into the development that changes are made. In a significant part, it is this boundary that presents itself as the demarcation for observing patterns, behaviors, and properties. And, it is this boundary that we find enigmatic, whether ascertained as a limit of concern or as determined by fiat. In the case of our universe it is surmised that "the boundary becomes indistinguishable from its contents" (Schiller 2009).

The arbitrariness of demarcation between things poses two problems for defining a system. First, the impacts of consequential bleed-throughs across demarcations may invalidate any preference or benefit imparted by an imposed isolation. That preference may be for convenience, expediency, or desirability (e.g., analysis may be easier). The consequences of a subjective or capricious decision about boundaries may introduce a randomness (or limitation) that increases the error in both the precision and accuracy of knowledge needed to design a product or service. Second, the ascertainment of patterns, behaviors, and properties is dependent on perspective. The perspective of the designer, the user, the customer, and other key stakeholders will at first be different. Within the boundaries of an entity (if such a concept has meaning), the observation of patterns will be different inside the domain of the product or service than that from an external vantage point. What may appear as dependencies that presage or present as patterns, behaviors, or properties may be dependent on other factors unobservable from the perspective of the observer. The result could be false data, erroneous information, misleading analysis, and theory in conflict with information. The perspective from which one examines an object or process with respect to its boundaries (*observer view*) will determine the kind and quality of data obtained and how the information is interpreted.

We can rationalize an idealized boundary that demarks platonic objects and process so as to avoid the problems of arbitrary demarcation by instead posing a condition by which the boundary might circumvent the problems of *observer view*. Consider a boundary that encompasses an entity's physical, functional, and behavioral incarnations throughout its lifecycle. Lifecycle covers the temporal domain for all events associated with an entity, for example, a product, service, or system. This situation also includes all influences on the bounded domain that are external to the presumed boundary. Moreover, all other entities in the universe (bounded or otherwise) are likewise isolated from the hypothesized bounded entity throughout that entity's lifecycle. Since all bounded entities over their lifecycle include all influences by other systems, the boundedness is never violated. There are no influences that have an effect on the hypothesized bounded system, as determined by those influences enacting across the boundaries of the hypothesized bounded system. Therefore, the *bounded* entity has not interacted with the rest of the universe. If such a hypothesized bounded entity existed, there would be no observables from which to know about it. As there would be no interacted with the hypothetical bounded entity, there would be no evidence of any kind that betrayed its invisible existence. Therefore, no entity can be formed or sustained in isolation from all else that is observable outside the *bounded* entity.*

For our purposes we can and do define boundaries, but that process need not be onerous or lend itself to problems for integration. With regard to integration, all types of boundaries need to be considered. Those types of

---

* If everything in the universe is made of the same fundamental elements (classically speaking, protons, neutrons, and electrons) that are fashioned in similar but perhaps circumstantially different ways, it seems likely there could exist such a hypothesized entity as a bounded isolated entity. While there can certainly be situations that defy direct observation of a region of space, we have limited means and tools to observe the influence of objects that are without such lifecycle boundaries as with the hypothesized entity. So either directly or indirectly we can detect many regions of space. However, to be blunt, for while it does seem likely that a bounded entity can exist in nature or otherwise for relatively brief lifecycles, long lifecycle entities (on the order of half the age of the universe would seem to be rather unlikely occurrences). The perceived continuum of discretely bounded entities (e.g., stellar systems, globular clusters, and galaxies) appear bounded in a fashion by gravitational forces that bare themselves to our knowledge of astrophysics. As we begin to explore in more meaningful ways to glean more information from phenomena at great distances from Earth, we notice the difference between what we expect locally from classical Newtonian physics versus that of quantum mechanics. Classically, measurements of an entity's position in time correspond to a referenced physical location. Our everyday experience reinforces this notion of seeing an object at a particular location and then if that object moves, observing it later at another location. However, that is not the case with quantum systems. When quantum systems interact, their local description challenges classical explanation (Bell 1965).

The quantum mechanical notion of "action at a distance" can be expressed through EMMI and the properties, traits, and attributes of objects. We can observe based on the limits of our technology and peer nearly 13.4 billion light years away. Is that the boundary of the universe? Our observations suggest a theory that we see a primordial broth of the beginnings of the universe, a prime example of integration that has taken time to nurture and mature. The concept of boundaries at the level of our universe or its constituent galaxies, stars, and EMMI is beyond the discussion in this introductory presentation of systems integration. Relying on the Parmenides' notion of unity is palliative only as we think little about the subject. The burden of naivety is shouldered by erudites.

boundaries that are important to integration relate to the physical object, they represent the extent of the uses of that object by other objects, and they relate to the behaviors of those users of the functions of the objects. How the boundaries interact is not always straightforward from the perspective taken by the observer. Different observers may perceive boundaries differently and in conflict with other observers.

From an integration perspective, the physical boundaries are the primary focus from which the functional and behavioral aspects are realized. Analysis and evaluation of boundaries should and can be performed before finalizing requirements for a product or service. Bringing together two objects exposes the functions of a product or service, whereas operating (or testing) those objects makes observable the behaviors of users (and other stakeholders). Any interaction that has a significant impact on the product or service needs to be considered and accommodated in the design, architecture, operation, and disposal. The lifecycle issues need to be considered in systems engineering and in systems integration. Each of these boundaries extends the EMMI transformed by the object's mechanism(s). From the perspective of an observer who is focused on establishing boundaries from an object (*an object-centric boundary perspective*) the most limiting of these boundaries is likely the physical boundary—the matter that comprises the object. That the physical boundary is considered limiting is meant to imply that the object minimum extent is deemed physical, and the functional and behavioral impacts carry on far beyond the physical boundary. The object's physical boundary (i.e., the physical boundary of the product or service) is most often extended by the user behaviors due to the object's functions. The functional boundaries are determined by the temporal and spatial relations between a "user" (object) and a product or service (object). The behaviors that result because of the objects or their function(s) (or in anticipation of same) are similarly bounded by the lifecycle of these kindred behaviors.

Scope

Products and services have scope; projects have scope. For many engineering projects, the scope of the product or service is defined spatially by a physical boundary. And in turn, sometimes the physical boundary is stated as the spatial extent of the product or service. For example, a hardback book is often considered spatially to be the extent of its physical dimensions (i.e., length, width, and depth). When the book is shipped, its weight is considered in addition to its physical dimensions. When the book is transported, its physical characteristics usually dominate the important boundary consideration. When the book is read, its physical dimensions play a role in the boundary considerations, as the book may need to be held and the pages will need to be turned. A book that is quite large may weigh too much to hold or may not facilitate easy handling to support the turning of pages. One of the functions of the book is to convey information. Should the book influence the reader through an interaction with the cognitive structures of the reader's perspective, the functional boundary of the book extends beyond the physical limitations of the words on each page. The publisher and author of the book have used the book (i.e., an object) to promote a form of learning. The scope of the work of the publisher is determined by the work that must be done to have the book accepted by the reader. Consistent with the *Project Management Body of Knowledge*, systems engineering parlance determines the scope of the book through the project's WBS (Turner 1993).* Work that is necessary to develop, integrate, and sell a book is the scope of the project. The scope of the work deals with the project as an enterprise rather than with the boundaries of the product or service when put into operations. Scope and boundaries are indeed quite different, but they are related. The casual definition that the boundary is or in some manner equal to the scope ignores the relation between the product's or service's boundary of operations and the effort that result in providing the user with a product or service.

Scope can be managed, whereas boundaries exist because of the design and use of products and services. When requirements change, scope changes. When requirements change, boundaries may or may not change. The processes that define a project's scope include developing a vision for the product or service, a roadmap that shows how the product will evolve over time, how technologies will mature and be replaced, when various upgrades will be released, what milestones will be delineated, and how the product or service criteria will be implemented.

For example, scope considers the contents of a book. When the reader considers the information gained from the book's contents and then integrates that information into their cognitive structures, the book has been used to convey information. When the reader applies that knowledge and thereby influences someone else, the boundary of the book has been extended by behaviors. The boundaries of the book can be said to encompass physical, functional, and behavioral aspects. The physical boundary is determined by the publisher and enabled by printing, cutting, and binding. The functional boundary is determined by the book contents and enabled by the cognitive structures of the reader. The behavioral boundary is determined by the reader and the listener(s). There certainly may be additional types of boundaries other than physical, functional, and behavioral, but these three are necessary and sufficient for systems engineering. These three boundary types capture the objective nature of the product both for its development and its use. Regardless of the boundaries, the project scope is defined in terms of procedures and events.

---

* Referenced in the *Project Management Body of Knowledge*.

Examples of scope include: work planned or accomplished; processes and procedures planned or accomplished; organizational entities involved or not involved; transactions to be completed or to be ignored; interactions or integrations scheduled or left unscheduled; items budgeted or not budgeted; information exchanged or kept confidential; ideas on an agenda to be discussed or left out; and concepts considered germane versus passed over or disregarded.

Carrying the metaphor of the chess match further, scoping refers to the rules of play. There are differences between tournament play (with penalties to player who distract their opponents, e.g., receiving calls on a cell phone, recording moves of pieces) and clocking time for beginning and ending play. The scoping of procedures and rules most often have only peripheral bearing on the boundaries of the game board. In essence, scoping guides the "work" (i.e., game-play) that is to be accomplished in connection with the game.

Defining the boundary of a product or service provides general guidance from which to scope the development and integration project. Scoping a project helps determine what should be included and not included within the limitations set by the product or service boundaries. Scoping a work effort assists the project team in making the myriad of decisions that constantly present "opportunities" to stray from the project's objectives. Scoping at the rule-level of carrying out project work reinforces the high-level policy and management guidance that is offered to show the path that is acceptable and presents the least expensive, most time-efficient way to complete the project.

Boundary Conditions

A condition is the circumstances that encompass an object; the factors that affect the manner and ways in which the object interacts; the situation in which the object operates; or the terms under which an object behaves. An object is influenced by its sensitivities to conditions. If an object is burdened by excess mass due to ice accumulation and adherence (i.e., sticking) to its exterior surface (e.g., wings of an aircraft), then ice is a factor for take-off and flight safety. The physical boundary of an aircraft's wings may interact with that of the atmosphere and result in ice "forming" on the wings. The boundary condition for the formation of ice may be temperature, humidity, and airflow over the wing's surfaces. The physics of icing metallic surfaces begins with nucleation centers that occur due to contaminants, edges, corners, and in general, surface roughness. Increasing the temperature, lowering the humidity, or increasing airflow may prevent ice formation on the wings. The boundary condition for de-icing of the aircraft wings is then potentially controllable by design and implementation of a capability that mitigates ice formation. Boundary conditions mediate the flow of EMMI across interfaces at boundaries.

Boundary conditions can be defined as mediation of capabilities that enact across boundaries. Consider a flow of EMMI between two objects, one defined within a bounded systems, the other left undefined (and therefore not included within the boundary of the bounded system). Describing the conditions which determine the interaction between these two objects are equally acceptable as boundary conditions. How those conditions exist, how those conditions apply, and how those conditions affect EMMI at the boundaries are boundary conditions. Boundary conditions are a way of limiting how EMMI affects a bounded object. It is most often boundary conditions that drive usage of bounded systems—how those systems can be operated, how they interact with objects and systems that are outside their boundaries, and how the users of such systems need to behave to accomplish their intended tasks.

Boundary Extenders

The interaction between two objects results in the enabling and use of function(s). Should that use be transferable from one object to another, the boundary for that function is extendable by that use. The condition to extend a boundary (or to not extend a boundary) is the boundary condition. To extend the boundary means to allow the object's EMMI to be transferred (or otherwise extended). The boundary of the object's functional domain is reached when the use that originated with the user and the object remains unused, that is, the end of the function's use (or appropriately, the end of the lifecycle of the function or the end of stability for the function). The behaviors of people (generally users and stakeholders) that result because of the objects or use(s) of the object are similarly bounded by the lifecycle of kindred behaviors.* If the originating object were to be destroyed, the function would necessarily cease, while the behaviors might or might not continue. The limit of influence of an object is reached when the receiving object's mechanism transforms input EMMI into an output that is related to but not distinguishable directly from the original object's output EMMI. In other words, the receiving object's behavior would appear indiscernible from that which would be expected given a variety of inputs. In essence, the transformation of input EMMI into output EMMI that occurs from object to object results in a change in the EMMI from object to object.

---

* More accurately, it is more than the mere uses of the object, but also behaviors that occur due to the object or in anticipation of the object.

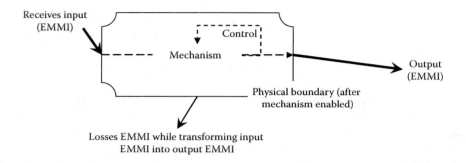

Receives input
(EMMI)

Control

Mechanism

Output
(EMMI)

Physical boundary (after
mechanism enabled)

Losses EMMI while transforming input
EMMI into output EMMI

FIGURE 2.5
Event.

One would suspect that this manner of changing EMMI would occur between every object, and that no two objects would maintain the exactitude of the original EMMI. Therefore, each event that represents a transformation of EMMI by an object records that event as an illustration of the causality associated with events. An event is depicted in Figure 2.5. An event only occurs *after* an object has received input EMMI. An event is observed *after* the enactment of the object's mechanism and an output EMMI has been released from the object. Events extend boundaries.

Objects and Boundaries

Objects are comprised of matter or energy in ways that manifest as physical properties, for example, an electron is an object. An object can be anything physical. Objects can also be anything symbolic, conceptual, relational, or "intellectual" as long as there is a physical manifestation of whatever form the *object* takes (Thurstone 1946). That two physical objects are touching means some of the constituent objects in one object have explicit physical contact with constituent objects in another object. It is currently beyond human skills to engineer two surfaces so that both have a continuous, uninterrupted area of surface contact along the entire length of the zone of touching. In some circumstances atoms or molecules from one object "drift," "migrate," or adhere to other objects. A mixing of physical constituent objects of one object with another object does not blur the physical boundary of either object (unless the objects develop changes in properties or traits). Alien objects intermingle with other constituent objects routinely (Langford 1971). If the properties of an object change due to these alien objects then there is integration. We loosely describe that area or volume of contact as the connection. When two objects are connected we mean that they are joined by such contact and not be impeded by those spots where no contact exists. The two objects have coupling and cohesion indicative of the type and kind of connection. If either object were to move, then the result of joining is to support the motion of both objects along their conterminous joint. For physical objects that are said to be connected, one can envision a boundary that signifies the end of one object and the beginning of another object. However, if the surfaces of the two objects were of the same composition and of the same structure, and were engineered to be *perfectly* contiguous throughout the touching zone, the physical interactions at that supposed boundary might not support the general notion of a boundary between the two objects. Were the boundaries to become chemically active (e.g., ionic or covalent), these previously independent surfaces might combine or fuse in some way. To fuse is to combine objects such that the identity of the results of the fused material is no more or no less than the constituent matter from both objects (Varzi 1997). Integration is different from fusing (e.g., data fusion) in that an integrated set of objects has characteristics that are different from either of the constituent matter. When two objects fuse they retain their individual characteristics except in the boundary regions of blending, which is a region of hybridization of both objects. If the two objects were identical in both their matter (in all respects, as if matter from one object were part of the matter from the other object, however physically separated the two objects were before their coming into physical contact) and in their constraints and limitations (ostensibly due to their local environments formed by their respective objects), then the fusion zone might take on related or similar characteristics as found in either of the two objects. The notion of a boundary incorporates the idea that there are other objects, distinct and distinguishable. Further, the notion of boundary provides a clear demarcation between the objects so objects retain some degree of complete or partial independence. The test for independence depends on detecting changes in characteristics *and* separability. These notions of boundary suggest dependence(s) that in some manner influence the action of objects and possibly change their respective freedom of self-control or undisputed or uninfluenced motion. Separability of the two objects requires that objects have no lasting effects as a consequence of their connections. That there is no lasting effect is to summarize the concept of interaction. Were the objects to change in ways that affected their abilities or capabilities and retain those changes after separation, they would have experienced integration for the time in which they were connected (and perhaps for some time later).

That objects have boundaries that come into contact is colloquially referred to as touching. When we touch something we do not expect to become part of what we touch, we do not expect to be changed in any way. Yet, some interactions are more than a casual brush with another object. If we were to breath in air that was contaminated with a virus that in some way resulted in us becoming ill, our bodies are said to have been unable to "defend" against the infectious agent. Sometimes we acquire an immunity (resistance as a consequence of antibodies or sensitized leucocytes (white blood cells)) to the infectious pathogenic agent. Immunity is the result of an interaction between the body (object) and the pathogen object (a protein-coated shred of deoxyribonucleic acid (DNA) or ribonucleic acid (RNA)) has now become part of the body's DNA or RNA structures. In the case of a virus touching the DNA or RNA in our body, the touching is likely an integration of proteins. The boundaries of both objects are presumed to be in physical contact. As we will determine, boundaries are more than physical. Functions and behaviors of objects can be juxtaposed (and even separated by great distances), yet still have causal effects.

Consider the case of a boundary between two people—one being the carrier of the virus (sender object), the other being the recipient of the virus (receiver object). Neither person comes into physical contact with the other, but a cough from the sender expels a pathogenically contaminated mix of air into the surrounding environment. The boundary of the sender's body is breached by the cough with the resultant spray extending into a volume outside the physical boundary of the contaminated object. The receiver moves through that contaminated region of air breathing in the virus, and later developing symptoms of the infection. The function of the sender 'to hack and expel air' from their lungs, throat, and mouth extends the consequences of the object, which is the same as their normal breathing—extending the consequences of the object by expelling matter (object). The function to 'to hack and expel air' from the sender has extended the boundary of the object. The physical boundary is therefore extendable by a function of the object. The functional boundary of the contaminated air has a lifecycle, that is, from the moment the pathogen has left the sender's body until the virus is no longer pathogenic or is physically unable to enter a host susceptible to infection by the pathogen.

Now we find that the habits of the infected receiver are unsanitary. The physical boundary of the sender's body has been extended by behaviors through the enactment of a function(s). The receiver incubates the virus for a period, develops symptoms, and becomes contagious. During this contagious time, the receiver shakes hands with another unsuspecting person and transmits the virus from hand to hand (physical boundary). The habits of the infected person are extending the boundary of the originally sick person, each person in turn becoming a vector for the pathogen. The physical boundary of an object is extensible by functions and by behaviors. The physical boundary has been referred to as the *bona fide* boundary and the extended boundaries (by function and behaviors) as the *fiat* boundaries (Smith 1994, 1995).

Objects can be physical or abstract (abstract in some incarnation, but transformed in the physical domain for the purpose of boundaries). Objects receive inputs, have mechanisms, transform inputs into outputs, and have losses. Their mechanisms transform inputs into outputs. The inputs are possible combinations of EMMI. Through their mechanisms, objects can transform the inputs into an output and send the results. Mechanisms are broadly defined to be that which operate in the context of forces.* Mechanisms have various controls that govern the transformations of inputs into outputs. Losses result from transforming input(s) into output(s). Outputs are measurable as performance(s). For reference: Figure 2.1 depicts the basic structure of an object.

Arrowheads indicate the direction of "movement" of EMMI. The boundaries of the object correspond to physical, behavioral, and functional influences on other objects. Boundaries are definable from either the object's view or that of another object. James Lake argued that "functional congruence must exist between phenomena that underlie a specified symptom pattern and phenomena operationalized as the mechanism of action ..." (Lake 2007).

Objects and Mechanism

Mechanisms are the means by which objects and processes change. Change means that an object or process is different at one instance from the previous or next instance. Change is precipitated by EMMI. For objects, EMMI exerts force(s) that influence(s) some or all the structures that comprise the object. The structures (Reed 2008) (e.g., mechanical, electrical, and chemical) *give way* to the influences of the forces derived from EMMI. Structures may *give way* to these influences depending on their susceptibility to such influence. When structures *give way*, they can have an impact on other structures influencing these other structures depending on the coupling and cohesion between the structures. Structures that *give way* in turn may influence other structures to *give way*. The properties of structures (those intrinsic characteristics that have resilience, i.e., relaxation and restoring action) have a semblance of reliability if their variations due to some range of force and that force's influence. For some structures, their intrinsic properties reinforce the similarity in nature of *giving way* based on like-kind influences on the structure(s). We refer to a mechanism as beginning with that which EMMI influences and ending with that which the mechanism produces as a result of that input EMMI. Mechanisms respond to EMMI and transform EMMI into a different EMMI. That transformation depends on the type of object and the circumstances and environment that provide that object with a context for transforming EMMI. In this way, mechanisms can be characterized by an enabling space, a transacting space, and an outcome space (Trockel 1999). That objects have structures

---

* Force is defined as the influences of EMMI on objects. Were there no influence on an object, there would be no force. That the influence should be measurable or detectable is of no consequence to this definition, as influence is relative. The test for influence is determined by the net of power (i.e., work done) on an object as observed by the outputs of that object's mechanism; changes in the object's properties, traits, or attributes; or other such changes in boundary, boundary conditions, physical issues, and functional or behavioral issues (Kocsis 2008).

imbues those structures with the mechanistic characteristics that result in *giving way* to influences from EMMI. The effect of a mechanism is to transform an input EMMI into an output EMMI. The output EMMI is parsed for convenience into two components: one describable as performance, the other as losses. Performance is measurable (with appropriate instruments with sufficient accuracy and precision). By definition, the major component(s) of the output EMMI are loosely considered to be the primary performance(s) of the object due to the input EMMI. All other output EMMI are said to result from achieving the primary performances attributable to the object for an input EMMI and context. These other output EMMI are defined as losses.

Introduction to Interaction

An action is the release or receipt of something due to the enactment of a mechanism. A mechanism is that which operates in the context of forces. An object sends (releases, sets free, or give ups) EMMI through the process(es) of an internal mechanism. Similarly, an object receives (accepts, collects, or takes up) EMMI. An interaction is defined by identifying the sending and receiving objects. Two objects are said to interact when the actions of both objects can be described as precipitating changes. The changes in objects due to an interaction are related to the effect(s) of the interaction. Those effects may include uniquely identifying the object that "sends" as well as the mechanism of the receiving object that is shown to be induced or stimulated to operate because of the action of receiving (that which was sent by the other object). However, an interaction does not by itself describe the causality of the interaction. Causality requires that the relation between two objects be modeled as the change in the sending object, the change in the receiving object, and the context of both the sending and receiving objects. Context is the situation or framework (Aerts et al. 2003) in which the interaction between two objects takes place. As such, interactions reveal a temporary dependency of the receiving object on the sending object within a context. Therefore, the context of the interaction may include multiple dependencies, for example, dependencies between the sending object and its context, the receiving object and its context, differences in contexts between the two objects, and the impact of context(s) on the object that is sent (and the object that is received, if the object sent is not what is received). It is through interactions that relations between objects are discernable. We will come back to causality in the discussion on emergence.

Interaction is characterized by the transfer of something from one object (sender) to another object (receiver). If the mechanism of the receiving object is enacted, then the input is transformed into something within the design limitations of the mechanism and the receiving object will attempt to use what is received. If what is received exceeds the design limits of its various mechanisms or is incompatible in some way, the receiving object may reject it, store it, or accommodate it. Regardless of the response by the receiving object, an interaction specifies that two events occur—one associated with the sender and one with the object. Whenever an interaction occurs, something is exchanged between the two objects, however one object may not have the capacity to discern the other object. A change in anything is associated with a change in energy. When something is taken away, energy is lost. When something is added, energy increases. And the actions of accepting or rejecting something are also events (which consume energy). Regardless of the type of action, energy is expended and lost. For every action there is a loss.

Energy, Material Wealth, Matter, and Information

The released energy, matter, material wealth, or data that is accepted by another object represents a limitation on the receiving object and constrains how the receiving object transforms input to output EMMI. An element interacting with another element is subject to time constraints (e.g., delay time for mechanism to operate and release energy, matter, information, or capital wealth, spatial distance between the elements, and types of *releases*). For example, energy is expended to perform the (service) processes of "moving wood" and "piling wood," together termed as "move wood." A small amount of that energy may have been exchanged with the pieces of wood through contact with the "mover of wood" (the entity that performed move wood). Move wood was not "free," that is, not without the expenditure of energy. To expend the requisite energy, mover of wood needed to have sufficient energy to perform move wood, replenish the energy expended, or live with a reduced capacity to expend energy. Energy comes in many forms, including from food and water. But no use of energy is free. The cost is always either in generating energy or in converting energy from one form to another. Continuing to expend energy without replenishment defines the nonreplenished lifecycle. For example, a nonrechargeable battery has an intended use and design lifecycle—one without replenishment.

Both generation and conversion involve losses. In general, no consumption of EMMI is free. Consumption, by the fact that a mechanism is active to enable consumption, means that EMMI is expended. And similarly, there is a loss in generating or converting EMMI. Interaction is defined as the transfer of EMMI. Therefore, interactions result in losses. Additional losses are incurred by generating or replenishing the transferred EMMI. Integration, like interaction, requires EMMI, albeit significantly more. This distinction between interaction and integration is seen in writings about interactions between entities through flows of material wealth, goods, and information in contrast to economic integration which includes various levels of cooperation (not seen with interactions), a wide range of benefits (not seen with interactions), a confluence of political and economic power (not seen with interactions), and a mechanistic change from that seen with interactions to as a removal of constraints with integrated structures. Interaction is broadly different than integration (Nieminen 2005).

Energy

Fundamentally, there appears to be only a few items that can be transferred between objects. EMMI are the primary items that appear to go from one object to another. Energy is transferable in many forms, including chemical, kinetic, and heat. The means of transfer can be related to the amount of force applied within the "sending" object. One model of this transfer relates to increasing the force applied to overcome the internal resistance of a mechanism that controls "sending" energy. When the resistance to the force by the mechanism is overcome, energy is expelled or "sent" from the object. Consider the case when the force is applied in an ever-increasing incremental fashion (such as if multiple send and receive interactions take place between two objects). Consider that in this case, the internal resistance of the mechanism is greater than some of the transferred energy. Each one of the instances of applied force is insufficient to stimulate the mechanism of the receiving object. Force is being applied, but the mechanism of the receiving object is not prompted to work. That mechanism could be the simple feat of detecting any energy that impinges on the receiving object. Energy is transferred but not detected by the receiving object. Force exists, but is below the threshold of "detection" by the receiving object and no energy is either gained or lost by the receiving object. No mechanism in the receiving object is enacted.

An object's resistance to change in location by an external force is related to the amount of mass of the object, the degree of coupling and cohesion* (Purao and Vaishnavi 2003; Darcy et al. 2005) of its bounded matter, and the impediments that restrict its movement. In the case of physical interaction, physical force is related to the rate of change of momentum of the object. We are concerned with the energy and force associated with the movement of an object, but not from the potential energy that might exist or from a force that may be applied (but is below the threshold of moving an object). From the perspective of interactions, if there is no movement, then there is no energy loss. Applying Newtonian physics, consider two pieces of wood, one piece resting on the other on the Earth's surface. In the Earth–wood–wood domain, each piece exerts a force on the Earth as well as a force on each other. The net of forces keep the top block from sliding "down" the bottom block due to the

---

* Drawing from the literature in software, coupling is the degree that one element influences another element "binding," and cohesion is the degree of relatedness between the cause and effect(s) observed.

horizontal components of the force of gravity opposed by the opposite force due to the peaks and valleys that characterizes the surface roughness that "locks" the two blocks together. Forces do not interact, masses interact (and therefore energy changes). The piece of wood on the bottom (i.e., sandwiched between the top piece of wood and the Earth) exerts a force on the Earth equal to its mass plus the mass of the piece of wood above multiplied by the acceleration of Earth's gravity (mean acceleration at the Earth's surface equals 9.83 m/s^2). In this static situation where there is no relative movement between any of the three objects, we observe no interaction. Even though the three objects (two pieces of wood and the Earth) are touching (and there is a force in the direction of the Earth's core), there is no movement. If there is no movement, there is no exchange of energy, and therefore no interaction. Physically, the surface fibers of the wood are mechanically locked together through their touching peaks and valley of surface roughness. Likewise, the bottom piece of wood does not move under the weight of the block above. The forces that are operative in the example Earth–wood–wood domain have not resulted in any movement. We are assuming there is no temperature differential between the Earth and the pieces of wood (i.e., no thermal energy transfers), and no other bonding between the Earth and the pieces of wood or between the pieces of wood. Simple contact between objects is not an interaction, unless there is an exchange of EMMI. For an interaction to occur there must be a transfer of energy, not the mere presence of a force.

Matter

Energy is also related to mass, where mass is the intrinsic property of matter (as measured by quantity of material). Matter is the structure of physical objects. An interaction based on transferring mass from one object to another results in a loss of mass (and energy) from the sending object and a possible net gain in mass (and energy) by the receiving object. The energy it takes for the receiving object to "take on" the mass may (or may not) require more energy than is the gain realized from the interaction. A net loss is possible. As with any EMMI interaction, the receiving object may or may not have a net positive gain due to the interaction, depending on the amount of EMMI expended through the interaction process.

Material Wealth

Matter is sometimes expressed as material wealth (De Marco 1960). Material wealth can be thought of as cash, investments (e.g., stocks, bonds, and marketable securities), and other equivalents (credit and debit cards). The rate of interaction of material wealth can be changed by enactments of law (e.g., taxes and encumbrances); interactions with individuals or institutions through investments, exchanges, financial transactions, and barter; general economic decisions that impact on the normal actions of living in society, and profligate lifestyle choices. Material wealth includes all that has the capacity to be converted into cash or cash equivalents. Therefore, one's time coupled with a means to generate material wealth during that time could be considered material wealth. Exchanging time, exchanging information, exchanging matter, and exchanging energy for remuneration in money are examples of the fungibility of material wealth. Broadly speaking, material wealth creates the financial capacity to perform work for money, similar to that of energy representing the capacity for work through mass. Material wealth has its place in both human endeavors as well as for natural processes. Material wealth is all that is referenced by mechanism coupled with abundance or plenitude. Some portions of the ground have an abundance of ground water, while others have a scarcity of ground water. For natural systems, material wealth can be thought of as a reserve from which to draw, given appropriately enabled mechanisms.

Information

Energy, matter (mass), and material wealth can all be thought of as information. The term *information* is used in the sense of data with a context. When information is combined with a model of relations and data flow, the implications of scaling and interpreting the data within a context forms the basis for knowledge. The word *information* is used widely in this book to refer to data, information, and knowledge.

Information is an inherent attribute of energy, matter, and material wealth. Information is carried by or in all three. The independence of information as a primary agent of interaction recommends its inclusion due to the inherent nature of objects interacting with objects. In other words, information is included in the nature of the interaction as well as in the essence of an object.

Whereas energy and matter are intrinsically related to mass, information and material wealth are constructs based on human behavior. Energy and matter are absolute in their embodiments as an object. In other words, only one object is necessary to describe energy or mass. Their role in interactions is to effect the changes that have occurred in two objects. Unlike energy and mass that have existence in one object, material wealth and information exist only as a result of transfer from one object to another. All the gold on Earth has no meaning other than as mass and energy unless another object provides the context of "need" or "want."* As a means of expressing an attribute ascribable to gold, an interaction between two objects is the only means of determining the "value" of the gold. The consequences of an interaction are predicated on those properties, traits, and attribute.

---

* *Need* is something you must believe will solve the problem, is possible, is affordable, can be provided when desired, and does not cause another problem of such significance that offsets the benefit of solving the original problem. A *want* is something that will solve the problem, but is not necessarily possible, affordable, deliverable, or acceptable. A need is absolute and unconditional. A want is a desire, as yet unfulfilled.

Information and material wealth are likewise embodied in energy and matter as the means for their interactions. For material wealth and information, the interaction process is a determinant process for the meaning of the material wealth and information.

Objects can be interpreted to have value and exhibit material wealth. Material wealth carries with it information. Both energy and matter also carry information, both as the conveyance of information that is specifically encoded within the energy or matter and as energy and matter individually. Among the many things that information signifies, for the purpose of deepening the appreciation for the types of mechanisms that are operative within objects, information can be gleaned from energy, matter, or material wealth by deriving data from the properties, traits, and attributes of the input EMMI, from the output EMMI, and from the lost EMMI. Due consideration must be given to provenance, congruence, and trust (Sztompka 1999).

Property, Trait, and Attribute

Property

Mass, m, is a unique property of matter. Property is tangible and physical (Hoppe 2004). Intellectual property is the property that can be shown to be tangible and represented in a physical form. Verbal communications is not a representation of intellectual property, but rather a process from which physical property can be used to represent both the intangible knowledge and the procedures by which that knowledge is transferred to something tangible, such as paper or recording on media. Properties have mechanisms. Objects can be physical or cognitive—both have mechanisms. Physical mechanisms can be related to energy and forces (traits of matter), while intellectual mechanisms can be related to procedures (e.g., legal methods, strategies, and steps—summarized as activities). Physical objects are the physical things we build. Intellectual objects are the processes that we use to build physical objects. In this context, services are considered to be physical objects with mechanisms describable as services.

Knowledge is an object. A mechanism of knowledge is the mental procedures to build cognitive frames and the enactment of procedures to carry out placing the knowledge in tangible form. The corporeal representation(s) of both the knowledge and the mental procedures and physical activities is shown as the piece of paper on which the knowledge is represented.

For integration, requirements, needs, design, and architecture are all objects of the intellectual kind—representable as knowledge (with cognitive structures, mechanisms (procedures), and representations (or models)).

Trait

Energy, E, is the confluence of matter and constant motion (i.e., motion that is constant in terms of the ratio of distance traveled by a mass, divided by the time to travel that distance, is termed as velocity, v). A trait is an object characterized by its boundary conditions. It is not the point that matter and motion may be different; rather that matter moving at a constant rate is different than matter not moving. Energy does not move matter. That difference between matter moving at a constant velocity and matter not moving is termed as the energy associated with the matter moving at a constant velocity, where $E = 1/2\ mv^2$ (kinetic energy). Energy, E, is also characterized by a mass moving at constant velocity vector (defined similarly as scalar speed and direction) as momentum, $P = mv$. For matter that is not moving, $E = 0$, and mass, m, has a defined magnitude that is not equal to zero. Compare the kinetic energies of a mass moving, first at constant velocity and then at a constant velocity that is three times greater than that first velocity. The kinetic energy difference is nine times (due to the squaring of the velocities—one velocity being three times greater than the other velocity), while the difference in momentum is three times. Motion can be considered to be limited by the amount of energy that is represented in the moving mass object, or conversely, the energy that is available for the moving mass object limits the motion of the mass. As such, motion (or energy) can be construed as reflective of a boundary condition. Motion of an object is suggestive of either a limit on the amount of energy (or movement) that is available for motion or a constraint that is imposed internal to the object that results in motion. In either interpretation, the mass stays the same. The mass object and the boundary conditions of the mass object moving have a trait termed as energy. Neither is energy a property of matter, nor is it an attribute of matter. Energy is a trait of the object called *moving (constant velocity) mass*. We can describe energy as a mechanism that can produce a force that causes motion. Force is also a trait of the object called *moving (accelerating) mass*. With a constant force, the momentum, mv, increases linearly. Consequently, force and motion of mass are comparable. Force moves matter.

Force, F, is characterized by an accelerating mass where the motion of acceleration, a, is constant in terms of the ratio of the distance traveled by a mass, divided by the time to travel that distance, divided again by that time to travel that same distance. This difference between an accelerating mass and a mass that is at rest is what we term as force, where $F = ma$. Force is the rate of change of momentum. Mass, as an object, at rest exerts no force due to motion (which is zero). Consider a mass positioned on a table that is situated normal to the force of gravity on the Earth's surface. The motion of the table is the same as the motion of the mass, both of which are connected to the Earth through their mutual attractions by the mechanism of gravity (inherent in all matter). The Earth, therefore, all objects, including the table and mass, are rotating. None of these objects are at rest (in an absolute sense).

The force of gravity is represented as $F = ma$, where the variable a is the acceleration of gravity, shown as $F = km_1m_2r^{-2}$, with $k = 6.673 \times 10^{-11}$ m^3 kg^{-1} s^{-2} and r the distance between the centers of mass of the object with mass m_1 and the object with mass m_2. The much greater force of gravity due to the Earth is pulling the mass toward the Earth while the table is resisting the Earth's force by countering with a force equal to that of the mass object. The result is a force of attraction that suggests that the mass is static, while in fact it is not. If the same mass-table object were situated normal to the force of gravity on the Moon's surface, that force of Moon's gravity would be approximately one-sixth that of the Earth's gravity. The reference frame for the measurement of movement has changed from the Earth to the Moon, with associated changes in the magnitudes of the gravitational attractions, but the force remains. In neither the Earth nor the Moon situation is the mass at rest. In both situations, the mass is experiencing boundary conditions; therefore, neither is force a property of matter nor is it an attribute of matter. We can describe force as a mechanism that can produce motion. Force is a trait of the object called *moving (accelerating) mass*. Energy and force are different than matter (Burgin 2003).

For integration, traits are matter with contexts (i.e., boundary conditions).

Attribute

In contrast, attributes are measures of properties and traits. Length is a measure of distance and size is a measure of volume or area. Attributes are objects characterized by their constraints. Distance traveled per amount of fuel (or energy) is one measure of piston wear or engine wear (Smith and Bahill 2010). Attributes do not have mechanisms. Kilometer per liter has no mechanism, nor does it have a boundary that limits it. The measurement of a kilometer or the number of liters is mechanistic by the nature of measurement. We know that the burning of a liter of fuel results in delivering a force to the tires that in turn interact with the road surface. Further, we know that burning a liter of fuel results in motions of an internal combustion engine-driven vehicle. For example, knowledge (an object with properties, traits, and attributes) is required to perform various activities. There are procedures (mechanisms) that when followed achieve a degree of precision and accuracy, and there are formalizations that are called for to represent the measured quantities in a tangible form (object). The tangible form can be conveyed to others as data or information. However, the data or information (e.g., distance traveled or distance with context, respectively data and information) are not mechanisms by themselves. There is no energy, no force, no mechanism, no boundary, and no movement; therefore, kilometer per liter is neither a property nor a trait. Attributes can be considered measures of performance. Measures of performance are testable.

If we were to limit the number of kilometers that were driven, we could determine to some degree the amount of fuel (or energy) that was used or could be consumed by knowing the fuel (or energy) consumption. However, to determine the fuel (or energy) consumption to a higher degree of accuracy and precision, we would further need to know the constraints imposed on the system. The limitations are given by the domain of the problem, while the constraints are a structural property of the solution. For example, limiting the size of the fuel tank or energy source places a maximum amount of fuel or energy that is available for use. The variability in the use of that fuel or energy is due to the heat of the engine, the weight of the vehicle, the altitude and slopes of the driving course, the rate of release of fuel, and the speeds of the vehicle (to mention a few factors).

Summary of Property, Trait, and Attribute

Systems have properties associated with the intrinsic nature of objects (corresponding to a mechanism); traits associated with conditions and mechanisms; and attributes that are imputed to intangibles that represent measures. From an integration perspective, we deal with objects, traits, and attributes. These are the three components of integration that need to be managed. Objects (both real and intellectual) are signified by their mechanisms, the engines that result in the interactions with EMMI. Traits illustrate the context (conditions, e.g., boundary conditions) in which the object is active. Attributes (measures) describe the constraints that are applied to the object's properties and traits.

Properties, traits, and attributes are testable. Properties are fundamental to objects and as such are not measures of performance, for example, mass is not a performance, although it can be verified as satisfying requirements through testing or analysis. Traits (properties in context) are related to the context of an object. A moving object can be measured for speed, relative to a standard of measurement and a reference point. We sometimes think of such standards as absolutes; however, they are relative in an absolute sense. Measuring to a set of standards requires close attention to the validation of that standard for its intended uses, for example, its fitness for use, given the particulars of the measurement circumstances. If the task is to measure the airspeed of a glider, then the standard for measurement could be terra firma (i.e., ground speed). However, the standard for measurement over the open ocean presumes a nonmoving surface (firm by all accounts), which is instead represented by a moving surface of water. Were the glider to attempt a water landing, the movement of the glider and water necessarily need to be matched in both speed and direction of movement for a very smooth touchdown. The direction of the water movement is potentially both horizontal and vertical (depending on the sea-state). In the horizontal direction, if the glider travels faster than the water movement and in the same direction as the movement of the waves, the glider will be tangled in waves—engulfed or inundated. In brief, smooth water landings are quite difficult to achieve. The standard for measuring relative velocity to a multidimensionally moving surface can be quite complex—deserving close consideration.

Epistemology of Systems Engineering Integration

The epistemology of systems engineering integration intends to provide a basis for the measurement of a system as a system (through its properties, traits, and attributes) (described by David Cropley as "characteristics, features, or properties" (Cropley 1998)). But that is rarely the case. Founded on the somewhat roguish testing regime followed for building products or services, we test what is convenient and expedient, not what is necessary and sufficient. In short, because it can be tested as part of development, we conspire to test and then presume that the test in some fashion reflects on what is both beneficial and pertinent for integration. This portrayal of testing should not be taken as an incisive rail against current practices in systems engineering. Indeed, it is not. The very nature of systems engineering is iterative; each pass at the work designed to improve on the previous incarnation, while systems engineering integration is recursive by design. It is expected, planned, and appropriate to test. However, testing to suggest improvements in work is not adequate to provide sufficient confidence in an artifact that it is ready for integration. Were the artifacts designed according to an idealized set of perfect requirements, and specified without fault, the integration work would then proceed with difficulties mired in different interpretations of what was done, wrapped in a medley of social behaviors. Epistemology provides the basis for certainty (Ferris 1997). But within epistemological thought, the reasons for believing that measurements are accurate and precise are mirrored by the reasons that data are marked by inaccuracies and imprecision. These issues of error are further complicated by trust or lack thereof in the data (or in the means of measuring or recording the data).

Metrics

The concern for integration is to develop a set of metrics for determining how well integration is proceeding. Those metrics could be in terms of the amount of time taken to achieve some specified level of integration; the amount of time that is remaining to achieve some specified level of integration; the trust (or lack of trust) that verification or validation will be satisfactory; the level (percentage or degree) of completeness (or lack thereof); the utility that can be ascribed to the integration that is accomplished (or remaining to be accomplished); and the utilization of resources to accomplish the level or the time to complete integration. In all, numerous metrics can be devised to determine how successful the integration efforts are compared to a standard or a measure in absentia. The concern for integration is to know what the leading (or lagging) indicators are to better gauge how well the work is progressing (or not). Management and systems engineers do not appear to have much control over integration. It is as if the entire integration effort is on autopilot with the integration effort taking on a life of its own—growing in time and swelling in its use of resources. Metrics include inputs from customers as to their satisfaction of the integration effort (Bahill and Briggs 2001). A word of caution is appropriate: customers and users (sometimes domestic, sometimes foreign) have unique metrics for their environment that may appear efficacious, but are inappropriate for an integration effort (Friedman 2010).

The disclaimer about metrics covers the pertinent issues of falsification through specious data or data collection; willingness to acknowledge problems based on metrics; knowing what to investigate to determine root cause; and determining the appropriateness of the metric as a determinant of interest. Having many metrics is usually the beginning of learning how integration is enabled and impacted on by the work, resources, and policies that invigorate it. Having the right metrics provides a clear focus on the issue; results in the correct decisions; and works toward the common goal.

The quintessential discovery, however, is to have focus on a single overall objective to coalesce and bind the spirit and interests of both the enterprise and the customers (and users). As such, the enterprise conspires to integrate the customer's decisions into the company's decisions, integrate the customer's needs with that of the project objectives, and integrate the customer's loyalty with the loyalty of that of the enterprise's workers.

The alignment of a single metric with the goals of the organization may not be as complicated or difficult as it may at first seem. In the case of an enterprise whose business is to deliver access to data (a service), the single metric of bits per second might adequately represent the totality of the business model. All business personnel would see their positions as fostering their support and focus on the company's metric, "provide individual users with greater than one megabit per second." If this metric accurately captures the goals and objectives of the enterprise as well as the customer's and user's view and uses of the enterprise's product or service, then the primary interest among the key stakeholders is broadly and generally in agreement. Such an enterprise might be an Internet service provider (Internet access—a service), Internet search engine (content search—a service), or a library (content provider—a service). No doubt there may be other metrics deemed important by the key stakeholders, but these other metrics will always seem less important than bits per second for users. The totality of services available to the customers and users, when combined in various ways, should focus and maintain the customer and user loyalties to the service-providing enterprise. Yet, even though a single metric can rally enterprise support and endear customers and users, other metrics are important. Figure 2.6 depicts a process flow diagram that relates metrics to customer, enterprise, and project.

Figure 2.6 illustrates the driving inputs for work to be accomplished by a project within an enterprise. From the customer, the project goals and objectives are laid out; the requirements delineated (as best known at the time); and the limitations of funding are specified along with the spending rate, milestones, reviews, and delivery schedule(s). In addition, any other resources that are available or are specified to support the project are included in the discussion. A contract between the customer and the enterprise is signed and the enterprise determines what resources will be made available to the project (consistent with their business policies and rules), and what constraints will be placed on the project (for revenue generation, profitability, and use of intellectual property). The project will be enabled by the business policies, governed by regulations and rules, and monitored according to enterprise metrics (also made known to the project). The project office will carry out its planning and decision-making processes to match the available resources (from the customer and the enterprise) to satisfy the goals and objectives set down by the customer. For planning purposes, a set of tasks will be developed in concert with the available

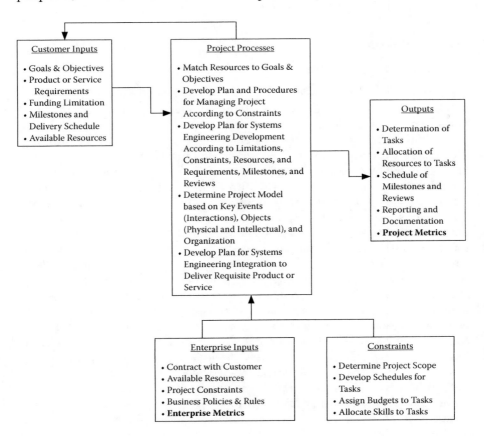

FIGURE 2.6
Process schematic showing metrics.

resources, skills, and limitations. A specific plan and set of procedures will be written to discuss project management given the constraints of project scope, task schedules, task budgets, and available (and planned) skills. Further, the customer and enterprise resources that are to be made available will be allocated to the tasks. Milestones, reviews, and schedules for the same will be proposed along with reporting and documentation requirements. Project metrics will be reflective of the enterprise metrics and tailored to meet the specific needs of the project (and the customer).

Metrics are not about trade-offs between what best to do versus what is expedient. Metrics are used to represent that state of being, the determinant of "how is it going?" both in the minds of the integrators and that of the systems engineer(s) and project management. Metrics are about the shared value of what the common goal needs to be. Never losing sight of that common goal is enlivening for customers, users, and the project team (the business or the enterprise).

General Nature of Objects

Capturing the full measure of actions that are enacted by the objects in economics, sociology, management, systems engineering, physics, and other disciplines requires that the properties are delineated, the traits are identified, and the attributes are settled on early in the analysis for either building a product or service or analyzing and evaluating an existing system. The binding of the objects with their traits (conditions) and modifiers (attributes) describe objects and interactions within the natural and human-built constructs; the determination of what is important to interactions (and therefore integrations) between objects must be broadened beyond the sending of energy and matter for the sake of only energy and matter, to include systems for matters involving human and social issues. Social interactions, knowledge, and information are the primary elements of exchange (Burgin 2003). Both material wealth (value) and information (patterns) are bounded (or constrained) by energy and matter, yet distinguishable from the energy and matter that carry the value and patterns for social interactions. The transfer of knowledge and information are enacted through physical matter and energy constructs, as well as representing material wealth. For economic transactions, material wealth is a primary element of exchange, with energy, matter, and information concomitantly involved. The embodiment of material wealth is likewise through physical and energy constructs.

For "forces" other than physical, various disciplines have developed constructs similar in nature to those most familiar in physics and engineering as being the impetus for change. Information processing acts as an economic force (Shaw 1990), the role of heads of families acts as a social force

(Kunovich 2009), and information (defined as data, context, and the model of interactions) is sometimes thought of as an intrinsic attribute of a message that precipitates human or computer (human surrogate) action (Shannon and Weaver 1963). The broader definition of "force" captures the general notion of overcoming resistance to change. For the purposes of interaction,* that change is internal to an object. Speaking only of an object's internal actions, the force imparted due to the interaction is opposed by the internal resistances of the object (e.g., mechanism and its control, and the environment of the object's internal structure, as distinct from the mechanism of conversion). In this sense, the object's mechanism reconciles the action of the object as a consequence of the interaction. That change in the object due to the interaction results in a change in the object's energy. Therefore, the energy represents the capacity of the changed object. Interaction results in a change in capacity.

The empirical form of interaction takes place through forces that act on mechanisms. Another metaphorical interpretation of interaction could be thought of as the exchange of capacity that results in actions of the objects. Since both objects have undergone change due to an interaction, there can be both individual object changes as well as change to the context in which the objects have interacted. The local environments may have received energy lost from the enactments of the object's internal mechanisms; the absorption of heat resultant from the physical heating of the atmosphere along the path of transmissions; or the reduction in the interpretable information due to a reduced signal-to-noise ratio due to volumetric dissipation of energy over distance. Both objects have experienced losses due to the interaction. The aggregate consequence of the metaphor of two-object interaction is not the simple sum of one action from the sending object with that of the action of the receiving object. The objects, their internal mechanisms, their actions, their losses, and the resulting changes they undergo are different. Systems are built up of objects that interact and can behave differently from one interaction to the next.

Examples of the transfer of energy, in the case of physical matter, include photons and electrons that collide with other physical objects. Transfers of energy also occur through forms of matter or ideas as representations of material wealth (including money, information, and knowledge). To reiterate, if there is no change in an object, there is no energy loss. Change can result from interaction that is internal or external. No change can occur without an interaction. Therefore, various constructs of energy could include building a spacecraft, designing a banking system, presenting materials to an audience, acting on plans to travel, paying for an item at the check-out counter, discussing philosophy, painting a house, or investing in education.

---

* For the purposes of integration, that change is an aggregation of the objects within a system.

Services and Products

Consider the action of piling wood, that is, that of moving the pieces of wood from one location to another. We call this action a service. An action (in support of a need) is called a service. We consider the pieces may have been moved several times (each time a service). Services are queues and accesses to facilitate the exchange and use of energy, matter, material wealth, or information. Products are usually thought of as physical items that have various functionalities. Products and services have functions. We could consider a pile of wood to be a product. Not all products are integrated. A cut piece of wood is a product that can be purchased at a hardware store or lumber yard. The wood is the result of many processes, including planning, cutting a tree, to milling, to grading (for quality), pricing, shipping, inventorying, stocking, to selling. Many products went into the pile of wood, individually considered as products. The aggregation of wood is a product of labor (move wood).

Products have only limited physical properties and attributes that can be used, but services are only constrained by access rather typically than limited. Limitations are conditions of boundaries, and once imposed they are immutable. The physical design of some products, for example, a single-hand instrument telephone, limits the number of people who can use the product, versus a product that is designed to accommodate a service with such restriction, for example, a speaker phone made of technology that supports communications with a small group of people. Technology would seem to be the limitation for carrying the communications beyond one to only a few people. Limitations can be thought of as the budget earmarked and the schedule determined for a project, whereas constraints are the apportionments of money distributed to the tasks along with its designated schedule. Constraints, however, are conditions of allocations, that once established are changeable, however vicissitudinous. Constraints are flexible within the overall limitations set for the project. Labor hours saved on one task may benefit another task, as dollars overspent on a task may force different behaviors on other tasks. Therefore, products have limited scalability in contrast to services. Services can be enabled (e.g., through technology) to be more scalable than products. Of course, a personal service (one-on-one) may not be scalable. Depending on the design, access to service functions can be made more scalable than access to product functions. Scalability is doing what is done with either more people performing the same activity or with one person being able to do more through some economy of numbers, technology, or process. Scalability is all about either doing what you do with more people doing the same thing or being able to do more with one person. Scalability in the first instance (more people doing the same thing with the same product) implies that each person requires a product, that is, scalability by single-user products. Scalability in the second instance (being able to do more with one person) is through efficiency by using a service. Scalability in this second instance implies perhaps a single product that (through services) provides a similar functionality as with multiple products. So, by either increasing the number or speeding up a service, scalability is achieved. Products and services can be thought of as different in this regard. For example, project management and systems engineering can be thought of as two distinct but related services, the end result of which is a product (or service).

While products and services are different, both provide for functionalities that can result in uses with various performances. The ability to perform or, as Taguchi (1986) prefers to think of it, the variance in the performance can be thought of as the quality of the product or service. In contrast to products that are built and then subsequently delivered, services are built and used simultaneously (Kotler and Keller 2007). The differences between products and services include how products and services are used and thought of, how they are delivered, how users behave once they have access or use, how users behave because of the product or service, how the management processes are planned and enacted, and how products and services are integrated into operations. Indeed, it serves the producers of products well to think of the services provided by their products with equal consideration to those intended uses as a product. What decisions will the customer make because they have purchased a piece of wood? The customer may be thinking about using the wood as means to support a bookcase. The service provided by the wood is to support a bookcase. The product function of 'to support' is not designed into the wood product per se, only envisioned by the customer who purchased and the user who used the wood to support a bookcase.

While our perception of integration may be skewed by the framing of a particularly vexing issue, the nature of integration does not change. The nature of integration is to result in the exchange of energy, matter, information, and capital wealth. The goal of integration should be to enable this exchange in an energy-efficient manner, consistent with the limitations imposed on a lifecycle basis and the constraints allocated in the architecture of the system.

Objects

We commonly think of an object as a fundamental element, entity, or representation. It may be atomic or an aggregation of entities. Objects can be physical or abstract. Objects may be conceptual, phenomenological, or ideological.

Some but not all objects are teleological. Animal objects can be said to be exhibit processes that are purposeful (Bainswanger 1990). Teleological processes explain phenomena by the purpose(s) or their end(s) they serve rather than by causes postulated to explain behaviors. Bainswanger (1990, p. 119) describes three conditions for teleological action: self-generation, significant value of the action to the object, and significant value ascribed to the cause. As we have observed, action is self-generated only if the object is a source kind. Human-built objects may be comprised of source objects, but the other kinds of objects are necessary for integration into a system (otherwise, the objects simply interact).

Objects may be comprised of other objects, each of which is related by interactions. Objects can be ordinary or elemental. Ordinary objects are macroscopic in size—perhaps represented as a piece of wood, a broom, a component of a subsystem of a large electrical grid, an eye or arm, or a human. Objects are distinguishable from other objects and nonobjects by their boundaries and boundary conditions, their mass and energy, their information and knowledge, their material wealth, their functions, and the behaviors they induce because they are available or anticipated.

Objects have boundaries. However, boundaries are neither fixed nor impassable. They simply pose a limit at which the endogenous operations of an object are expected to be* or are substantially diminished. External actions may affect an object regardless of the boundaries of that object. These boundaries can be of several types: behavioral, functional, physical, abstract, process, and representational. Boundaries of an organization can be determined by behaviors, physical objects (entities), functions, abstract concepts, processes, and representations (or models). Boundaries are not always easily recognized. Behaviors of various types form boundaries by which some people in an organization will not cross. These might be legal or ethical boundaries. Different organizations may have varying degrees of adherence to ethical boundaries, some visible some not. Behaviors are describable in terms of observed reactions to influences of energy, matter, material wealth, or information. Functional boundaries illustrate the limits on the use for a product or service. Physical entities (objects) may have a visible corporeal boundary, for example, the physical edge of a dustpan that when placed on a flat surface such as a floor provides a function of 'to ramp to' so dust can be swept into the dustpan. Functions are identifiable at boundaries of objects. In essence, a function denotes a boundary and a boundary condition—the circumstances in which the function can be used by a user (act as a ramp and repository for dust, in this case). By simple analogy, the boundary conditions for all macroscopic objects are describable in terms of observed reactions to influences of energy, matter, material wealth, or information on the boundary of an object. Abstract concepts hold that unrelated items can be juxtaposed and bounded. A photograph of a red chess piece with a black and yellow car battery on a stark white background may not offer any recognizable pattern to a group of people. However, the mere physical presentation of the photograph is recognizable as an abstraction bounded by the edge of the visible imagery. The photograph can be thought of as a limiting measure of bits of information to

---

* Intentions (i.e., the cognitive conceptions) are ascribed to the fulfillment of needs through uses of human-built artifacts.

be transmitted in a noisy communication channel (Shannon 1948a, 1948b; Cover and Thomas 1991). Representational boundaries might be models of what a person considers being the extreme limit of endurance, or the absolute maximum number of pounds of ice cream that are possible for one person to eat in a 20-min sitting.* These boundaries represent models by which a person acts. Representations are the results of processes.

An object is distinguishable as a microscopic object, indicative of subatomic particles, atoms, and molecules. Similarly, the six types of boundaries (behavioral, functional, physical, abstract, process, and representational) apply. Living systems have boundaries as listed in the six types. For objects, behaviors are describable in terms of observed reactions to influences of energy or matter. While the literal application of material wealth (e.g., money) or information would not seem to have an effect on elemental objects, both are convertible into energy and matter which do have such influence. For this reason, objects are said to be subject to influence (i.e., exhibit behaviors) from transfers of energy, matter, material wealth, and data.

Object Types

Objects are differentiable by their input and output characteristics. Type 0 interactions are self-induced through internal mechanisms. The Sun's internal mechanism to radiate is an example of a Type 0 interaction. Type 0 interactions are the results of stored energy used to drive an internal mechanism. Type 0 interactions are typical of "sources" of energy. Type 0 interactions are one type of energy source to enable or sustain the interactions necessary for a system. The outputs of EMMI can be received by other objects and interact with those objects.

Type 1 interactions result from the complete absorption of EMMI. By the nature and enactment of its internal mechanism, the receiving object remains anonymous or unacknowledged. A Type 1 interaction is potentially receivable by objects, but is not received, is received and not recognizable as an accurate representation of the sending object, or is received and the receiving object does not respond to the sending object. Type 1 interactions are inhibited or masked by physical, functional, or behavioral reasons (internal or external to the receiving object). A human yells for help believing someone will hear the clarion call. But no sound is heard by a rescue team. No one responds to the calls for help. It is the nature of a Type 1 interaction to be the interaction that could have been (but was not for some explainable reason). All things natural and human/animal-made can interact in a Type 1 manner. Type 2 interactions are sent and received. The Sun releases energy that intersects and collides with Earth. Type 2 interactions include an electron moving exoatmospherically toward Earth under the influence of Earth's gravitational field (EaG-field), experiencing a "collision," being captured by

---

* Strict rules of safety apply; not meant as a challenge.

a hydrogen atom, and releasing energy, or an autonomous robot that automatically reroutes internal electrical energy to recharge batteries without an indication of the remaining charge needed to maintain a minimum threshold for operations of all its subsystems. The yelling for assistance is a general request directed to anyone, regardless of language. The autonomous robot's rerouting of energy may be based on elapsed time since last recharge, or a software algorithm that relies on internal sensor inputs to estimate reserve capacities of its rechargeable batteries. A Type 1 interaction is initiated from within. In contrast, a Type 2 interaction eliminates (or discharges or "sends") EMMI due to some external receipt of EMMI. Examples of Type 2 interactions include the person responding to the yelling for help. Type 1 interactions reflect the internal needs or intentions of an entity, for example, the self-initiated requirements for survival. Type 1 interactions are in response to internal processes, the mechanically induced self-regulation for fulfilling basic needs. Type 2 interactions are the responses to external stimuli, the simultaneous or reflexive reactions based on are capabilities within the entity's structure. Regardless of the type of interaction, the mechanical processes that carry out the actions of the "send" and "receive" functions are limited by the entity's capacity to initiate a "send" or respond to a "receive." Further, the mechanics of interaction also preserve the constraints of the entities. The architecture of the entity and the mechanism for interaction are constrained by their design and implementation. Therefore, interacting entities are subject to limitations, conditions, and constraints.

Limitations describe the extremes of operability of an entity at its boundaries (the physical extend of an entity). Limitations are methodological or procedural schemas that either define or signify intended extremes. Limitations can be organization or mechanistic, procedural (rules and policies), and social (customary and acceptable behaviors). Limitations are sometimes described as conditions of boundaries (i.e., boundary conditions). Once the limitations are instantiated in the entity, they form an immutable structure. For example, the physical design of some products, for example, a single-handset telephone, is optimized for a single user. The limitations are imposed through both the design and technology which combine to provide a distance at which a voice can be heard (from the perspective of both the person speaking and the person listening). Limitations can also be thought of as the budget earmarked and the schedule determined for a project. The project costs shall not exceed $10 million and the deliverables are due no later than 2 years from the start of a fully executed contract. These are limitations agreed to by the parties, stated in the contract between the parties, and enforced by penalties. The parties to the contract are limited by the agreement. Limits apply to what can be done versus boundaries that apply to the physical extent of entities.

Within the limitations of the contract, constraints are the apportionments of money distributed to the tasks along with its designated schedule. Allocations impose conditions on objects and processes. The concept of interaction captures the observation that an entity or agent (for another entity) initiates or responds to an entity or agent subject to limitations, conditions, and constraints.

We distinguish between objects who have Type 1 or Type 2 interactions with Type 1 objects or Type 2 objects. Type 1 objects produce Type 1 interactions (internally initiated) and Type 2 objects produce Type 1 or Type 2 interactions depending on conditions and context. In other words, Type 2 objects can elicit a response and respond to an input, whereas Type 1 objects eliminate energy, matter, material wealth, or data only due to an internal process. Both Type 1 and Type 2 objects can interact. Examples of Type 1 objects are uranium ore, a uranium-enriched nuclear reactor core, or the Sun. Examples of Type 2 objects are an electronic resistor, a car, a building, or a piece of wood. The piece of wood interacts only after experiencing an input from another object, for example, friction due to touching another piece of wood. If the force of friction between two stacked pieces is sufficient to resist the force of gravity (that would "pull" one block downward), then the pieces of wood do not move. No energy is transferred. However, if the piece on top is piled high with more pieces, the friction at the boundary between the lower pieces may become insufficient to resist the force of gravity and the upper pieces slide down the lower pieces. When the movement occurs, energy is transferred between the moving pieces. Type 2 objects require an input of EMMI or they do not interact. In the case of pieces of wood, touching is not interacting. Only when the pieces move is there interaction. Type 1 objects do not require an input to eliminate EMMI. This elimination of EMMI from Type 1 objects may result in interaction with a Type 2 object if conditions permit. Both Type 1 and Type 2 objects change when eliminating EMMI. The extent of activities of Type 1 and Type 2 objects is limited to interactions. Interaction between Type 1 and Type 2 objects is a necessary condition for integration. Both Type 1 and Type 2 objects are required for integration. Interaction that involves two objects "sending" and "receiving" energy, matter, material wealth, or data (in an informational sense) is required for integration. Integration implies a system.

For example, consider placing a piece of wood (object) on top of another piece of wood (object). Being careful to place the wood so there is either some overlap or some measure of stability in their placements, add another piece of wood to the "pile." If each piece of wood that was placed in the pile stayed exactly where it was placed, then it is not interacting (as there is no movement). However, since the friction between each piece of randomly placed wood is probably insufficient to resist the effects of Earth's gravity, most likely the blocks of wood settle and move as they are placed (or thrown) onto the pile. Consequently, the sliding blocks of wood interact with other pieces of wood in the pile. The interaction of a piece of wood is a Type 2 interaction as energy is transferred due to the movements. The interaction only takes place once another block of wood is placed so that it touches and moves. As the pile grows, the pieces of wood are touching on any one or more of their sides or edges. The action of one piece of wood on another represents the physical boundary that inhibits the "free" movement of the wood and the basis for energy transfer if the blocks of wood move. The action of board-on-board due to friction is a property of the wood surfaces and a trait of the circumstances surrounding the movements of those surfaces. The action of movement is due to the mechanism enabled by the force of gravity. The result of such a mechanism is observed in the ability to add wood to make and grow a pile in height and breadth. However, if there was no friction between the wood pieces and between the wood pieces and the surrounding terrain, the wood pieces would just slide from their placement on another piece of wood (also presumably sliding unabated) and continue their motion without constraint according to the local topography. On level ground (again assuming no friction and no other impediments), no pile is possible. Without friction no energy is transferred from block to block or from block to the Earth's surface. Motion would continue unabated, undamped. One of the conditions for integration is that the constituent parts must interact. No integration and no system are possible without interaction. In this case, the piling of wood requires interaction due to friction to build a pile. If a piece of wood retains its original (prepile properties and attributes) and does not change as a result of interaction with another object (e.g., a piece of wood), then the pile of wood will remain just that—individual pieces of wood in a pile (every piece of wood remaining exactly where it was placed). Regardless of the force of friction, only Type 2 objects are included in the pile of blocks of wood. No integration is possible by the process of "piling the wood" (Figure 2.7).

Contrast the rather random acts of piling wood versus stacking lumber. Stacked lumber (assuming that it was carefully placed so the stack had lateral and longitudinal stability) would not move due to the Earth's gravity. The stability of the stack is in fact ensured in part due to Earth's gravity. As there is no relative motion between the stacked lumber, there is no transfer of

FIGURE 2.7
Wood pile. (Courtesy of Timothy L.J. Ferris, photo taken in Pebble Beach, California.)

energy between the boards. Again, only Type 2 objects are involved—no integration is possible with the wood.

The whole of either the pile or the stack will be equal to the simple sum of its parts. Consequently, the pile of wood built through interactions with other pieces of wood (and the ground) is not a system as there is nothing more or less than the individual pieces of wood regardless of their being piled or stacked. Interaction per se does not always result in a system—an integrated whole. The pile or stack of wood could be taken apart without changing the properties or attributes of any piece of wood. The parts are as distinct before helping comprise the pile as when the parts are conceived of as a pile. There is no integration, merely concomitant action where each part remains a part but never becomes a whole by parts or by itself. Systems can be conceptualized as self-reliant entities that are simultaneously wholes and parts (Koestler 1968). In other words, neither the pile nor the stack of wood would be considered as a piece of wood, and neither would a piece of wood be considered as a pile. But a piece of wood would be considered as part of the pile. Neither the pile nor the stack of wood is a system, although there can be interaction between the pieces of wood. The pieces retain their exact identity, properties, and attributes before being placed into the pile as well as after being piled. Knowing something about part of a system portends some insight into the interactions and objects of another part of the system (Kuhn 1974).

All interaction is point to point (Buede 2009), object to object. People interact with a book when they look at the book. If light reflects from the book and impinges onto the retina, the contents of the book can be resolved and with proper training, the book can be read. The brain collects the light energy and interprets the "encoded" data (the reading process). If the reader puts in no effort in interpreting the information (by simply staring at the book with no focused cognitive activity), the contents may be unrecognized and leave no "footprint" or impression on the reader. The interaction with the book is that of "looking," but not of reading. Page after page of looking without reading does not result in anything gained from the interaction with the book. Yet the person has perhaps spent considerable time with the book, only to later realize there was no retention, no cognitive awareness of the book's content, and no lasting effects of the interaction with the book. No knowledge was gained—there has been no integration of the reader's cognitive structures with the contents of the book. Interaction without integration occurs all too often (in the case of education). Of course there are other times when we may watch a movie, be totally engrossed in its content, and later wish the nightmares would go away. We would probably prefer a natural immunity to certain diseases rather than submit to the integration of viruses and the resultant compromise of our immune system. Integration is not always desirable.

Interactions between objects may or may not result in integration. But integrated systems are by definition interacting. We build structures that are capable of interacting. But until they do interact in various ways, those structures are not systems.

Constraint

The tapestry of relations between objects and processes and their properties, traits, and attributes constitute the domains where constraints are present. "A constraint is a relationship that is maintained or enforced in a given context" (Mayer et al. 1995). For this presentation we use the term *relation** which is meant to be the general term for the degrees of connectivity, coupling, and cohesion that affiliate or bind objects and processes. A constraint exists according to the conditions that govern an object or process. These conditions are specified by the amount of EMMI that is available for an object or process. The mechanism of the object acts as based on the amount and kind of EMMI. The output of the object is a transformation of that input EMMI and the effects of that EMMI on the object's mechanism. Similarly, the conditions under which a process is enabled and driven by input EMMI through its procedures (a similar notion as mechanism for objects) and then carried out by objects is likewise constraining for the process as well as for the processes that interact with the constrained process and with the objects that interact with the constrained process. An object or process under such conditions is said to be *constrained*. Interactions with a constrained process or object is constraining on all objects or processes that interact therewith. The governing condition that initially established a constraint propagates that constraint through interactions with other processes or objects. There may be some processes or objects that are unaffected by the constraint as their mechanisms are either not sensitive to the constraints imposed by the interacting process or objects or in some manner unaffected by the conditions imposed by the EMMI. This lack of sensitivity could be due in part to a very low coupling or cohesiveness or a mechanism whose response is longer than the lifecycle of the constraining interactions (or the lifecycle of the propagation of those constraining interactions).

Frameworks

We interpret observations, postulate principles, and derive laws starting from theory. Our interpretation of observations is confounded by our perspective, measurement, and biases. Theories can be thought of as consigned principles, that is, fundamental statements that are comprehensive in their applicability and generality through their agreement with observation. Empiricists rely on observable regularities from which to build a formal model. This aim of this model is to characterize the causal links that result in the observed regulari-

---

* We use the term *relationship* to signify an association or dependence of people with other people.

ties (Klepper 1996). This structure and narrative that embodies and conditions the model is termed as a theory frame or (referred to a framework). A framework aids in formulating hypotheses and identifying the "... kinds of causal conditions and process patterns that seem relevant for a given range of issues, they offer concepts that correspond to these identifications, and they give reasons for the choices made" (Rueschemeyer 2009).

The essence of models and theory can be explained through frameworks. A framework is characterized by consistency of logic, and the continuity of method. However, to be pertinent for systems integration that involves many disciplines and fields, a framework must also be applicable across disciplines and scalable from micro- to macrosituations. Frameworks can be used to analyze situations. A situation is defined as the constrained and limited characteristics of a set of properties, traits, and attributes.

The goal of the framework should focus on describing a general model of integration—one that applies equally well to systems, system of systems, and nonsystems. The allure of general systems theory (sometimes referred to as "systems theory") strived to point out the similarities between the disciplines with regard to their construction, while over time the desire was to develop models that would more inextricably link at least two different fields of study (Boulding 1956). The systems theory movement captured the good intentions that recognized the (1) striking similarities in the system likenesses covered by the subject matter across various disciplines and fields, and (2) the oftentimes equivalent thinking, knowledge structures, and processes, and models of various phenomena. But beyond a few intriguing correspondences, on further investigation, nothing conclusive seemed to coalesce into a definitive theory for any two fields. Two quotations seem relevant.

> The voyage of discovery is not in seeking new landscapes but in having new eyes.
>
> **Marcel Proust (1871–1922)**

> There is a soul of truth in things erroneous.
>
> **Herbert Spencer (1820–1903)**

Although these two authors were contemporaries, they may not have known each other; the context of their remarks is not readily apparent, but their striking correlation in general feeling suggests that there might be more to what we know than in the way we express what we see. The systems theorists spoke often and kindly about the concept of integration, but not with regard to bringing ideas together. System theorists observed supposed patterns and suggested principles that indeed are important. Those principles, once placed in the context of a proper framework, become the gems of discovery that so eluded these early thinkers.

A framework of integration should be operative regardless of the type of objects that are to be integrated. The concepts of integration should apply equally well to all objects, systems, the discussion on systems, and systems engineering, and every other discipline and field that bring things together with the purpose of relating the whole as more than its parts. Such a legitimate framework would help capture the essential ingredients of the systems thinking. Yet, it is neither a multitude nor a family of frameworks that is required for this role. A single, all-inclusive framework can be constructed in which consistency and capaciousness are both conveyed and comprehensible. However, if the framework was purely theoretical, it would tend to (1) cover only a limited number of object classes and variability, or presumably lose its relevance; (2) not necessarily enlighten the practice of systems thinking (or systems engineering); and (3) ignore that which it could not explain. And if the framework were particularly practical, (1) observables would appear without implicating any tenet of causality; (2) there may be no context for what one knows with regard to what one does not know (without reference to pre-existing conditions); and (3) there may be no reliable means to plan actions for that which one does not know in advance (assumptions must be all encompassing, therefore predetermining the answer by trivializing the problem).

The approach to developing a consistent, utilitarian framework is to apply a few general principles (some from systems theory) to explain empirical phenomena and then predict new behaviors that should be observable in all systems. To that end, the focus is on descriptive measures that confirm our normal thinking about concepts that help guide us through investigation and interpretation of our actions. The commonly thought-of notions of investigation, management, research, construction, decision making, and analysis should be enlightened by this framework that we tat. This framework should make apparent the rationality, inference, or discrimination that is commensurate with our innate ability to explain, convey meaning, and carry out our everyday tasks. The desired framework will reflect a combined theoretical and practical presentation.

Frameworks are comprised of frames, each frame representing a set of concepts that together are a different perspective of the domains that make up integration, that is, the defining objects and the interactions between objects. For integration, those domains are (1) the processes that when enacted provide the guidance needed to bring objects together, and (2) the results of those processes—the products or services that we desire as satisfaction of our goal. The process domain governs the building of objects while the product (or service) domain is the result of those efforts to put parts together.

Process Frame

The model of the roles played by cognition, procedures, and representations of cognition and procedures (herein referred to as models or representations) within the process frame is representative of the management processes. The model of the roles played by the physical, functional, and behavioral characteristics (of the user) within the product (or service) frame is the result of the work to build the product or service. Both the process and product (or service) frames are comprised of a recurring collection of variables (termed the key variables) that are always seen to be essential to the observed effects of the work. For example, the typical management planning, procedures that are laid down and enacted, and the results of that planning and worker's undertaking are captured in the three classes—cognitive, procedures, and representations. The effects of management processes are demonstrated and replicated by every project or experiment without exception. While the details may be different, the basic processes of management are carried out by the process frame.

Object Frame

The model of the physical incarnation of products and services is exemplified by objects. Objects are or represent material structures, material wealth, and information. From these physical entities come energy or matter. We interpret energy and matter as energy or matter, and in certain instances we interpret energy or matter as material wealth or information, or both material wealth and information. The Sun (object) is a physical structure, made of mass and emitting energy. That the energy conveys information about the make-up of the exterior of the sun is discernable through investigation with spectroscopes and analysis of the spectrum derived, therefrom. The energy has conveyed data, which when collected and organized by a sensor, result in information. The sun's energy impinges on the Earth, activates cellular structures (plants as objects) to grow (assuming nutrients, soil, protections, water, etc. are also provided). Those plants have value, are marketable, and can be sold. The barter of goods for plants (objects for objects), or plants for money, are reflective of a value system for material wealth. The effects of EMMI from an object and from the object itself are demonstrated when we build or use a product or service.

Key Variables

If one of the key variables in either frame is omitted, similar to an experiment (and approach) that fails to replicate the result of another experiment (that is thought to be very similar by its nature and specifics), then the results achieved when all the key variables are incorporated exposes either a change in experimental (or project) conditions or is suggestive that another key variable is operative and has been omitted in one of the experiments. This is *not* to say all of the variables that conspire to realize a given experimental result are included as key variables. Most certainly there are other variables with some import. But the omission of a key variable is essential to bringing about the desired result. Missing a key variable points out a deficiency in the formulation of what is necessary to give the framework its predictive worth. Frames capture the key variables, that is, the minimum set of variables that are necessary and sufficient to explain the resultant experimental results in a fashion that the experiment can be shown to be replicated, that is, similar results are consistently achieved. Other variables may exist that have effects (perhaps noticeable under certain conditions) that might refine the primary results of an experiment that is predicated on the key variables. But the principal correlations and patterns due to the key variables are not changed substantially, except in those cases when specific conditions apply. Research often focuses on new conditions that may be suggestive of new key variables, or perhaps a new theory that explains the totality of variables in a manner consistent with previous theories and also predicts new phenomena.

Essence of a Framework

A framework that is all inclusive of the subjective direction to accomplish a task is needed in conjunction with the objective results of those accomplishments. The hallmark of a framework for integration is its consistency to reproduce similar results from each use. A framework that is characterized by consistency has logic, continuity of method, applicability across disciplines and fields, scalability from the micro to macro (and vice versa), and the flexibility to accommodate a variety of differences across and within its classifications. Most importantly, the definitive framework for integration should focus on the eventual prospects of at least not inhibiting the development of a definitive theory of systems integration. To investigate the essential elements of systems integration, a framework is developed and presented that reflects causality in a system—that which derives from cause to effect. An integration framework provides the basis for identifying principles that have substantial theoretical foundation(s). Systems integration can be thought of as having fundamental provenances, a few of which are listed: (1) engineering principles (which are interdisciplinary), (2) systems thinking (multidisciplinary), (3) economics (determination of value, risk, and consequences), (4) acquisition (the catalyst that moves a concept through development into operations), (5) social science (the mechanisms of human behavior), and (6) management (the processes that govern the direction, control, communications, planning, organization, and team-building). The field of systems integration applies principles from science, nature, and sociology to build desirable and worthwhile systems. Systems thinking extends this paradigm by attempting to encompass metalessons from all disciplines. Metalessons signify the maturation of the discipline through considered discussions about the philosophical bases for theories, the provenances of models, the efficacy of frameworks, and the operative frames that capture the essences of an experiment. Systems thinking continues to contribute to the development of systems theory through the discovery of universal principles that transcend discrete disciplines. It is more than pattern recognition that drives these discoveries. It is the recognition that frameworks not only clarify knowledge what is known but also point to missing elements. Systems thinking and systems integration together empower systems engineers to consider the whole problem and possible solutions in the context of lifecycle issues, including costs and time constraints, and achievement of system performance. So, it is also incumbent on those who analyze systems to discern the mechanisms and behaviors that are represented through certain variables. Therefore, the charter of systems thinkers is to create ideas, build products and services, and analyze behaviors and other clues that suggest mechanisms that are often incomprehensible or unachievable by other means. The systems integration framework originates from this merger of systems thinking, systems theory, fundaments, and principles.

The intent of this discussion is to develop an integration framework that reflects theory and best practices in systems fields—engineering, sociology, psychology, biology, cybernetics, computer science, economics, management, and the like.

Building on and expanding the work of Schlager and Blomquist (Schlager and Blomquist 1999), frameworks should be compared based on nine factors: scope and boundary of inquiry; underlying model; impact of decisions; role of stakeholders; conceptualizations and explanation of action; measures; and metrics of quantification. If the framework was particularly theoretical, it would tend to (1) cover a limited number of object classes and variability; (2) not necessarily contribute to the practice of systems engineering; and (3) overlook what it could not explain. If the framework was overly practical, observations would be made in the absence of implications; there would be no context for defining facts in the context of what remains unknown; and there would be no reliable method to plan contingencies around unknown variables.

Causality

If interaction is the cement of systems, then by this presentation, the causality of events is caused by the concoction of objects, mechanisms, and behaviors that have conspired or happened. The three driving determinants that form the scientific foundation of the causal mechanical perspective are (1) the sufficiency of the EMMI that activates a receiving object mechanism and was transformed into performance (and losses) that in turn carries on similarly with another object, and so forth (termed as the modal causality); (2) the local circumstances surrounding a specific event (termed as the proximate causality); and (3) the conditional causality that related modal causality to proximate causality. As a group, these three types of causality are termed as the objective causalities.

Modal causality is the root cause of all events. Modal causality is the basic source of events (the historical provenance) that provides the foundational causes from which local circumstances (proximate causality) and the apparent most direct event (conditional causality) arise. Combined objective causalities are the *sine qua non* of causes. The modal causality provides the historical trails of events, that is, limits what is causally possible; the proximate causality focuses through localization in time and space, that is, further limits the likelihood of an event, whereas the conditional causality completes the triad of objective causalities by constraining the context and circumstances surrounding the sequences and trails of events. The number of modal trails is literally countless; some of the extraordinarily high number of proximate events are perhaps identifiable, while the conditional constraints are usually readily observable just prior to an event.

Objective causalities are posited to be both necessary and sufficient to render a complete explanation of an event—substantiating the causal connection.

That the necessary and sufficient EMMI is received by an object transformed into performance, to, in turn, activate another object, which in turn combines in some way with other objects, leads to a "trail" of connection by interactions. A receipt of EMMI by an object (object to object) represents an event. An event is defined as the enactment of a mechanism by input EMMI transformed into output EMMI (performance).

Events transpire as a result of every enactment of a mechanism, whereas functions occur at the interface between two objects. Similar to objects and their enactments at the atomic level scaled to enactments at the galactic level, each object that receives EMMI's describable as events according to the same form and structure of input and output.

A building suffers damage in an earthquake, then fails structurally and collapses. The events that lead up to the building's destruction are imminently describable. Beginning at some point in mid-sequence, the land became available for use as a building site; permits and permission were obtained; the architecture was submitted, reviewed, and approved; construction followed established practices; routine and special inspections were accomplished according to regulations; people then occupied the building; the building shook and then collapsed. There were countless events that occurred in these much abbreviated trails of events leading up to the collapse. Another trail of events transpired within the Earth's crust. The crustal movements that occurred leading up to the precipitous release of energy was defined as the earthquake that shook the building. As the ground shook and the building collapsed, a cause and effect were determined. One trail of events might have involved an inspector who missed an important set of weld joints during a routine inspection in the early days of construction. Another trail of events might involve an earthquake engineer who contacted the geological survey regarding the geotechnical specifics associated with the building site. Perhaps the geology data were incomplete or the earthquake engineer underestimated the potential impact of an earthquake. Another trail of events might be the Earth's geographical properties that were changed due to a volcanic eruption, half-way around the Earth. While the connectivity of these trails most certainly lead to the collapse, the risk of the building collapse can be only partly determined. The event in question (the "focus" event—the earthquake) is the event that one wants to discuss in terms of the details of causality. That the focus event is preceded by objectively measurable events is complicated because these events are mostly countless, with a few being identifiable. The level of delectability of proximate events challenges technology and human cognition. If any one of these objective events is implicated in a proximate event, then it is usually termed as a "direct" causal event (insurance companies may associate an earthquake with an increase in seasonal rain (as was the case in the year prior to the 1906 San Francisco earthquake), but that correlation may not be known to be causal). Direct causal events are those events that share the responsibility for the event along with its direct consequences. Should a building inspector not observe or by circumstance ignore a safety item designated as critical, the action of the inspector may be implicated in the collapse of the building. As such, the inspector may bear some responsibility for the collapse, being associated with a proximate cause.

Causality in its most general sense means to not be random. There is always a pertinent and identifiable relation between an object and an event, or an object and a process, or a process and a process, or an object and an object. The notion of randomness is rejected for this presentation. Fundamentally, *if* our knowledge was sufficient to know the meaning of all events, all objective causes, and all circumstances, then (and only then) could the notion of randomness be rejected. As the aim of this book is to explore boundaries, identify constraints, and posit relations between objects and processes for the purposes of interaction and integration, the deeper the exploration, the greater the information, and (perhaps) the better match with our perceived realities. Certainly, the greater our acceptance of the results, the "deeper our debt" to the like-kind guiding notions.

Objective causalities exactly imply the sequence of events, one instant at a time. The determination of the boundaries of the objects spans any number of events until the output EMMI of the sending object (to be here considered as the cause, i.e., the cause-object) is transformed by another object. It is reasonably arguable that every object at least has the potential to transform their input EMMI into something that is different, but perhaps not so different that the change is discernable following examination by an outside observer. So, at best, an outside observer might perceive only a few events, misexamine some, not detect changes, and maybe not even recognize a change as having taken place. The result is a blurring of objects (their physical boundaries, their functions, and behaviors) and their respective boundaries. It is suggestive that identifying mechanisms may be difficult at the elemental object level and that only after some degree of integration will mechanisms be discernable and examinable.

But only an arbitrariness contrived by our will limits the extent of the phenomena about which we inquire. Human consciousness implicitly limits perceptions in a way that may seem arbitrary and capricious (to a different observer), resulting in arbitrary categories of objects, and by extension, arbitrary boundaries between them.* Therefore, the unanswerable question remains, "What is a system?" By inference, without being able to define a system, the objective to discuss, extend, or build a system would appear problematic. We cannot consider that which we cannot conceptualize or define. The rationalities of inference are based on induction (Mill 1882; Holland 1962; Newell and Simon 1972; Holland et al. 1986; Hutter 2007), abduction (Peirce 1934), deduction, comparative (Przeworski and Teune 1970), and systemic thinking (Francois 1999). And, inferring a system is neither knowing nor contemplating the interaction and integration that must take place. Yet we presume to answer the question of "what is a system" when we contemplate systems as such things as people, families, planets, oscillating gadgets, trees, cities, networks, ships, and insects. This is a question that is remarkable by its innocent fundamental nature, yet unanswerable as presented. The question is intractable, having either an infinite number of answers or none. Like the universe such questions may have no bounds.

The domain of knowledge that has accompanied our inquiry over the past 2500 years does not have the requisite features or power to example all that is required to provide an answer. But we are not without choices. Continuing to invoke philosophical notions that have maneuvered our thinking and enquiry has helped humankind with both conceptual and material progress. This choice is available, adopted, and widely practiced. I pose another choice: consider applying the Parmenides method—What is it that is? What is it that is not? What is it that cannot be?—not to the question of what is a system, but as what processes are involved in making a system. The answers might be suggestive of processes that portend *systemness* rather than a literal answer to the question: What is a system? With some contemplation on how to think about the problem of determining how integration works and does not work (i.e., not beginning with a definition, but instead investigating the nature of integration from objects that are not integrated, objects that could be integrated, and objects that we believe are integrated, and then having more luck than we deserve when developing constructs to try out) the results just might happen to turn out to be robust and offer insights. At the very least, the past five years of research on integration has shown me too many false hopes and exposed a paltry few nuggets of insight. Fortunately, those nuggets served as guideposts to develop a reasonably robust consistency of ideas. These are

---

* Private communication with Dr. James H. Lake, board-certified psychiatrist, clinical assistant professor, Department of Psychiatry and Behavioral Sciences, Stanford University Hospital, Stanford, California, September 13, 2011. Dr. Lake asks us to "consider the concept of an inherent human/psychological 'need' to think about the universe in an ordered way, which inevitably leads to attempts to categorize things in terms of hierarchies or systems? Kant discusses this concept, i.e. there is no 'pure seeing' (or other perception), but only 'seeing aspect' which is the imposition of the mind's order onto the world." Dr. Lake continued, "Would it be helpful to include discussion of time in your ontology, as causality in relation to time was Parmenides' major concern?"

offered for your consideration and comment. Our challenge is to begin unraveling the enigmas of integration that brings us systems.

Causality, Mechanisms, and Correlation

A foundational element of scientific investigations and theories is an appreciation of the importance of mechanisms. Mechanisms illustrate the empirical causalities of events. This view of causality suggests that events are precipitated by mechanisms from which we infer causality. Events are the results of actions through mechanisms. We term the sequence of events causality—event by event. Causality is not correlation and correlation is not causality. Causality is formed from the modal threads of events leading to the proximate events (nearby in space and time) from which the conditions are stipulated to select the necessary and sufficient events. Causal events have both provenance and pertinent specificity. Correlated events have nexus, without satisfying the three types of causal events required for strict demands of recognizing cause(s) and effect(s). Correlated events may provide a clue to indicate a causal chain, but correlation by itself fails to identify key variables that feign causality. This difference between correlated and causal events suggesting simple probabilistic occurrences is an inadequate test for causality (Sage and Armstrong 2000).

In all situations, where dense threads and networks of events intertwine, there may appear to be a limit to one's ability to identify an adequate test for causality. This effect of integrated objects (i.e., systems) is suggested in the social sciences and the natural sciences. Perhaps the difficulty resides in our inability to identify the event chains leading up to a particularly targeted event or miss characterizing correlation and causality. We are left again with determining cause and effect based on probabilistic effects of one object caused by another object. While this mathematical approach is appealing and generally reasonable tractable, the question arises—what amount of historical events are necessary to establish modal causality. Often, there is due attention to the proximate events and conditional events, but the causal provenance may be either unknown or quantifiably small. Before the proximate events, the usefulness of modal causality is to establish the event lineage that leads to the determination that an event is indeed possible. Were there no detectable lineage, then one would need to account for and explain spontaneous events. Spontaneous events would arise from essentially no outside source of EMMI, only from their internal mechanism(s) within their physical boundaries. We may be somewhat limited in our appreciation for events that occur without input EMMI and that limitation we can deal with through a scientific approach to causality. But even the reliance of one object forming spontaneously suggests the question—"how" while it is difficult to fathom the notion of a beginning in a physical sense, the question "how" is answered simply as there is no beginning and no ending, only a continuum of object and action. In this matter, the issue of spontaneous events is dealt with without reference to other equally interesting questions, such as why, where, who, when, and what.

Determining the number of modal chains of events is important for analyzing proximate events, with the objective of identifying conditional events. Missing a modal event may prevent us from uncovering a conditional event. The impact of missing a conditional event is to suggest the existence of a spontaneous event—which has been disregarded as previously described. Not allowing spontaneous events simplifies the analysis of proximate events to identify conditional events.

Model for Objective Causalities

The earthquake-building failure example points to several components of a model that can be built for objective causalities (modal, proximate, and conditional). A model shapes what we see and how we act (Senge 2006).

The aim is to advance a means of capturing the key relations that are fundamental to objective causes. A model that gives justice to the sensitivities of the variables, yet fairly captures all that is reasonably germane. The test of sensitivity should concern all that enlightens the conditional causes, that is, those that are constraining for integration. Limiting the purview of the model, we add worth to understanding what we observe (it is conditional on our actions and relatively straightforward to analyze) and viability to its use as a tool from which to analyze and evaluate for planning and predicting. The test of reasonableness is built into the model's ability to support theory which facilitates confirming these predictions. However, a model without an adequate framework may not capture the essence of how to think about the subject, its premises, and its nuances—in this case, perhaps the modal and proximate causes. Therefore, the model must consider the limitations that are imposed on the work, the confounding and proximate causes that affect the work, and the conditional issues that are seemingly under our control. For a project, the modal causes include the funding limits (total and incremental) and the schedule limits (total and by stage or deliverable); the proximate causes include the influences of family members on the work habits of the project team (either individually or collectively); while the conditional causes include project management's allocation of resources and skills.

A framework structures the sequences of relations shown within the dimensions of the model and expands on relations by applying various principles (Garud and Kumaraswamy 1995). Frameworks facilitate managing of concepts.

The model for objective causalities must deal with objective behavioral issues and objects as well as their mechanical properties (the objective component), and the processes to provide the objective component through cognition or rules, procedures or activities, and models or representations of cognition or procedures (the subjective part). The model for objective causalities is rationally consistent with a core interpretation (Luhmann 1982) of the Luhmannian perspective of systems theory (Luhmann 1995a, b).* These two parts form the model from which a framework is devised to guide thinking and management of relations. The objective part encapsulates the resultant product or service that is built by the systems engineer while the subjective part is describable as the management process used by the system engineer to satisfy the requirements of the new product or service (i.e., perhaps a system). The objective part can also be construed as nature's objects interacting with EMMI, with the subject part construed as the set of allowable interaction, the procedures of those interactions, and the manner of representing those means that are recognizable by the objects to be integrated.

Objective Causalities Framework

A framework for objective causalities applicable to interaction and integration must reconcile the sociological aspects of systems integration, integration method, processes, activities, and acts (referred to as the subjective factors) with the physical and functional aspects of the product and service during development (and integration) and when operational (referred to as the objective factors). The blending of these social and engineering aspects results in a nexus of cognitive and literal activities that build products and services. Figure 2.8 depicts a diagram showing events (customer, developer's enterprise, and developer's project). The causal events that impact on the project (the customer and the enterprise) are initiated in the past. The modal causalities reflect these historical modes from which near-term events are resultant. Budgets and schedules are limited during the modal causality phase by the source of funding for a capability. The customer determines a set of requirements that are consistent with the limited funding and schedule, then begins the processes to either purchase the product or service or have the requisite requirements satisfied by a custom development and integration. A contract

---

* Within the ongoing debates of sociological theory, two basic views have emerged—the view of social processes as systemic or as consensual decision making. Since systems engineering is as much a social science as it is an engineering discipline, the debate is relevant to the discussion of integration. Systems engineering is about building products and services. The social aspect of management, economics, and team activities are important and germane. The engineering aspects of bringing technologies and structures together in a meaningful way reveal the efficacy of the resultant product or service (Bausch 1997).

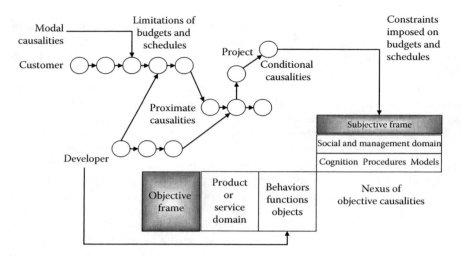

FIGURE 2.8
Framework for objective causalities.

is initiated and the enterprise begins to define a solution based on the initial set of requirements from the customer. The formation of a team for the project represents the proximate causalities that determine that the key directions and decisions will guide the project. Interactions between the project team members and the customer reveal a series of problems that will strain the limited funds and most likely delay the delivery schedule(s). One of the project workers is determined to take time off to rest but is told that the work is too critical to even rest. The worker quits (one of the conditional causalities). A difficult situation was then compounded and the customer canceled the project. The subjective frame intersects the objective frame in a framework that combines the three different causalities. The only caveat is that the causalities must have objective (quantifiable) measures.

The modal causalities define all that influences the framework for integration. Those influences include all that led customers to become customers, users to become users, developers to become developers, companies and regulations to organize the work environment, and future stakeholders to be impacted upon by the results of project work (i.e., products or services that are developed and integrated). The modal causalities are those that make up who and why we are. Modal causalities determine the relations between the past and the future for all things. Either someone is a participant or they are not; either resources of a certain type will be used or they will not. The existence of the customer, the project, the product, or the service is determined by modal causalities. Therefore, the existence of the objective causalities framework for a particular project is deterministic due to modal causalities.

Proximate causalities are determined by the customer's requirements; the project's initiation, formation, and organization; the bringing together of a team of workers; the allocation of resources; the development of an overall schedule and budget; and the organizing of work tasks. The focus of the proximate causalities is to define and enable the project work. The previously determined specifics of the work packages, the assignment of personnel, the interim due dates, the allocations of specific resources, and allotment of facility usage are all indicators of proximate causalities. The customer exists, the project exists, the work exists (all due to modal causalities), but the particulars are coming into existence as decisions are made (the work of proximate causalities). Proximate causalities were most often conditional causalities, showing the historical traces of decisions that have resulted in the present. However, all proximate causalities are not known (nor knowable).

The conditional causalities are very local, indeed. Going to work on time, staying late, taking vacation, using or not using a particular method or tool, having a conversation (or not), and checking your work all point to conditional causalities. Conditional causalities are controllable to an extent.

The management processes for systems engineering (i.e., planning, communicating, directing, controlling, team building, and organizing) are formulated and carried out through processes. The results of the systems engineer's work are the physical object and their connections that provide a user with function and result in various behavior. The relation between the physical object and the function that it enables it at the interface with another physical object is referred to as the consistency (or integrity) between an object's function and the structure formed by the objects (Clark and Fujimoto 1990). The management processes that guide the building of products and services result in the construction of objects and their resultant functions. Therefore, for functional integration to occur, there must be a frame of planning, organizing, directing, communicating, controlling, and working with other people (teamness), referencing Garud (1995) for the work of Astley and Brahm (1989) who states "the functional integration of modules as part of a coherent system, an overarching 'framework' of planning and coordination would be necessary." The systems engineering activities and procedures are defined in the 3×3 matrix of tasks for the framework of objective causalities.

As Glennan points out in "Rethinking Mechanistic Explanation" (Glennan 2002), there appears to be a seemingly irresolvable difference between the approaches to interpreting the fundamental nature of scientific explanations. Salmon (1984) proposed a causal approach to unification via mechanisms, and Kitcher (1989) proposed unification based on patterns that explain unification. Unity is a sought-after concept that illustrates the reuse of various social mechanisms (some of which may be considered fundamental) (Hedstrom and Swedberg 1998; Gross 2009). That social scientists (Torres 2008), biologists, chemists, physicists, economists (Williamson 2009), and systems engineers should adopt a mechanistic view attests to the pervasive nature of proof that appears required to pose arguments that are by some means convincing. It would seem that if we can appreciate a mechanistic way of transforming one thing into another we have comfort in being able to describe the result of the transformation of an input into an output. With a plethora of mechanisms that appear to be active ubiquitously, the pleasure of characterizing all of them may be daunting. But this monumental task does not obviate the power of mechanisms as unifying transformers of EMMI. Regardless of these apparent difficulties, causality can be modeled using mechanisms and produce results acceptable for systems engineering and systems engineering integration. The further development of a theory of integration relies on a mathematical formulation (following the mereology proposed by Stanislaw Leśniewski (Stanford University 2007)), more experimentation across a wider set of disciplines and fields, the discovery of more principles and fundaments, and extension of current cognitive structures.

The global perspective of mechanisms (Glennan 2002) is only partly described by these two mentioned approaches for unification. Other approaches should be assumed to exist. In fact, mechanisms may be reexamined, refined, and replaced, but mechanisms will be a fundamental part of the model and integral to the framework. The same is to be said about the frames. The objective frame will follow the objective user behaviors, the product's functions, and the physical entities. The subjective frame will represent the cognitive part (however that is conceived or formulated), the procedures to carry out the cognition, and the thinking about the models or representations of the cognitions and procedures. The general nature of a project for people to build or change ideas into a product or service is the essential character, the essence of the objective causalities framework. For convenience, the framework is broken into two regimes, each representing a frame—one subjective, the other objective. Subjective and objective measures are discussed in Chapter 3.

Objective Frame

The objective causalities framework also translates into the natural regime (i.e., without human interventions) in the same manner, retaining its inviolate character. The phases of the lifecycle may change or be redefined or eliminated. However, the general nature of nature is to exist within the paradigms of low energy usage to achieve stability, and once past the threshold(s) of stability, transition through instability with catastrophic releases of energy. The complete transformation of energy to matter and matter to energy is representative of the observations and laws of nature that we are beginning to recognize in greater detail.

The objective frame of the framework of objective causalities is the final result of the work efforts managed under processes—the physical objects, the product or service functions, and the objective behaviors that were determined by the development team to result from the use of the product or service or in anticipation of the product or service.

The importance of the model and framework for objective causalities is in its general utility as a descriptive means to gauge the general nature of causality. As such, systems engineering can be replaced with any discipline or field of work—the framework be adapted and applied appropriately regardless of the type of product or service that is desired.

From a systems engineering perspective, the processes take a lifecycle perspective that incorporates: conceptualization, planning, design, development, testing, and delivery of the product or service, then continuing with acceptance, operations, support, maintenance, upgrades, and ending with disposal. From a systems integration perspective, the process takes on an event-driven perspective that endeavors to provide stability in operations for products and services.

Subjective Frame

Social processes form the subjective frame of the framework of objective causalities. The role of the social process for systems engineering is to describe the cognition and cognitive structures that deal with objects. In the case of people, the subjective frame deals with thoughts that guide people, inspire them, support them, provide needed resources, and encourage, cajole, or compel them. The management of the project for systems engineering is a social process (Yasui 2010). The set of factors that make up this social thinking process (subjective frame) are consistent with scientific sociological investigations that divide the factors into planning, procedure (mechanisms), and models of the plans and procedures (mechanisms). In this sense, procedures are termed as mechanisms. For clarity and to avoid confusion with dual use of the term mechanism, "mechanism" is used in the engineering and science sense to describe the inner workings of an object, while "procedures" are used instead of social mechanisms in describe processes.

For natural objects, the results of interaction may be sporadic interaction or aggregations of a temporary nature or integrations that are metastable or stable. A natural process has an input (EMMI) that responds according to a set of rules and is enacted through a set of procedures and activities which are exemplified by accurate representations of the enactment of the mechanistic rules and the mechanistic procedures and activities. The output of the process is the performance of the process, activities, or acts.

The result of a set of processes (referred to as the result of a procedure (i.e., social mechanism) or procedure) is called an event. One process could be referred to as an event, a set of processes can be referred to as an event (or set of events), or an abstraction of processes (or a single process) can be referred to as an event. An event implies the activation of a mechanism (i.e., physical mechanism) that is embodied within an object.

Summary of Objective Causalities

The frame of objective causalities is applicable to building cars (when engineers formulate plans, follow procedures, and model expected results), to delivering cars with requisite physical characteristics, required functions (with acceptable performance(s) and acceptable loss(es) to achieve those per-formances); predicated on their design, architecture, development, and inte-gration work to achieve the requisite desired behaviors that accompany the physical object and accommodate the other objects that are (or deemed to be) part of the race event. Beyond the objects of cars and people, the framework can be used to construct the objects that comprise the totality of the "race event." The "race event" could include the "crash" event, or not, depending on the objective of constructing the "race event." This grand ensemble would include objects of the race, formulated by planning and carrying out process(es) for procedures to manage the organizing of which cars compete in which race (assuming more than one race), where the cars will be positioned before and after the race, where the winner will be stationed or presented with the results of winning, and where the car will be serviced before leaving. Every aspect of the abstraction referred to as "the race" is definable with the framework of objective causalities. The bringing together of the "race event" constituent objects depends on definable interactions that are either observed or deemed reasonable given the before and after events. No doubt there will be emergent attributes that result from the "race event." As we will find in Chapter 3, emergence is commonplace. Temporary emergence(s) arises dur-ing the event, for example, a mechanic using a wrench tightens a lug nut, scrapes the fender, and rips a shirt sleeve (the rate of ripping a shirt is an attri-bute of the interaction between the shirt and the means of ripping; whereas the rip in a shirt is an attribute of the shirt). Temporary emergence can be sustained as long as the appropriate EMMI is available and the context and circumstances support the emergent attribute(s). This reversible circumstance is metastable, with the mechanic recovering fully, the car repainted, and the rip in the shirt sleeve mended. Unless there are irreversible actions (such as could result from a devastating car crash), the only interactions that occur are included in the "race event." The notion of time that in some manner links events can be investigated by dividing up "race event" into its causal trails, its proximate conditions, and its local circumstances.

If all interactions were stopped at the same instant, all objects would cease their behaviors, and all motion would cease simultaneously. In the next instant of infinitesimally small time duration, all interactions, behaviors, and motion would be allowed to continue for an infinitesimally small time duration, but then frozen again an instant later. Captured in the method is the time localization of an event, and at what spatial position (e.g., location) it occurred. Comparing that moment of occurrence with that of a measure-ment standard (such as time in seconds, or increments thereof), the event can be said to have occurred at a specific time. The onset of this infinitesi-mally small period of activity to the beginning of the next infinitesimally small period of activity is referred to as the duration of the event. Events concatenate and intertwine coextensively ad infinitum, one instant in a sequence trail of instances. Objective causalities are defined as the relation between two events—that one event uniquely determines the next event because the period of separation is infinitesimally small, that EMMI from one object was indeed the determinant factor for the activation of an object's mechanism, and that the output of an object is causally enabled by a single-input EMMI.* An event is theoretically the result of all objects, all interactions, and all EMMI in the universe, some interactions being directly involved at a specific instance. The expression of a single event is immediately cogged into a trail of chained events that proceeds to interact and perhaps integrate with other objects. Each object through its properties, traits, and, attributes can either retain its independence or develop various forms of dependencies in an integrated structure with various degree(s) of stability. Events concatenate and intertwine coextensively ad infinitum, as one instant in a sequence. Causality is the relation between two events—with one event uniquely determining the next event.

Objective causalities are exceedingly problematic to follow event by event. It is only after some measurable grouping of events has occurred that we may be capable of detecting and discerning some sibilance of causality (or at best correlation). The functions of objective causalities are those that occur between grossly detectable objects or by their influences of EMMI.

For human-made objects, the functions of objective causalities (1) relate the management or policies and rules of development with the sustainment of the objects; (2) support analyses of the events that result from the process domain; and (3) identify the considerations from which to discern the expected (or anticipated) behaviors of the users with those physical entities, functions, performances, and losses (Garud 1995). The framework of objective causalities relates the management of development and the sustainment of human-made objects (products and services) to the activities and events that must be completed. These activities (or project tasks in the case of systems engineering or systems integration) are defined in a 3 × 3 matrix that is the nexus of the cross-framed correlations between processes and the objects, that is, between the subjective and the objective frames, respectively. The objective causalities framework for integration is shown below. The 3 × 3 matrix that relates the subjective frame to the objective frame is laid out in blank rows and columns. For interaction and integration we refer to the framework as the integrative systems framework.† Figure 2.9 illustrates the flow of work through the systems framework that integrates the subjective and objective frames of information and data.

---

* Of course, this discussion opens up the issues of simultaneously received EMMI, an insufficiency of one EMMI that requires additional EMMI to activate the mechanism, other causal factors within the object or the environment, and so forth. For the purposes of this book, we ignore all other such effects. These other effects are certainly interesting to ponder and will be dealt with using the mathematical representations of mereology following (Stanford University 2007).

† Given that the operands for process (abstractions, procedures, and models) and the operands for objects (behaviors, functions, and physical entities) act in a 3 × 3 matrix, the integrative systems framework is also referred to as the "3 × 3 matrix."

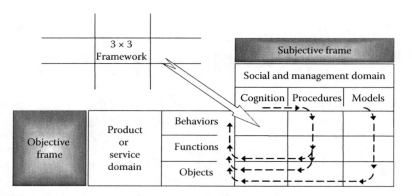

FIGURE 2.9
Integrative systems framework 3 × 3.

Figures 2.8 and 2.9 are similar. The explicit depiction of causalities is shown in Figure 2.8, while the procedural work-up based on those causalities is show in Figure 2.9.

This match-up of the subjective and objective frames occurs in the sequence outlined below, always beginning with cognitive structures, completing a nexus of one subjective–objective item, then moving on to the next nexus. Progressing through the 3 × 3 matrix begins with a work-up beginning with the subjective frame. Depending on your style, manner, and preferences for thinking about objects and processes, it may be more comfortable to begin with one or the other of the subjective or objective domains. Different people can also participate, proving their thinking styles better fit for one domain or the other. That said, it is advantageous to first focus on the social and management issues (cognition, procedures, models) if you think like a manager; on the product or service issues (e.g., objects) if you think like an engineer; on user behaviors if you think like a systems engineer; on functions if you think like a systems engineering integrator, on boundaries and user behaviors if you are an eclectic-systems integrator-thinker. Additional roles can be readily mapped to the 3 × 3 matrix. For convenience, a typical sequence is delineated beginning with cognition, progressing through each item in its domain that matches with the objective frame; then next to all items indicated for procedures; finally covering all items indicated for models.

- Cognition–objects; cognition–functions; cognition–behaviors
- Procedures–objects; procedures–functions; procedures–behaviors
- Models–objects; models–functions; models–behaviors

Detailing each of these conjunctions of domain items indicates the scope that is considered when mapping the frames to enlighten the integration effort. Beginning with the subjective frame each domain item is described.

Cognitive Domain

The cognitive domain involves the abstractions and reasoning that take place when thinking about a particular subject. All types and modes of thinking are involved with the cognitive domain, including conceptualization and interpretation. Another view of the integrative framework is shown in Figure 2.10. The specific topics covered for the nexus of processes (subjective) and product or service issues (objective) frames are outlined in the 3 × 3 matrix.

For the cognitive domain, the relations between concepts that are important to the user and reflective of the user's intentions should be represented in the product or service through the design of objects, the enactment(s) of functions that reflect the uses as well as the decisions that will be made with or as a consequence of the product or service.

The cognitive domain spans many disciplines and skills, including psychology, sociology, psychiatry, creative thinking, problem solving, management, and integration. The cognitive domain encapsulates the cognitive structures of questions, knowledge, information, and data. Beginning with the cognitive structures (the subject frame) places the context for objects. Context is most important for determining the utility of processes that build, integrate, or enable objects. The following listing spans many of cognitive structures for the

			Integration method		
			Processes		
			Abstractions (and reasoning)	Mechanisms, procedures, activities	Models, representations
O · B	P r o d	User behaviors (associated with or due to product*service)	Conceptualization pertinent to user behaviors due to product*service	Process and mechanisms describing user behaviors due to product*service	Models or representations of the user behaviors
J E	u c t *	Functions (associated with or because of objects that comprise product*service)	Conceptualization delineating uses provided by product*service	Process and mechanisms achieving complete portroyal of product*service functions	Models or representations showing all functions
C T S	S e r v i c e	Physical entities (associated with or because of objects that comprise product*service)	Identifying and interpreting the product*service physical artifacts, and ascribing meaning	Process and mechanisms resulting in the development of all physical elements	Models or representations of all physical elements

FIGURE 2.10
Integrative framework–nexus of processes and objects.

process of "to plan." Cognition is a vast field of study. When particularizing the 3×3 for a specific integration project, a complete process decomposition is necessary to identify the objective and subjective causalities that are relevant to the project. This decomposition should be completed before starting to outline the objective causalities within the framework of integration.

1. Cognitive structures (i.e., "to plan physical entities") take into account the physical entities that will be delivered. Typical questions include: How will the user use, support, and in general deal with the physical objects? What decisions will the user make because of the physical objects? Where will the user place the physical objects? How many users will be needed for the physical object(s)?

2. Cognitive structures (i.e., "to plan function") take into account the functions, functional performances, and losses that result from achieving those functional performances. How will the user enable and use the functions? Given the functional performance(s), what decisions will the user make? How will those decisions change the user's uses of the physical objects or functions? What other functions might the user require to carry out their work? What processes are necessary to assist the user to make full use of the delivered object(s).

3. Cognitive structure (i.e., "to plan behaviors") takes into account the behaviors of the users before the physical entities are used, during their use, and after their use. When the physical objects are not in use, what should the users anticipate?

4. Cognitive structures (i.e., "to organize physical entities") take into account how the physical entities will be organized for use; how they will be organized by functions(s); and how those organizations impact on how the users make decisions and behave organizationally before, during, and after use. How do users organize their work, with and without the objects?

5. Cognitive structures (i.e., "to direct physical entities") take into account how the objects are commanded or placed into operations; how the functions are sequenced for various types of operation; and how the users orchestrate their work—guiding their uses from one task to the next.

6. Cognitive structures (i.e., "to control function") account for the feedback required by the physical objects (e.g., user interface); the knowledge that a function is being performed properly; the confirmation (and adjustment, if necessary) to achieve the desired performance; the recognition and adjustment of the losses resulting from the system performances to satisfy a constraint or a limitation; and the changes required in behavior to complete the tasks.

7. Cognitive structures ("to communicate physical entities") account for the physical expressions through physical entities of communications;

the physical means of carrying out communications; enactments of the functions that facilitate communications; the thinking (pre-, during, and postcommunications) that is required to carry out, plan for, and enable communications.

Procedural Domain

1. Procedural structures (i.e., "define") take into account the wasted effort resulting from duplication of effort associated with overlapping roles and responsibilities; the sometimes contentious sparring between fiefdoms when resources are constrained, or power trumps corroboration. Alternatively, the spirit-increased productivity in a highly competitive environment may encourage duplication, inspire aggressive developments in strong ozone environments,* and foster good sense decisions based on strong questioning and demanding reasonable levels of proof for statements and work efforts.

2. Procedural structures (i.e., "accommodate change") take into account the facts of change and the rate of change. At stake is the reliability of the procedures to result in the desired outcomes. The credibility of the work is suspect when the credibility of the procedures does not show resilience to change. Often the rate of change is overwhelming if the premise of sound procedures is long-term stability.

3. Procedural structures (i.e., "architecture for power") take into account the need to assign roles and delegate responsibility and accountability as a means to achieve and sustain good governance.

4. Procedural structures (i.e., "paradoxy"†) take into account the stakeholder requirements for differentiating products and services from "other" products and services. Especially important for new product or service development is the demand to distinguish the product or service from competitive offerings. User behaviors often reflect the novelty in products and services in their feelings toward their work, their colleagues, and their involvement in teamness. Procedures that were developed in 1816, will necessarily have cultural, social, and stereotypical baggage reflecting a management style and manner. Procedures should reflect the enterprise metrics, the business ethics, and the project specifics.

---

* High energy dissociates air molecules leading to higher levels of ozone in the atmosphere. One who has high energy and does not hold back on sharing it is said to have "high ozone," or affectionately referred to as "ozone."
† Even if the product or service is new inside, the look and feel must also be "new." New has cachet, represents "cool," "state of the art," and "leading edge" (Luhmann 1995b).

Model and Representation Domain

1. Model representations (i.e., "show meaning") take into account the need to convey the concepts and procedures that were conjured and worked out to give to others. But until the ideas and activities are analyzed and evaluated, recorded and documented, and referenced and controlled, concepts remain concepts, and ideas for work remain just that, ideas.

2. Model representations (i.e., "support meaning") take into account the uncertainty that remains after the documents have been read and more questions remain. Were it possible to write all that is required for everyone to understand and acquire the information they need to accumulate appropriate knowledge, there would be no need for "support meaning." But only a fortunate, few people take away the meanings from only corporeal, material presentations. Representations of functions, or objects, of user behaviors, and of processes provide the paradigm for systems thinking, systems engineering, and systems integration. But the paradigm for products and services is different. Theirs is the interactions with all else in their operational environments; in other words all but the previous list. By example, representations of a function or object are the product or service. However, the product or service should not be thought of as its representations, rather it should be thought of in its environment and contexts as it interacts with all else. This is neither a mission nor application-centric view nor a data-centric view. It is not focused on its internal architecture or its user's intentions. "Support meaning" is the view of interoperability at the highest level of abstraction. Model representation "support meaning" is the understanding of operations from the viewpoint of "others." Model representation "support meaning" is the ultimate determinant for integrative success. Did the product or service do what it was supposed to do? If yes, success. Every stakeholder did the job intended—system designers adequately captured requirements; system architects adequately provided what was need when and where it was needed; developers overcame the adversities of project foibles and engineered reliable objects; system integrators produced value without pretense; systems engineers thought appropriately in systems; and users made the right decisions. Support meaning is the physical realizations on which success is built.

3. Model representations (i.e., "defend meaning") take into account a broad range of subjective issues that trump all things thought. Within the province of a project, people working with people often approach their use of time in different ways, ways that are sometimes counterproductive to efficient execution of work tasks. Having

communicated verbally (procedure) the results of a decision (abstraction and cognition), the workers are free to interpret, modify, and defend what they heard. Without the physical documentation, the decision can be changed by behaviors and circumstances. The number of workers, the number of requirements, the number of changes, and the number of decisions stack to beyond what anyone person can sort and make sense out of. Leaving a trail of information that models and represents what has happened and what is planned synchronize a lock-step team response.

Function

Objects have boundaries that extend beyond the physical entity. Objects that physically connect with other objects (i.e., touching) have a defined boundary at the exact location and instant of contact. That contact may be of some permanence, episodic, periodic, temporary, occasional, or one-time. Regardless of the temporal nature of the contact, the physical connection of two objects results in an interface. That interface is the boundary between the two objects. Specifically, the exchange of EMMI between two objects and satisfaction of the interface boundary conditions creates a function* that did not exist before the *connection*. Interfaces between physical objects act as mechanisms that transfer EMMI, that is, the influence of one object's

---

* Several definitions of functions and functional frameworks have been posed and described in the literature (Van Wie et al. 2005), but these have not been tailored explicitly to the issues of interaction and integration. A common theme in these definitions is to relate functions to behaviors and structure. Sorting through the differences in meaning for the same word across multiple disciplines is daunting, and trying to reconcile the multitude of definitions (each supported by varying degrees of scholarship) is overwhelming. Rather than approach the essence of integration in that manner of simply defining terms, a few observations guided by introspection were foundational ideas for many of the definitions in this book. Consider a wooden block (object) sliding on a hard surface (object). This scenario demonstrates two objects interacting to perform the function of "to move." From a discussion of this experiment, the nature of interaction can be exhibited in great detail. Taking care to reflect on the variables and their relations in this experiment, then applying the basic premises of systems begins to reveal the nature of integration. The word "function," for example, has multiple meanings spanning mathematics and sociology, while the word "system" is defined conveniently for biology, sociology, cybernetics, engineering, and systems engineering. In the case of the work "system," there is notable deference given to political motivations by systems engineers for defining "system" just beyond the evolving capability of the engineer. In the mid-1950s engineers recognized the need for systems engineering and differentiated systems engineering thinking as broader than the focused disciplines of engineering. Over time these engineering domains broadened to ply the practices in a broader (more system's fashion). Similarly with systems engineering, the evolving marketplace has extended systems engineering from its beginnings to systems, system of systems, and systems of systems. It is a bold ascertain that the differences between engineering and systems engineering could be interpreted as having an inkling of political interpretation (Hall and Fagen 1956; de Souza 2008).

structure on another object's structure acts as a surrogate mechanism. Therefore, the interface exposes a mechanism on the common boundary of the two objects. That surrogate mechanism vanishes when the objects lose their connection. Since mechanisms provide a means of exploiting the transfer of EMMI between two objects, there is a potential advantage from the perspective of another object that can in some way manipulate that "exposed" mechanism. That advantage reveals itself an as additional capability not seen within the internal structure of an object. For example, a user (object) may pick up a hammer (object). That the hammer rests on a workbench offers the user no advantage in driving a nail into a piece of wood. The hammer is potentially available for use. That the user comes in contact with the hammer (through the activity, "pick-up hammer") takes advantage of the interface that joins the user's hand and the hammer. The process of the "pick-up hammer" is only achievable if the user's hand (or logical extensions by some other means) comes in contact with the hammer. The capability that is not available to either of the two objects is enacted at the moment of making a connection. It is the connectivity of two objects through EMMI that results in a new capability. The capability exists because of the interaction between the hammer and the user, while the capability is said to be enabled by the act of connecting the hammer and the user. That there is a connection reduces the degrees of freedom of both objects. Reducing a degree of freedom is termed as a limitation, if the reduction in the degrees of freedom is out of the control of the receiving object. When an object is *used* (exchanging EMMI with another object), both objects have a constraint imposed on them that was not operable (in existence) before the interface was established. The limitations are different for each object. From the perspective of the send object, its internal allocation of EMMI (e.g., output versus losses) is deemed a constraint. That constraint results in output EMMI that is received by another object. The receive object is limited by the input EMMI and in turn allocates the EMMI according to its internal constraints. The primary difference between a limitation and a constraint is the perspective from which the EMMI is either input (limitation) or is allocated (constraint). The constraints are different for each object. When an object is constrained, a function results. When there are multiple constraints, there are multiple functions.

From a functional perspective, a hammer and a piece of wood are dramatically similar. The difference between a solid metal hammer and a piece of wood is related to the differences in the internal mechanism(s) of the two objects. These differences result from their individual properties and traits. For example, one object's internal mechanism absorbs incident radiation (EMMI) and converts it to heat, while another internal mechanism reradiates that converted heat to beyond the physical boundary of that object. The other object turns out to have very low thermal conductivity and reradiates very little. Other differences in properties for the two objects, for example, are the mechanisms of absorbing mechanical shock and the temperatures for melting and evaporation. The hammer and the piece of wood are similar in a functional manner, but their differences in properties result in different performances for those functions. As objects, the piece of wood and hammer have many common uses for humans (i.e., the wood and hammer present nearly the same functions during interactions with people). If the shapes of the hammer and piece of wood were similar, then the wood object could be used to drive nails into wood. The density and hardness of the hammer would seem to be greater than that of the wood, so driving nails would dent the wood, whereas the hammer might not experience denting. The function of 'to hammer' is carried out by a person with either the wood or the hammer, resulting in the nail being driven into a wallboard. The performance of the function 'to hammer' is different for the hammer and wood.

From a user's perspective, the object's performances (due to properties, traits, configuration, shape, size, etc.) distinguish the hammer and the piece of wood. Many users recognize this distinction and will employ a 2 × 4* to drive a nail, remove another piece of wood that is hinged by a nail, pry open a door that is jammed, or prop open a box lid. Both the hammer and the piece of wood offer mechanical properties supportive of these and many other such functions. The enactment of an object's function at the interface between one object and another object signifies the functional boundary of two objects. However, unlike the physical boundary which is the limit of the interaction of an object's matter, the functional boundary begins at the physical boundary and extends by the object's EMMI to the furthest extent of the influence of its EMMI. Integration is object to object, with emphasis on demonstration of functions.

From the perspective of a user, functions are expressed in part through their use(s) and their performance(s). Uses are often thought of as performances, so uses and performances are dependent factors. Generally, there is a mix of functional requirements and performance requirements that drive a project's development and integration processes. To achieve functional performance(s), the sending object's mechanism results in loss of EMMI and the receiving object's mechanism experiences a loss. Should that loss exceed the output performance, the object's function is said to "have degraded," "does not function," or "does not have the function." If the function's performance ceases to be produced by either the sending or receiving objects' mechanism, the function can also be said to not exist or to be degraded such that the function is unavailable.

From an integration perspective, objects are connected to demonstrate functions. Testing for functions is usually accomplished by measuring the performance(s) that can be achieved by the connection between two objects.

Should an automobile not move because of not having a source of energy from gasoline, diesel, electrically charged battery, compressed air,

---

* A "2 × 4" is an unspecified length of lumber that is nominally 4 cm thick by 9 cm length, having been reduced in size from a historical sizing of 2 inches by 4 inches.

hydrogen, or by being pushed by another object, for example, the function's performance of the 'to move' function is zero. If the source of power is operative and has sufficient EMMI for the automobile's mechanism to function as designed and fulfill its tasks, then the object's performance is said to be measurable. Should there be a nominal value, m, that is expected as the measure of a function's performance (under various conditions and circumstances) the performance is said to have a *target* performance value. If there is a variation in performance that is characterized as distributed over a small range of performances centered on the value for the *target* performance, then the *target* performance is regulated or controlled by a mechanism. By pressing down on the accelerator in the automobile, the performance output of the car's movement can be controlled according to the constraints imposed by the design of the automobile and the allocations that result in the various constraints. The variance in the output of the mechanism(s) (termed as the variability in performance) for a given input (in this case from the action of the driver (user) of the automobile driving the automobile) can be considered to be the quality of the function "to drive." A function is describable in terms of its performances and its variability in performance (quality) (Taguchi 1986). The functional boundaries of an object are measurable by its performance(s) and its losses that result in achieving those performances.

In addition to an object being recognizable by its physical boundaries and its functional boundaries, the behavioral boundary of an object is determinable by either the behaviors exhibited by the receiving object due to the influences of receiving EMMI or in anticipation of the receiving EMMI. The behavioral boundary of the sending object results from its influences due to either its physical boundary or its functional boundary. This is to say that the sending object has responsibility for its action(s) and that determining the results of physical and functional boundaries is an important activity for integration. Knowing the boundaries of the objects to be integrated helps delineate the sequencing of objects for demonstration of functions. Recognizing that some objects require multiple interfaces to demonstrate a top-level function demands great care in building up all the subfunctions during the integration process.

No single object results in a function. A single object has no physical use without some sort of connection (based on Type 1 or Type 2 interactions) with another object. In general, that connection is through EMMI. The connection can be made physically with matter or remotely with energy, material wealth, or information. Each physical boundary is porous to energy, material wealth, and information. The boundary is determined by various connections between objects. The boundary conditions at an interface are describable in terms of coupling, cohesion, abstraction, and granularity (size of the increments of EMMI) of the EMMI that is transferred.

An automobile has the function of "travel" when the wheels are in contact with road and they have sufficient traction with the surface of a road to the car to move without the interface between the wheels and the road.

The term "function" includes what something is used for (i.e., user intent), what it could be used for (i.e., design intent), in addition to what it was accidentally used for (Ariew 2002) (i.e., opportunistic intent). Function can be understood as belonging to one or more of these categories. Function is defined relative to a particular stakeholder's perspective. For example, from the point of view of a developer, the function of a subsystem A might be the interface between two subsystems B and C. But from the point of view of a user who wants to encourage the adoption of a new interface protocol, subsystem A may be their only means to guarantee payment of license fees for use of their protocol. Both functions (interfacing subsystems B and C and enforcing payment for using protocol) are ascribed to the same set of objects and are adequately defined, although they may differ in performance and quality based on the stakeholder's perspective. The set of objects for one function should be considered from the various stakeholder perspectives and represented as dependent representations of overlapping functions mapped to the same set of objects.

A system function may have any number of performance parameters and likewise several quality requirements associated with each performance. Functions can be added, deleted, or changed. The output EMMI from a single object that interacts with a second object to produce another output, the totality of outputs is called performance. Specifically, this performance is the performance of the function. As a means of accountability, performance measures aid in understanding how well functions achieve their goals. Performance measures allow for comparative ratings from which to determine similarities between functions.

A function occurs because of the interaction between two objects, commensurably at the physical interface between two objects. The interaction of EMMI between two objects enables an output (performance) that is dependent on the mechanisms of two objects acting across their common interface. If provision is made (designed and architected into the product or service) then a user may access that interface and be enabled to "use" the function, but not the interface. It is important to realize that while being respectful to the prevalent emphasis on interface specifications, the key issues for integration reside with the enactment of functions (their connectivity, coupling, and cohesion) and not solely with the transfer of data between objects. Functions can be compared by their combined performances of the two objects. For example, two cars racing on a dirt tract have various measures one of which is determinable by the winner of the race, that one crosses the finish line before the other (assuming both cars started at just after the signal to begin the race). The race event involves two cars, each performing their individual functions of 'to race.' The objects of the two cars and the starting line from the start of the race; and the function of 'to race' for each car is formed by a car's interaction with the dirt tract. The performance of that interaction helps determine the change in speed of each car.

The performance of the process of racing is comparable to other processes. The performance of the process of one car racing is to either win or lose. The performance of the function of one car is in their instantaneous speed or its acceleration. In this example of cars racing, the aggregate human-construed object of "the two-car race" has a lifecycle. By chance if the two cars were to collide with each other, they would interact through EMMI. And further, if by chance the two cars ejected their drivers unharmed, then burst into flames and burned as an aggregated mass, parts of two cars might disintegrate, or fuse, or retain an individual similitude to their precrash structures. Again, this event referred to as "crash" can be described as an interaction, or should the resultant object take on properties, traits, and attributes different than those of the precombined objects, then the "crash" can be described as an integration.

Quality

That functions may have a measurable performance and a loss attributable to achieving that performance is inherent in the structure of objects. Quality can be associated with the loss (Taguchi 1986, 1990; Taguchi et al. 2005). In this manner, quality refers to the consistency of performance, or alternatively, the deviation from a target value (i.e., the performance requirement). Quality indicates how well a function is accomplished by the system and is a measure of the loss due to the performance of that function. Quality can be represented as a loss function (Taguchi 1986; Taguchi et al. 2005) (see Chapter 6 and Appendix 2). The greater the loss resulting from the deviation from a performance that is nominally the best, the poorer the quality. Overall, the quality imputed to a set of objects characterizes the stability of the performance(s) and the function(s) ascribed to that set of objects. The implications of poor stability relates to (1) nondelivery of the set of object's functionality, (2) delivery of the set of object's functionality (within the range of performance tolerance), or (3) delivery of performance beyond the range of specified performance tolerances. Functions describe the intentions of the design. For functions, integration is the relationship between the mechanistic intentions expressed through the design and the performance of objects through their EMMI.

The losses to achieve various performances are of two types—controllable losses and uncontrollable losses. Controllable losses can be measured and fairly determined to be associated with specific events. The losses due to variation in performance are controllable to the point that variations due to stochastic noise are controllable. Losses due to conversion efficiencies for EMMI are controllable by the mechanisms of transformation of input EMMI into output EMMI. If the mechanism for transforming EMMI is of a certain type that results in a conversion efficiency (input/output) of say 80%, then there is a 20% loss that does not translate into output performance. That 20%

may be reduced by changing the mechanism of conversion. For example, changing from incandescent light bulbs to florescent lighting improves the conversion efficiency of electrical energy to lumens outputted. The result is a 13-W florescent bulb that has the equivalent steady-state output of a 60-W incandescent light bulb. There are slight differences in the "color" wavelength of the emitted photons and the lifetimes of the photon generation mechanisms are also different.

Take for example, cooking a package of dried noodles. Undercooking the noodles by 90% of the suggested cooking time of 4 min results in 24 s of heating water with only a negligible amount of heat transfer to soften (i.e., "cook") the noodles. These barely warm, wetted noodles are unlikely to be eaten. The result is that such undercooked noodles will be thrown away with a loss of the purchase price, water for cooking, 24 s of energy consumption, the energy expended by the person(s) cooking to prepare the noodles for cooking, placing water in a pan, cooking, removing, and tasting the "cooked" noodles. These tasks can be translated into costs and attributed to undercooking noodles. In contrast, cook the same noodles for 24 min (700%) longer than the suggested cooking time. The noodles are hydrolyzed, substantially devoid of flavor, without physical substance (as desired from noodles cooked according to suggested cooking instructions). The losses are considered in the same manner as with the undercooking. In both cooking sequences, the "cooked" products were discarded (full loss of purchase prices). The result is more spent (i.e., more loss) from cooking the overcooked noodles than with that of the undercooked noodles. The loses due to variation about a performance (due to regulation of enactment(s) of mechanism(s)) results in an inefficiency in achieving a desired output for losses due to the mechanism transforms are the total losses of the functional mechanism.

References

Aerts, D., Broekaert, J., and Gabora, L. (Eds) 2003. A case for applying an abstracted quantum formalism to cognition. *Mind in Interaction*. Amsterdam: John Benjamins.

Ariew, A., Cummins, R., and Perlman, M. (Eds) 2002. *Functions: New Essays in the Philosophy of Psychology and Biology*. Oxford: Oxford University Press.

Astley, W. G. and Brahm, R. 1989. *Organizational Designs for Post-Industrial Strategies: The Role of Interorganizational Collaboration*. Greenwich: JAI Press.

Bahill, T. A. and Briggs, C. 2001. The systems engineering started in the middle process: A consensus of systems engineers and project managers. *Systems Engineering* 4(2): 156–166.

Bainswanger, H. 1990. *The Biological Basis of Teleological Concepts*. Irvine, California: Ayn Rand Institute Press.

Bausch, K. C. 1997. The Habermas/Luhmann debate and subsequent habermasian perspectives on systems theory. *Systems Research Behavior Science* 14: 315–330.

Bell, J. S. 1965. On the Einstein-Podolsky-Rosen paradox. *Physics* **1**: 195–200.

Blanchard, B. S. and Fabrycky, W. J. 2011. *Systems Engineering and Analysis*. Upper Saddle River: Prentice Hall.

Boulding, K. 1956. General systems theory—The skeleton of science. *Management Science* **April**: 661–671.

Buede, D. M. 2009. *The Engineering Design of Systems: Models and Methods*. Hoboken: John Wiley & Sons, Inc.

Burgin, M. 2003. Information: Problems, paradoxes, and solutions. *Triple C (Cognition, Communication, Co-operation)* **1**(1): 53–70.

Clark, K. and Fujimoto, T. 1990. The power of product integrity. *Harvard Business Review* **68**(6): 107–118.

Cover, T. M. and Thomas, J. A. 1991. *Elements of Information Theory*. New York: John Wiley & Sons.

Cropley, D. H. 1998. Measurement, theory, and information. *Information and Control* **41**: 275–304.

Darcy, D. P., Kremerer, C. F., Slaughter, S. A., and Tomayko, J. E. 2005. The structural complexity of software: An experimental test. *IEEE Transaction of Software Engineering* **31**(11): 982–996.

De Marco, E. P. 1960. *Engineering Economy*. New York: The Macmillan Company.

de Souza, R. A. 2008. Maturity curve of systems engineering. *Systems Engineering*. MS thesis, Monterey: The Naval Postgraduate School, 114pp.

Ferris, T. L. 1997. Foundation for medical diagnosis and measurement. *School of Physics and Electronic Systems Engineering*. PhD thesis, University of South Australia, 350pp.

Feyerabend, P. 1993. *Against Method*, Third edition. New York: Verso.

Francois, C. 1999. Systemics and cybernetics in a historical perspective. *Systems Research and Behavioral Science* **16**: 203–219.

Friedman, A. H. 2010. Improving enterprise decision-making: The benefits of metric commonality. *Aeronautics and Astronautics*. MS thesis, Boston: Massachusetts Institute of Technology, 102pp.

Garud, R. and Kumaraswamy, A. 1995. Technological and organizational designs for realizing economies of substitution. *Strategic Management Journal* **16**: 93–110.

Glennan, S. S. 2002. Rethinking mechanistic explanation. *Philosophy of Science* **69**(September): S343–S353.

Gross, N. 2009. A pragmatist theory of social mechanisms. *American Sociological Review* **74**: 358–379.

Hall, A. D. and Fagen, R. E. 1956. Definition of system. *General Systems* **1**: 18–28.

Hedstrom, P. and Swedberg, R. 1998. *Social Mechanism: An Analytical Approach to Social Theory*. Cambridge: Cambridge University Press.

Holland, J. H. 1962. Outline for a logical theory of adaptive systems. *Journal of the Association for Computing Machinery* **9**(3): 279–314.

Holland, J. H., Holyoak, K. J., Nisbett, R. E., and Thagard, P. R. 1986. *Induction: Processes of Inference, Learning, and Discovery*. Cambridge, MA: The MIT Press.

Hoppe, H. H. 2004. Property, causality, and liability. *The Quarterly Journal of Austrian Economics* **7**(4): 87–95.

Hutter, M. 2007. Universal algorithmic intelligence: A mathematical top-down approach. Manno, Switzerland, Dalle Molle Institute for Artificial Intelligence (IDSIA), Galleria 2, CH-6928 Manno-Lugano, 70pp.

Kirk, G. S., Raven, J. E., and Schofield, M. 2009. *The Prescocratic Philosophers.* Cambridge: Cambridge University Press.

Kitcher, P. 1989. *Scientific Explanation: Minnesota Studies in the Philosophy of Science.* Minneapolis: University of Minnesota Press.

Klepper, S. 1996. Entry, exit, growth, and innovation over the product life cycle. *The American Economic Review* **86**(3): 562–583.

Kocsis, J. G. 2008. *Determining Success for the Naval Systems Engineering Resource Center.* Department of Systems Engineering. MS thesis, Monterey, CA: United States Naval Postgraduate School, 101pp.

Koestler, A. 1968. *The Ghost in the Machine.* New York: Macmillan.

Kotler, P. and Keller, L. K. 2007. *A Framework for Marketing Management.* New Jersey: Prentice Hall.

Kuhn, A. 1974. *The Logic of Social Systems.* San Francisco: Jossey-Bass.

Kunovich, R. M. 2009. The sources and consequences of national identification. *American Sociological Review* **74**: 573–593.

Lakatos, I. and Feyerabend, P. 1978. For and against method: Including Lakatos's lectures on scientific method and Lakatos-Feyerabend correspondence. Chicago: University of Chicago Press.

Lake, J. 2007. *Textbook of Integrative Mental Health Care.* New York: Thieme Medical Publishers, Inc.

Langford, G. O. 1971. *Experimentally Obtained Metastable Atom Excitation Functions for Helium, Methane, and Ammonia.* MS thesis, Department of Physics. Hayward, CA: California State College.

Luhmann, N. 1982. *The Differentiation of Society.* New York: Columbia University Press.

Luhmann, N. 1995a. *Social Systems.* Stanford, CA: Stanford University Press.

Luhmann, N. 1995b. The paradoxy of observing systems. *Cultural Critique* **Fall**: 37–55.

Mayer, R. J., Crump IV, J. W., Fernandes, R., Keen, A., and Painter, M. K. 1995. Toward a business constraint discovery method (IDEF9), p. 11. *Information Integration for Concurrent Engineering (IICE) Compendium of Methods Report.* College Station, Texas, Knowledge Based Systems, Inc. and the United States Air Force Materiel Command, Wright-Patterson Air Force Base, 149pp.

Mill, J. S. 1882. *A System of Logic, Ratiocinative and Inductive, Being a Connected View of the Principles of Evidence, and the Methods of Scientific Investigation.* New York: Harpter & Brothers.

Newell, A. and Simon, H. A. 1972. *Human Problem Solving.* Englewood Cliffs, NJ: Prentice Hall.

Nieminen, A. 2005. *Towards a European Society: Integration and Regulation of Capitalism.* Helsinki: University of Helsinki, 465pp.

Peirce, C. S. 1934. *Collected Papers: Volume V. Pragmatism and Pragmaticism.* Cambridge, MA: Harvard University Press.

Przeworski, A. and Teune, H. 1970. *Logic of Comparative Social Inquiry.* Malabar, FL: John Wiley & Sons.

Purao, S. and Vaishnavi, V. 2003. Product metrics for object-oriented systems. *ACM Computing Surveys* **35**(2): 191–221.

Reed, I. 2008. Justifying sociological knowledge: From realism to interpretation. *Sociological Theory* **26**(2): 101–129.

Rueschemeyer, D. 2009. *Usable Theory: Analytic Tools for Social and Political Research.* Princeton: Princeton University Press.

Sage, A. P. and Armstrong, J. E., 2000. *Introduction to Systems Engineering.* New York: John Wiley & Sons, Inc.

Salmon, W. 1984. *Scientific Explanation and the Causal Structure of the World.* Princeton: Princeton University Press.

Schiller, C. 2009. *The Adventure of Physics—Vol. VI: A Speculation on Unification.* Motion Mountain. ISBN: 978-300-021946-7. Retrieved July 29, 2011, from creativecommons.org/licenses/by-nc-nd/3.0/de

Schlager, E. and Blomquist, W. 1999. A comparison of three emerging theories of the policy process. *Political Research Quarterly* **49**: 651–672.

Senge, P. M. 2006. *The Fifth Discipline: The Art & Practice of the Learning Organization.* New York: Doubleday Currency.

Shannon, C. (1948a). A mathematical theory of communications. *The Bell System Technical Journal* **27**: 379–423.

Shannon, C. (1948b). A mathematical theory of communication. *The Bell System Technical Journal* **27**: 623–656.

Shannon, C. F. and Weaver, W. 1963. *The Mathematical Theory of Communication.* Chicago: University of Illinois Press.

Shaw, M. 1990. Prospects for a discipline of software. *IEEE Software* **7**(6): 15–24.

Smith, B. 1994. *Fiat Objects.* ECA194 Workshop, Amsterdam: ECCAI.

Smith, B. 1995. Formal ontology, common sense, and cognitive science. *International Journal of Human Computer Studies* **43**: 641–667.

Smith, E. D. and Bahill, A. T. 2010. Attribute substitution in systems engineering. *Systems Engineering.* **13**(2): 130–148.

Stanford University 2007. *Stanford Encyclopedia of Philosophy.* Stanislaw Leśniewski. Published online, http://plato.stanford.edu/entries/lesniewski/. Retrieved August 13, 2011.

Sztompka, P. 1999. *Trust: A Sociological Theory.* Cambridge: Cambridge University Press.

Taguchi, G. 1986. *Introduction to Quality Engineering: Designing Quality into Products and Processes.* Tokyo, Japan: Asian Productivity Organization.

Taguchi, G. 1990. *Introduction to Quality Engineering.* Tokyo, Japan: Asian Productivity Organization.

Taguchi, G., Chowdhury, S., and Wu, Y. 2005. *Taguchi's Quality Engineering Handbook.* Hoboken, NJ: John Wiley & Sons, Inc.

Thurstone, L. L. 1946. Comment. *American Journal of Sociology* **52**: 39–50.

Torres, P. J. 2008. *A Modified Conception of Mechanisms.* Heidelberg: Springer Science + Business Media B.V.

Trockel, W. 1999. *Integrating the Nash Program into Mechanism Theory.* Los Angeles: Department of Economics, University of California and Bielefeld University, 13pp.

Turner, J. R. 1993. *The Handbook of Project-Based Management.* London: McGraw-Hill.

Van Wie, M., Bryant, C. R., Bohm, M. R., McAdams, D. A., and Stone, R. B. 2005. A model of function-based representations. *Artificial Intelligence for Engineering Design, Analysis and Manufacturing* **19**(2): 89–111.

Varzi, A. C. 1997. Boundaries, continuity, and contact. *Nous* **31**: 26–58.

Williamson, J. 2009. *Probabilistic Theories of Causality.* Oxford: Oxford University.

Yasui, T. 2010. A new systems engineering approach for a socio-critical system: A case study of claims-payment failures of Japan's insurance industry. *Systems Engineering* **14**(4): 349–363.

3

Foundations in Systems Integration

Introduction

One could ask why the concept of integration has not been considered a subject requiring thorough examination of joining parts to make a whole, or involving profound contemplation of the whole as constituent parts. The process of joining things together in some manner that involves interfaces and transfers of various types is perhaps the most widely talked about, least understood subject in any discipline. Individuals not versed in systems engineering often seem unable to appreciate the befuddlements indicative of performing systems integration, or more insidiously, system of systems integration. Systems engineers and systems integration engineers often find integration vexing and problematic, but forge on with the work because it appears tractable (especially given enough time and money). In general, well-intentioned people know what they want to integrate and why. If the technology is mature and in some form of use, then integration may seem possible (buying into the platitudes that off-the-shelf works so why not use it). A few of the torments of integration center on dealing with unreliably performing technology that has been designed and architected into objects; having rushed objects from iterative thinking that underpins development into the processes of integration which benefit more from recursive thinking; reducing budgets for integration tasks; and shortening schedules to delivery. The usual desire for integration is for interoperability of objects and processes to achieve some effect in their intended operational environment.

Indeed, the typical guidance for integration planning offers such platitudes as "human factors considerations must be included in every solution" and "do no harm." The nuts and bolts of planning are in defining a sequence of activities that will bring together the objects into objects, and then into being as the system (object). Given the uncertainties of developing new objects that will be integrated into a system, planning for integration would seem to be problematic at the outset. Typically, the inputs to the integration plan are the system design and architecture products, the objects that will be developed, schedules and resources for the work tasks, and the plans for various types and levels of testing. Systems integration planning is usually completed early in the lifetime of the project to develop a new product or service.* Integration should not be relegated to that effort which results in a *whole* by following some set of best practices.† The principles discussed in Chapter 1 offer specific guidance on how to better perceive integration and therefore how to apply appropriate practices. Best practices for one type of work may not work as best practices for a different kind of work. Systems engineering is quite different from systems integration. The planning is different, the structures are different, and the thinking is different. Systems engineers (as others) do systems integration planning and object integration itself. For example, integration of uncannily interactive objects is not amenable to cook-book implementation. The best chefs improve their recipes each time they prepare a dish of food, sometimes trying new ingredients or increasing or decreasing their amounts, and revising the cooking times. Perfection is reached when the recipe in the chef's head (intellectual object) is written down (physical object) and is shown to be scalable from small to large portions, when the customer feedback is strongest, and when the processes and procedures are time-efficient and cost-effective. Once the recipe is worked through, tested, mapped and synchronized with kitchen processes, and integrated with the procedures, skills, and habits of the kitchen staff, then (and only then) will the recipe be considered a success. Unlike developing a new product or service that is (by definition) unlike the previous project, the project team must instead rely on principles that will be applied to the specific circumstances of the new project. For integration of a new product or service, waiting for the component objects to be developed likely leads to missed opportunities to demonstrate early many low-level subfunctionalities. The essential issue is recognizing when an object is ready for integration at various levels of functionality. The common, but erroneous, perception is to "perform integration when the hardware and software components are developed and delivered by the development team."‡ Citing principles from Chapter 1, Principle 5: The Principle of Forethought suggests developing a plan that includes early identification and testing of object subfunctionalities, that is, integration begins as soon as two objects can be integrated. Implementation of activities supporting Principle 5 can be derived from Principle 2: The Principle of Partitioning. Identifying the partitionable subfunctions early lowers the risk of development that results by waiting until a subsystem or equivalent object has completed development and unit testing.

---

* The emphasis in this book is on developing new products and services. A few notes have been added to deal with the differences for upgrading or integrating an existing product or service with another product or service. These notes are by no means comprehensive or complete. They only serve to point out a few differences with that of new product development.
† For systems engineering, a best practice is iterative development and improvement. For systems engineering integration, a best practice is successive approximation based on recursive thinking.
‡ Care has been taken to not cite references for such faux pas statements.

The subtleties of integration are wrapped in a puzzling ontology that has as yet defied simple, sustained scrutiny. This book suggests that integration is more than bringing together different parts (objects) in a manner to be merely combined or summed. Fundamentally, integration is *not* the act of combining two or more objects into a whole. Nor is the discipline of integration a process that orchestrates acts of aggregation and combination. Rather, integration is a method of dealing with the logic and rigor necessary to know what and when to bring parts together, parts that might already be within wholes and parts that are barely parts (speaking to definitize objects in this case). Integration is nominally part and wholes, but the wholes are different than the individual parts. Further, the boundaries of the objects suggest what should be partitioned and integrated, rather than some other delineation or mapping onto design or architecture of the future product or service. The difference of thinking in integration* results in a system.

General Systems Thinking

Systems integration builds on an assemblage of principles that are relevant for general systems theory (Boulding 1956; von Bertalanffy 1968; Miller 1978; Jain 1981; Rapoport 1986; Klir 2001), pertinent to systems engineering, as well as the practices of every discipline and field—all work. Without a broad appreciation of general systems theory, which abstracts domain-specific knowledge to a set of high-level criteria that reflect discipline knowledge, the attempts to perform integration (of thoughts, or data, for example) from other disciplines may be done in the absence of proper foundation and context. The difference between data and information is context. The difference between information and an integrated form that relates the totality of experience subjective and objective experiences in a cogent, relevant, and useable form is knowledge. Consider human systems integration and the question of potential or actual losses incurred at the system level for two design strategies: just-in-time versus on-demand information. At issue is the timeliness and quality of flow of information, that is, the losses incurred due to poor integration and ineffective use of humans in a computer-enabled environment. The work of systems theory explicates the relations, dimensionalities of boundaries, interfaces to objects and processes, forms of interaction (EMMI), and the functions and behaviors that determine and evoke action. This fabric of concepts weaves general systems theory. An all-inclusive fabric is needed—a fabric characterized by consistency of logic, continuity of method, applicability across disciplines, scalability from the interdomain's micro- to macrostructures, and capaciousness. Most importantly, the optimal fabric needs to focus on the eventual goal of describing a definitive theory of systems integration. Systems integration contributes to the development of general systems theory through universal principles and laws

---

* Specifically in terms of partitioning objects according to their three types of boundaries.

that transcend discrete partitions of thinking, for example, disciplines and fields. For the purposes of thinking in systems, partitioning must be considered with great care. Boundary confounds, partitions constrain, and inferences drawn need to be generalizable. Together, thinking in systems and thinking in integration empower systems engineers and systems integrators to consider the problem space as a whole and therefore the possible solution sets in the context of lifecycle issues or discrete events. The fabric of general systems theory rightly originates from this merger of thinking in systems and thinking in integration. To suggest a fabric that improves descriptive and empirical results and accounts for unobserved phenomenon, a set of best thinking* for systems theory and the guiding principles from systems integration need to be blended. Together, systems thinking and systems integration form this theoretic fabric. The balance between theory and principle can be maintained by committing to two conditions: For every theoretical construct there shall be a corresponding principle that typifies the duality of their applications, and for every principle there shall be a corresponding theory that embodies its use and relates to the fabric through its context.

A consensus on general system theory (von Bertalanffy 1928, 1968), its integration into systems engineering (Boldyreff 1954), the development of cybernetics (Wiener 1948; Ashby 1957), the dynamic behaviors of complex systems (Forrester 1958), the relevance of chaos theory (Lorenz 1963), the maturation of sociology (Buckley 1967), the considerations of living systems (Miller 1978), the structures of information systems (Lewis 1994), and the acquisition, building, and integrating of complex systems (e.g., military customers (Schilling 2005)), focus on spatial and temporal conformities, forces, mechanisms, control, and hierarchical levels of entangled interactions.

The argument generally posed in support of general systems theory focuses attention on spatial and temporal conformity, forces, mechanisms, control, and hierarchical levels related through complex relationships. That focus presumes that spatial and temporal constructs determine the perspectives of contexts, properties, and states. Rather, we determine to focus on events rather than the typical spatial and temporal constructs; we develop a mereology of objects and processes. Therefore, we revise the traditional thinking that systems are referred to as hierarchical or multilevel (Simon 1962, 1973) and characterize these appearances as hypothetical rather than derived from a natural orderliness or empirical data set. For example, hierarchy can be represented in linear Hilbert space (Gabora 2002) where states are defined as mathematical objects reflecting the properties of measurements. These measurements are grounded in a definable physical reality. Combinations of number, type, and state(s) of elements also include forces. Forces can be further differentiated into mechanisms and controls. Therefore, number, type, state(s), and forces are the essential components of complex systems. In this way, Gabora avoids the need to define explicitly the term

---

* Rules of thumb, rules of dumb.

"complexity" which has evaded attempts to operationalize—in formal mathematical terms. Further, systems can be constructed through their recursive (Schiemenz 2002) and recombinant conceptions, the natural result could be the appearance of hierarchy and can be modeled as such. While some notion of hierarchy is relevant and perhaps inevitable, we have concern about commensurability of levels across fields. Cross-field abstractions are different and identifying corresponding levels of abstractions is both contrived and problematic. Further, systems theory was predicated on inductive rationales and the observations of nonlinear actions (von Bertalanffy 1968). The supporting structure for inductive thinking applied conforms to the adaptations and conveniences of the authors (Holland 1986). We take away the generalized form that includes eight characteristics of theory and principle for systems integration:

- Knowledge can be represented by rules of condition to achieve action (initial, operating, and output conditions are made up of force and mechanism).
- Rules are based on current and future events (rather than states), suggesting that causality is related to multiple kinds of inputs and partitioning of objects and processes.
- Rules regarding automata are similarly defined and enacted in lower-level and higher-level structures.
- Superordinate relations value the metasystems with more structure than those objectively associated with lower levels.
- Synchronic and diachronic rules promote superordinate relations.
- Multiple sources of weak interactions as initiators result in measurable outputs, the result of which is integration.
- Two classes of mechanism are possible—those that revise parameters (for objects) and those that generate plausibility useful rules (for processes).
- Mechanisms require initiators that fall within bounds of the required thresholds to result in measurable outputs (and losses).

We build on these characteristics to develop the framework for objective causalities and to investigate the workings of interaction and integration. We promote inductive thinking to explore the depth and breadth of knowledge associated with actions and events (regardless of discipline), the Gestalt principles of similarity, and the notional construct of models of how interaction and integration show their effects. The one and the many, that is, the considerations of integration, are describable from quantum mechanics (Aerts 1983); to cosmogony, referencing Parmenides (Fairbanks 1898); and cosmology (envisioning grand organizations) (Senge 2006). That description is systemness.

Determining Systemness

Consider placing two pieces of wood together, one on top of the other, so that their individual intrinsic properties and traits resist movement through the forces of friction and gravity. Each piece of wood is a part (object). Stacking the wood makes two parts (object), yet these two parts do not make a whole that is different than the individual pieces of wood. The wood parts interact as a result of their physical contact, but they are not integrated. Each piece of wood retains its individuality. Now, nail these two pieces of wood together— the three parts (two pieces of wood and a nail) combined to make a whole. The interactions are between the wood (part 1) and the wood (part 2), the wood (part 1) and the nail (one end), and the wood (part 2) and the nail (the other end). Additionally, one end of the nail interacts with both pieces of wood (during the "drive the nail" procedure). But the wood is still the wood and the nail is still the nail. Neither the wood parts nor the nail have changed because of the combining; they are still individually the same pieces of wood and the nail. Yes, there is now a hole in both pieces of wood, but the pieces retain their substantial identities and characteristics. Were the nail to be removed from the pieces of wood, the hole would probably remain depending on the age of the wood, the moisture content within the fibers of the wood, and the size of the nail (i.e., the hole). Interaction—yes, but no integration has occurred. The wood–wood–nail aggregation is not a system.

Combining the two pieces of wood with a nail provides some additional benefits (and different functionality) over that of a nail, or one piece of wood, or the placement of one piece of wood on another piece of wood, or the placement of a nail on top of the two stacked pieces of wood, or the placement of two stacked pieces of wood on top of a nail. If the wood were nailed so that one plane face of the wood was along the x-axis and the other plane face of wood was along the y-axis, then the wood–wood–nail combination would form a right angle. Figure 3.1 depicts the configuration of wood–wood–nail.

A piece of wood (object $e_{1.1.2}$) is joined to another piece of wood (object $e_{1.1.3}$) by a nail (object $e_{1.1.1}$). This new object (object e_2) can be used, for example, as a bookend to provide lateral support for edge-standing books or as a bracket to hold a shelf.

Stability

In some sense, stability can be thought of as a change in a "state" which has certain properties that are desirable to maintain. We can think of "trigger events" that change the stability and therefore the "state"; the desirable state is lost. In this case, stability is defined as the ability to apply restoring forces to mitigate events that trigger changes in the status quo. One example of a stable system that can turn unstable is a supply chain. In the world of providing goods to people, there are oftentimes a number of organizational entities involved in moving goods that fall betwixt the completion of a product or

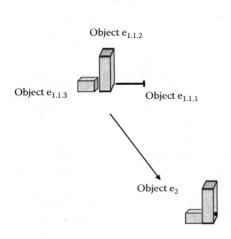

Object $e_{1.1.1}$ interacts with object $e_{1.1.2}$
Object $e_{1.1.2}$ interacts with object $e_{1.1.1}$
Object $e_{1.1.2}$ interacts with object $e_{1.1.3}$
Object $e_{1.1.3}$ interacts with object $e_{1.1.2}$
Object $e_{1.1.1}$ interacts with object $e_{1.1.3}$
Object $e_{1.1.3}$ interacts with object $e_{1.1.1}$
Object $e_{1.1.2}$ interacts with object $e_{1.1.1}$

FIGURE 3.1
Objects interacting.

service and operational employment by the users. Moreover, products and services are sometimes moved from one organizational entity to another as the products and services make their way to customers and users. Supply chains can be architected to be customer-centric, being driven by the buyer's timeline for purchasing their desired quantities or amounts. Other supply chains are architected to be supplier-centric, instead accounting for the specific needs of the suppliers in preference to the buyers. Supply chains are appropriately viewed as systems (Haskins 2007). Products* and services (objects and EMMI) *flow* from one organizational entity to another (objects). Information (EMMI) regarding what is needed to satisfy perceived supply or demand signals from suppliers and customers are referenced to what objects are available in the *chain of flows* between organizational entities. Money (EMMI) also *flows* between objects. From the perspective of interaction and integration, supply chains are representable as objects, EMMI, and processes (Kleindorfer and Van Wassenhove 2004). The stability of the supply chain (which is normally thought of as a single sequencing of events) is dependent on both the supply process and the demand process. All objects in the chain act as supply and demand operators. In other words, thinking of supply chains from only the perspective of the buyer or seller of a product or service can lead to instabilities in movement of objects, information that is delayed or distorted, and ineffectual satisfaction of the needs of buyers or sellers. When an instability in supply or demand is propagated through the supply chain, wildly unstable operations ("the bullwhip effect") are observed. The reliance on delayed or distorted information in a chain of decisions can lead to an increase in the volatility of a supply-driven system (Sousa 2004). The

---

* Sometimes referred to as "boxes."

volatility increases geometrically as the information ripples through the system, each object attempting to over- or undercompensate for the abnormal input. Return to stability can require a "reset" of the inputs or a change in the controls that govern each mechanism in the chain of events.*

From an integration perspective, the stability of the supply chain is describable in terms of structural asymmetric flows of objects and EMMI. Satisfying supply for highly varying demands for products or services requires the supply system to be broadly scalable in several dimensions—in essence a network supply chain having high levels of capacity achieved in a single chain or in a web of chains. Systems engineers might refer to such a web as having back-up or redundant traits. Systems integrators might think in terms of redundant functions. Stability in a system is achievable and sustainable in multiple ways.

For a system, stability is essential to maintain its system-like characteristics. From this example of "the bullwhip effect" an errant input causes a chain of events that can ripple, interaction by interaction, leading to unstable operations or configurations of objects. Considering both the example of a supply chain and the example of the bookend, object e_2 in Figure 2.1, stability is an important facet of both a system and groupings of objects. In the case of the supply chain, instability might be an over- or undersupply or demand; therefore, the supply chain can be thought of as always being stable in its flows (to a first order), but instable in its supply or demand. As such, the stability of a supply chain is dictated by the people who drive the supply and who make demands for products or services. Extrapolating these causes for instability for the supplier or the customer or user, an effective means of stabilizing supply and demand to meet highly variable needs is to develop a network of suppliers that supply through a network of chains to respond effectively to highly variable demand. Effectiveness is determined by the quality of the delivery of the product or service, that is, the deviation from the desired quantity and timeliness of delivery. In the case of the bookend, instability might be a loosely connected set of wood blocks, a nail that detaches easily, or wood that splits apart due to cracks generated from driving the nail. The implications of these examples are that often just grouping objects is sometimes as desirable as systems.† Therefore, stability in the bookend is achieved by reducing the variation in physical connections from that of "firmly" connected.

Not all groupings or integrations increase value or utility from the perspective of the buyer or seller. Likewise, integration is not necessarily better than aggregations. Integration is more than achieving certain effects from a system or system of systems through it operations. Integration requires appropriate

---

* Interactions can be thought of as resulting in a single, one-time event, as a sequence of interactions, each precipitated singularly by the previous interaction (chain of events) or multiple interactions, each precipitating multiple interactions (cascade of events).
† While there are many aspects of systems that are desirable, simple aggregations or conglomerations of objects are equally important and should be justifiably as important as systems.

interaction (as in the example of the supply chain) to establish and maintain stability. And, stability for a grouping depends on the uses of the final product or service.

Metastability

For a system to exist and sustain itself as a system, it requires a semblance of metastability or stability to continue as a system. While stability is achievable through interaction, system properties and traits are only sustainable as an integrated system.* We define metastability of a system based on the naturally occurring metastabilities in galactic nebulae (Langford 1971). Metastability is the intrinsic property of a group of objects that persists in an apparent equilibrium of interactions between objects where only a small disturbance in the established interaction can dramatically change (reduce or increase) the system's lifetime. Such an aggregation that is sometimes not-a-system and at other times a system seems *precariously* close to existing or not existing as a system. We referred to such an entity as a ProtaSystem—one that is transitional between acting as an aggregate of objects and a system of objects. Changes to a ProtaSystem could result in complete loss of some measure of presumed stability or, alternatively, through a progression of changes a substantial increase in the group's lifetime (and might now be referred to as a system if system-like properties were sustained). ProtaSystems are metastable, while Systems and NotaSystems are stable. Figure 3.2 depicts these differences.

Even with stability, systems have varying lifetimes, so ProtaSystems are characteristically sufficiently unusual that their anomalous behaviors tend to take considerable investigation to ply a measure of appreciation for the cause(s) and effects of the transitional characteristics. The importance of metastable systems in the natural world cannot be understated. These transitional ProtaSystems are incubators for stars and planets and breeding grounds for sea and land species. An example of a metastable system is galactic nebulae where stars and planets are birthed. Interactions between atoms and molecules are sufficiently long so as to produce spectral lines that are not observed in the Earth's atmosphere (and therefore perplexing for many years). In the case of a metastable system with an unequivocal loss in stability or dramatic improvement in stability, the objects show changes that are observable in their properties as well as in their traits.

---

* The statement that system properties and traits are only sustainable as an integrated system would seem to be a tautology. The degree to which the system properties and traits are stable depends on both the spatial extent of those properties and traits and the degree of stability. Some systems are quite stable (i.e., have very long lifetimes, but are metastable for long periods). For example, animals and plants living in tide pools are not always under water and not always out of water. Yet, tidal pools are teaming with life that would seem to cling precariously to a very small area of existence.

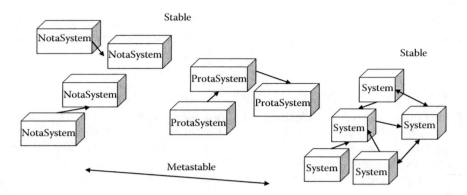

FIGURE 3.2
Metastability and stability.

Instability

Instability is not the opposite of stability. Instability results in loss of functionality or performance. The consequences of instability are generally correlated with loss of value. Aggregations of objects generally have only a few EMMI events with determinable losses, should an object fail to perform as expected. A system comprised of objects has a multiplicity of functions with an intricate set of variables that determine performances. There is a higher level of dependencies built into systems rather than with an aggregation of objects. Losses are higher when systems cease to be systems. Our dependencies on systems are higher than on aggregations of objects. We depend on the Sun–Earth* system for life. Whether objects are grouped and actively sending EMMI as aggregates (NotaSystem), protasystemic aggregates (MetaSystem), or systemic aggregates (System), stability is determined by their ability to provide steady, reliable, and durable functions and performances. From an interaction and integration perspective, stability is only achieved and measurable by functions, their performances, and related quality.

Integration Perspective

Integration may seem explicitly fundamental to all structures, but it is more appropriate to say that it is interaction that is explicitly fundamental to all structures, with integration being different than interaction. Systems

---

* Probably should include the Moon, since tidal and weather effects are strong determinants for moving and mixing the atmosphere, providing the dynamics for orbital stability, and facilitating crustal movements that cause magma to flow, soil to rejuvenate, plant life to flourish over eons.

do not exist everywhere and do not exist in stable configuration at all times. Therefore, integration neither occurs everywhere nor at all times. Why and where does integration occur? Why and when does integration not occur? While it may seem intuitive that interaction is inherent throughout the lifetime of a system (implying stability), it is more accurate to consider the lifetime as a continuum of integration. Rather than thinking of lifecycle as conception to disposal, it is more appropriate from a product or service's perspective to think of various stages of interaction and integration. The lifecycle paradigm is from the perspective of the customer, developer, and user. The integration perspective is from the perspective of the product or service. Systems are opportunistic, by their ability to adapt to changes in EMMI.

The system is the result of integration, or conversely, integration is the result of achieving system behaviors. Lifecycle can be thought of as merely the temporal interpretation of integration, whereas the events of integration are interpretable in terms of lifecycle stages. Posing lifecycle as the primary orchestration and organization of systems engineering in itself is not new. And in the recent past, several have suggested an integration process model (Project Management Institute 2000; Booher 2003; Jain et al. 2010; Tvaryanas 2010) suggesting that there is a distinguishable difference between lifecycle process models and integration process models. The paper by Rashmi Jain specifically proposes a systems integration process model (Jain et al. 2010) that is based on a lifecycle view of systems integration. The systems engineering process models do in fact regiment the portal to the fundamentals of integration, but without sufficiency in detail and interactiveness with other processes and with a dearth of discussion concerning objects and processes.* Without integration, no process would integrate with another process, no object would combine with objects to make a system, and no product or service functions would be group-governing and reflective of the whole. Every process, every function, and even the physical space that encompass the domain of interest exist only because of interaction and integration, and both necessarily for a system. Integration requires interaction, but interaction

---

* I confess I am unable to appreciate the writings of many authors who profess to discuss integration. To be clear, their reasonings seem to apply the word integration aptly as if I already knew what it meant and how to do it. Were it merely to bring two parts together to form a whole, then the difficulty of integration would be rather uninteresting and perhaps easy. Yet, I have found integration to be extremely interesting, but most certainly not easy. Integration seems very fundamental, perhaps at the level of existence. Integration silently poses as an invisible cement that permeates, binds, and forms all that we are and much of what we do. Humans are systems that live and work in systems. We also build and work with objects that are not systems. What is a system and what is not a system? Systems interact with nonsystems and systems. It is difficult not to interact with either. Somehow the prevailing views of integration are logically premised or defined as simply putting parts together to form a whole—or at least simple stated by many accounts. Integration will only be simple when we know how to recognize it, how to describe it, and how to do it.

does not imply integration. Lifecycle is merely the temporal interpretation of integration, not the impetus.

Rather than thinking of integration in a lifecycle sense or in a process sense (as is typical through the use of systems engineering process models), this presentation focuses on the progression and results of interaction and integration for a set of processes and objects (the mereology of integration). Thinking in systems engineering terms for integration means addressing and answering three questions: Is the concept solution effective in solving the impetus problem such that the needs of the stakeholders are satisfied? Is the design and architecture effective in enabling the appropriate functions with their requisite performances and quality to implement the proto-solutions? Is the system of objects and EMMI realized in such a manner as to verify the effectiveness of the concept through testing? Figure 3.3 depicts a view of a systems engineering for integration.

A high-level summary of the systems engineering process model (i.e., development and integration) is mapped to an integration systems model which depicts the process model view in terms of an integration systems model. The three questions are laid out as factors and issues that conspire to show concept effectiveness, proto effectiveness, and system effectiveness. For each stage in the integration systems model, the risk is reduced due to integration as a means to answer the three questions.

There is a noticeable change in the type of thinking required to answer the questions as the product or service progresses through development and into integration. Concept effectiveness is strongly interactive with proto effectiveness through iterations. The result of that interactive (iterative) relation is

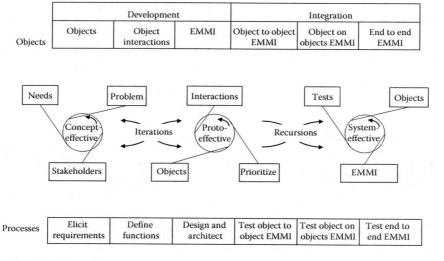

FIGURE 3.3
Integration systems model.

an integration of concepts and ideas that signify whether the planned integration approach will be effective. Effectiveness is determinable by both the subjective measures of the team's work (Giachetti and Rojas 2007) and the objective measures of the results of the team's work (Rahming 2009). The transition from proto effectiveness to systems effectiveness is indicated by interaction that favors forward-looking coordination to demonstrate functionality as a consequence of integration over iterations and rework to correct problems with objects and their EMMI.

Beginning with needs, problems, and stakeholders, a set of requirements is laid down, revised, and agreement reached by the key stakeholders. Based on this initial set of requirements, the project establishes direction, pacing, skill sets, resource requirements, project metrics, and work packages. A concept of operations is developed, which, when combined with the creativeness of a designer and architect, results in a set of specifications. These specifications are coordinated through tasking for the development team to make a product or service (build and test, and integrate and test). Objects and interactions with EMMI are defined to conceptualize a solution that is consistent with the budget, schedule, and available resources, and that when implemented will be both sufficient and necessary for solving the defined problem. Objects and their interactions with EMMI are delineated, organized, and prioritized to lay out the blueprints for the task of building and testing, and integrating and testing. When finally combined, the objects and interactions produce the set of EMMI that meets the performance and quality requirements for the resultant product and service. The transformation of a proto-effective set of objects and their EMMI into a system of objects and system EMMI marks the essence of an integrated product or service.

Essence of Integration

Integration can be described as occurring naturally through mechanisms that work according to various processes to transform subatomic particles into atoms, atoms into molecules, molecules into structures, structures (e.g., stars and planets, products and services) into superstructures (e.g., galaxies and clusters of galaxies).* If we think of integration as transforming parts into a whole (as is commonly viewed), integration has been variously defined for enterprise applications (Gold-Bernstein and Marca 1998); local applications, data, functionality, and processes (Dueker and Vrana 1995; Giachetti 2004);

---

* Subatomic particles appear to be systems, as do atoms, molecules, living things, stars and planets, and some products and services.

organization (Ashby 1962; Galbraith 1977); and connections of elementary-level subcomponents to higher-level assemblies, subsystem, or system (Ducamp and Lagarrigue 2007). Systems engineers view integration as that means of building a system (INCOSE 2008), rather than thinking of integration as the consequences of what happens when individual objects are put together. A systems-level view of integration emphasizes end-to-end functionality enforced through system design and architecture. For human-built systems, a function can generally be thought of as how objects are used to accomplish something, that is, a capability of a system. For naturally occurring systems, a function can be generally thought of as how objects interact with other objects to work within their mechanisms to gain or sustain stability, that is, showing some degree of preference for low-energy expenditures (Katifori et al. 2010). A function is the essence of interaction between two objects, and for integration, a function is a structural property of the relations between objects. The essence of integration is providing functions that are unachievable by any individual object; performances that surpass the totality of individual functions; and quality that engenders stability for availability of those functions, and stability in the performance of those functions.

A natural system is often thought of as having capacity—capacity to withstand environmental "hardships." However, a predominant theme for systems engineers who build and "integrate" systems is to consider integration in terms of recursive or progressive approaches while working with data and interfaces (Buede 2009). Typically, one scheme or a combination of several schemes is imposed to carry out the mechanics of software integration through data and interfaces. These schemes include top-down, bottom-up, thread, mixed, and various combinations of top-down and bottom-up. The *International Council on Systems Engineering (INCOSE) Systems Engineering Handbook* (INCOSE 2010) recommends a bottom-up approach to integration, beginning with the atomic-level* data items (Broersen 2003) that enable the most detailed functionality (the most detailed functionality in the system hierarchy) to build from the individual objects into subsystems, with the final output enabling top-level system functionalities. This bottom-up view of integration deals with the specifics of each object through interfaces, data items, and timing—the details of ensuring the fundamental connectivity and flow of EMMI between objects.

There is no correct way to integrate objects. There are, however, ways more correct than others, depending on the circumstances surrounding development and integration. An example of a propitious means of integration builds processes based on the principles developed in Chapter 1. Consider a project that is behind schedule and over budget, with objects still in development, and no integration work yet started. Let us assume that we have a strong alignment of strategies (Principle 1: The Principle of Alignment), and further let us assume we have followed good practices in planning (Principle

---

* The smallest discernable level of operation.

6: The Principle of Planning). Yet, even with the key stakeholders in violent agreement with regard to strategies and good recognition of patterns for planning and scheduling, the project development work lags. If the key stakeholders agree on strategy, then the problems in development may be due to requirements to engineer with less mature, less reliable technology in order to achieve advancements over existing systems. A very bad decision would be to push unreliable technology in engineered objects into an integration environment. A very good decision would be to insert a suitable surrogate technology into the engineered objects to serve as a "placeholder" during integration. Since only a portion of the engineering work is predicated on less reliable technology, and a primary *objective of integration* is to provide for a stable system (builds on Principle 2: The Principle of Partitioning), integration can proceed along with development. The key to integrating with surrogates is modularity in functionality (Stone 1997). Industry recognizes the need for modular functionality, but resists making the investments required to realize modularity. Modularity of functionality means that all perceptions, uses, interactions, and embodiments of functionality are equivalent. The difference is the physical delivery of the function. In other words, the function is provide, but the performance(s) and qualities of those performances are not equivalent to the same function provided by different physical objects. The form (physical representation) and the fit (dimensionalities) may be different. The performance(s) may be different. And the deviations from the performance target(s) may be different, but the function is present and reliable. As form, fit, and function are generally described in the product baseline, the process model would need to be amended to allow for functional equivalency through modularity. For existing products and services, the modularity of form, fit, and function is a preeminent expectation and requirement (Herald et al. 2007). Thinking in integration is more about making the product or service a reality. Thinking in systems engineering is a focus on making components comply with specifications that reflect requirements, that is, what product or service should do. Should-do thinking is systems engineering development, whereas will-do thinking is systems integration.

The emphasis on functionality (as defined in this book) for integration (as defined in this book) is different than how the term modularity is used in other contexts; in fact there is a wide disparity in definitions (Gershenson et al. 1999). A typical view of modularity expresses the dissimilar nature and independences of module components from other aggregations of components, along with simplicity as a hallmark of improved lifecycle issues (Stryker and Jacques 2009).

Whether the objects are elementary particles or artifacts within subsystems, integration is a deterministic factor in achieving wholeness (that of systems wholeness). Integration can be observed and appreciated through human endeavors by design, architecting, and implementation. As a natural phenomena, objects act through their energy and matter mechanism, and follow similar logic that has been applied to products and services. By this commensurability of logic, examples, and structures, the intent is to lay out a general view of integration—one that transcends the particulars of any one discipline, trade, or practice. Systems integration applies Principle 3: The Principle of Induction to develop recursive thinking that presupposes the patterns that must be followed in integrating a product or service given circumstances.

Purpose of Systems Integration

As the primary objective of integration is to provide for a stable system, the purpose of systems integration is to make provisions for achieving system performances that meet the requirements of the system. By all intentions, integration realizes the functions that offer performance and quality increases over that of human-centric operations.

Automation

As is often the case, automation replaces people. The corollary is that automation can save money sans people-intensive operations. Historically, labor costs increase at a faster rate than lifecycle costs associated with highly reliable machines employing mature, proven technology. Institutionalized technology produces fewer wasted resources than directly attributable to humans through human activities. Evidence of automation "improvements" creep into our daily lives. Automated teller machines have eliminated bank walk-in deposits and nominal money withdrawals. Check-out from stores (e.g., grocery, hardware, and apparel) are piloting and implementing self-check equipment and procedures. Systems integration applies technology to bring together systems that offer advantages over individual actions using NotaSystem objects.

Technology

Technology* is the scientific, mechanical, electronic, or chemical means of improving people's performances or by providing or enhancing their indigenous functions. These improvements provide for (1) making better decisions, (2) doing more work faster, and (3) doing work that could not be accomplished

---

* Institutionalized technology is defined as technology that is congruously integrated into an enterprise, organization, or operational setting.

before by any one individual. As such, engineering is an enabler to bring technology to people. And systems engineering facilitates lifecycle thinking to not only make improvements but also to achieve improvements taking into account lifecycle issues, reducing impacts on stakeholders (including the environment), and mitigating unintended consequences due to the building, operations, or disposal of a product or service. The systems engineer fights with technology and knows its cost; the project manager knows the value of technology and is fearful of its nature; and the systems integrator knows the price of technology and provides value in products and services.

Improvements

Many of the improvements due to integration may not be readily apparent at first glance. Even people working on a project assume many integrations of technology, often taking for granted the "invisible" work needed to integrate. Accounting of integration work is mixed with development and testing activities. Obtaining a clear perspective of what integration is and is not is blurred from this perspective, as is the general feeling for how much cost is included in integration efforts. Broadly defining integration as plans, methods, and tools (Hassellbring and Reichert 2004), integration covers a great percentage of work done on a project. Therefore, improvements due to integration occur naturally throughout the spectrum of project work. Narrowly defining integration as only that period designated for integration of the system and not for individual system objects is commonplace and widely accepted for products developed for the U.S. Department of Defense (Haskins 2007). With this thinking about integration (as a stage in which a product or service goes through), improvements due to integration are quantifiable in terms of time and money. But broadly speaking, integration is the removal of all impediments that inhibit system stability. Improvements due to integration then occur as a consequence of method and processes (Giachetti 2004).

Automations and improvements due to integration materialize after developing the inherent properties or traits of an object. These properties and traits are especially noticeable when they are singled out as essential (either by requirements or by recognition of significance) during an attempt at integration with an existing object or system. For example, the properties of the Hubble Telescope mirror were acknowledged early when developing the key requirements for low weight and good thermal properties. When integrating a new technology that calls for changing test procedures, problems can arise when old ideas and plans for testing are assumed to be applicable to new technologies. Integration looks at processes as well as objects to achieve requisite performances. Particularly careful and considered analysis is required when upgrading or changing an existing product or service. Upgrades and changes deserve specific discussions as a great percentage of systems engineering work is focused on improving, sustaining, or extending the life of systems already in operations.* These improvements increase the effectiveness and efficiency of integration, provide for more suitable capability for the users, improve the performances of the overall system and particularly for certain functions, and help the users make better decisions because of the delivered service or product.

Tasks of Systems Integration

The major (top-level) tasks of systems integration are (1) characterizing and providing the requisite objects in the form of products or services, and (2) defining the interactions and integrations that are conducive to providing operational effectiveness (e.g., process integration, functional integration, physical integration, and behavioral integration). These tasks are carried out in a step-wise fashion.

Systems integration combines an engineering approach to building a proto-product or proto-service from disparate objects with a systems engineering approach to thinking in systems about the product or service. There are 12 basic steps for carrying out integration. These steps are listed as sequential, but in fact they required both iterative and recursive ordering and thinking.

The 12 steps of integration for new product development cover the general planning and scheduling of events as well as the specifics of the events (including objects, interactions between objects, and processes).

1. Integration planning
   a. Identify the key events that demonstrate subsystem and system functionalities.
   b. Coordinate the team's work and activity flows to accommodate the key events.

---

* The topic of upgrades and integration is beyond the scope of this book. However, a brief discussion is warranted to point out the key issues from an integrator's perspective. Integration for upgrades or changes is neither bottom-up nor top-down, but rather at the level of interactions between objects. The information needed resides in design and architecture of the existing system, the operational concept, the key requirements, and the expected results from the upgrade or change. An analysis of the physical, functional, and behavioral baseline (existing product or service) is completed to identify and evaluate the impacts of an upgrade of change in terms of the systems operations, the user interactions, the user environment (including procedures, social, and economic issues). Scenarios should be developed to posit the range of actions possible, the conditions under which those actions become causal to systems operations, and when the futurity of events warrant a further investigation and evaluation. The investigation would include the kinds, types, and frequencies of decisions that are made due to the upgraded or changed system, or due to the anticipation of the upgraded or changed system.

    c. Establish the sequencing of events that need to occur to demonstrate the bottom-up development of subfunctions.

    d. Determine the degree of work concurrency that can be supported by the project resources and satisfy the needs for development of the various subfunctions.

2. Integration scheduling

    a. Schedule the dates of the events identified in Step 1.

    b. Plan for and schedule the durations of both the integration work and the events.

3. Identifying two objects that are to be integrated (object A) and (object B)

    a. Begin by first specifying the lowest level of integration for a defined object A, where object A will become a minimum component necessary to demonstrate an identified subfunction (multiple subfunctions may be possible with the defined object A, suggesting that the integration team must decide on the integration method and approach to demonstrate single or multiple subfunctions).

    b. Based on the subfunction associated with defined object A, define an object B that will become the complementary object that when combined with object A will provide for the demonstration and testing of the identified subfunction. Depending on the decision made regarding object A and single or multiple subfunctions, object B will follow with that decision as its guidance for single or multiple subfunctions. However, only object A and object B should be considered as the objects for that decisions, meaning no additional objects should be allowed to enter into the subfunction demonstration. Should the level of detail of the defined object A (or object B) offer more than one "opportunity" for integration, the level of detail is too high and a lower level of detail needs to be defined and used as object A (or object B).*

---

* For upgrading or changing existing products or services, the level of detail for object A is already specified and built, existing as the "as-is" determinant of object A or object B. In the case of an upgrade or change to an existing product or service, the object added to the existing product or service (object B) should be integrated one subfunction at a time. Whether the integration should be as a completed object B with all subfunctions included depends on the method and approach for integration testing and the desires of the stakeholders of the existing product or service (object A). Should the operations of object A be critical, real-time, involve issues of safety, then a not-in-service (but fully operational) product or service should be used for integration. Preferences of the integration team and users (of object A) should determine the approach for integration. However, if object A cannot be taken out of service or a not-in-service object A is unavailable, then the subfunction should be integrated one at a time.

4. Determining the boundaries of object A
   a. Physical
   b. Functional
   c. Behavioral

5. Determining the boundaries of object B
   a. Physical
   b. Functional
   c. Behavioral

6. Specifying the interactions that characterize the relation(s) between object A and object B
   a. Boundary conditions
   b. EMMI

7. Describing the interaction(s) between object A and object B
   a. Inputs
   b. Outputs
      i. Performance(s)
      ii. Loss(es)

8. Describing the characteristics of the interaction(s) that serve as a function for the user (verification)
   a. Conditions for the function (test planning)
   b. Boundaries to enable the function (test planning)
   c. Limits of the function (test planning)

9. Attempting to demonstrate the function
   a. Test setup
   b. Measurement approach and meaning

10. Evaluating the conditions for the functions* of
    a. Reliability
    b. Availability
    c. Vulnerability
    d. Susceptibility

11. Analyzing the boundaries that enable the function

---

* For the purposes of this book, we define requirements only in terms of functional requirements. In other words, there is no accommodation for "nonfunctional requirements." As such, reliability, availability, vulnerability, susceptibility (and the like genre of "ilities") are deemed as functional requirements with target performance(s) and quality requirements representing deviations from those requirements.

12. Determining the limits of the function (validation)
   a. Scenarios and vignettes*
   b. Environmental limitations†
   c. Use limitations‡

Defining Terms

The style of presentation in this book is to develop a reasonably consistent set of defined terminologies that will clarify the concepts and usage of terms related to integration work. This lexical exercise (the semantic circumambiency) does not attempt to develop nor cannot potentially achieve a complete ontology for use with integration, systems engineering, or systems engineering management (Quine 1960). Appropriately then, the author takes a first step and makes an exiguous effort to formulate a set of quasi-defined terminology. Words have meaning. While that meaning is often ambiguous in both written and verbal usage, it is important to agree on consistent usage of terminology in order to express the nuances of meaning uncovered by research.

The terms defined in this book by no means stand up to rigorous standards demanded by scholarship. They only purport to elucidate sufficiently enough to assist the reader in differentiating between the customarily used terms in a manner that is meaningful for the purposes of integration. For example, it is quite common in systems engineering to refer to functional decomposition and then proceed to decompose processes (as if functions were the same as processes); lump verification and validation together (as if

---

* Scenarios are the futurity of possible outcomes of event; vignettes are more detailed sequences of events that highlight particulars about a scenario (a possible set of circumstances, conditions, and constraints, i.e., the environment of the future).
† Scenarios limit the concept of operations and the architecture to the environment of the future. Vignettes work within the structured future environment to determine how a system's architecture will respond to various interactions with objects through EMMI. Scenarios and vignettes are useful tools to discover missing or influential stakeholders, uncover possible sequences of events that are particularly interesting (causal), and to develop strategies to explore consequences and uncertainties that follow from the consequences of a decision or actions taken.
‡ Validation of current uses of a product or service involves determining what the product or service is capable of doing. Validation of future uses of a product or service (aside from being problematic because of the unknowns associated with the scenarios and vignettes) requires people other than today's users to posit those future uses. Today's users are steeped in today's product and service, and oftentimes find difficulties extracting themselves from today's thinking. A group of potential users of a future product or service should be presented with a paucity of information, only appropriately and sufficiently linked to the general notion of today's product or service so as not be encumbered thinking about tomorrow's product or service in its future environment.

they were not only similar, but cognitively the same methods enacted by the same people at the same time*; or view requirements analyses and specifications as interchangeable). They are not interchangeable. The general use of some terms in this book, such as methodology, method, processes, and mechanisms, is detailed where they are explicitly used in the context of integration.

Then, briefly, methodology indicates a set of methods that are consistent with a theoretical foundation that is based on principles and philosophical assumptions. Method connotes a systematic, step-wise logical way of carrying out processes. Approach are the specific steps, that is, the way. Processes are a set of goal-oriented activities carried out according to an intended theme or approach. Processes are driven by cognitive activities that result in procedures (referred to as social mechanisms in sociology) whose output are models (or representations) of what was thought of as modified by what was done to make happen what was envisioned. As regards social mechanisms, Neil Gross (University of British Columbia) provides a review of social mechanisms as sometimes unobservable structures or processes, sometimes observable and thereby with apparent motive(s), as low-level building blocks, as causal, as means to transform (events) by mechanisms, as an intermediary process, and as chains of actors dealing with chains of problems—the processes by which cause and effect are founded (Gross 2009). Procedures are carried out through activities. Activities are sets of behaviors expressed in an orderly array of acts (many of which might have some relevance to the activity). Acts relate to activities (through an approach) that in turn relate to a process, consistent with a method that might be part of a methodology that is harmonious with a theory. The causality of an interaction is also forced through the lens of the method of discovery and the theory of its interpretation.†

General Ontology and Mereology of Integration

The kind of universe that can support the existence and observation of physical objects allows for the creation of an ontology of "things."‡ Things can be restrictive or inclusive based on philosophy and principles. This kind of

---

* Clear distinction is made in most presentations (books, papers, and talks); however, the conceptualization of "V&V" remains prevalent.
† There is much to say about causality in regard to integration. Determined by objects (their mechanisms) and EMMI, integration presumes causal relations between objects and processes. That our knowledge of systems and engineering is most likely flawed in many respects, our presumption for being able to integrate objects, let alone predict what might happen when we attempt to integrate objects is particularly impudent. Yet, that is the very essence of advancing our knowledge. I say, forgive the brazen ones, their mistakes are tomorrow's advancements. In the words of a major Russian poet, F. Tyutchev, "An idea once expressed is a lie" (Turchin 1977).
‡ "Things" are defined as contrivances that justify cognitive structures for thinking.

universe is not "aware" of these contrivances, nor is this kind of universe cognizant of rules governing such contrivances. Our worldviews afford us a great deal of freedom, while this kind of universe most likely has much less. By delimiting properties of EMMI, we proffer definitions and explanations; we "justify" theories and pretend laws of nature; and we add to wisdoms and inklings of "great" thinkers. Were the reasonings of the "great" thinkers by some measure "correct?" If so, how would we know? The ontology for integration posited by the author presumes the kind of universe where objects (both physical and intellectual) and their EMMI are secondary to space–time. Space–time does seem to prefigure kinds of relations that are possible. This kind of universe might support* the ontology and mereology posed in this book.

The way in which integration is viewed determines what should be considered as the variables of integration and how those variables should be grouped. The totality of these variables and their grouping(s) is the ontology of integration (these theoretical worldviews—classical, relational, and mechanistic—are described in Section "Nature of Physical Objects"). The author suggests that there are at least three theories of integration that are consistent with thoughtful reasoning about the lifecycle of products or services, in which integration has a relation to lifecycle stages. The difference between the three theories can be summarized by the following example. Consider an object that is moving and emitting (or conveying) EMMI. Movement is relative to a reference frame and EMMI is conditioned on the environment, the inputs and outputs, and the mechanism of the object. Further, movement and EMMI may be related (as is the case for decelerating charged particles, i.e., electrons or protons, emitting x-rays, termed bremsstrahlung†). Therefore, the object's movement may be related in some ways to the object's properties, traits, and attributes.

Nature of Physical Objects

From a perspective of classical mechanics, that is, the worldview most frequently adopted when engineering a product or service, objects are thought of as having various performances, all of which are measurable to a degree, sufficient to satisfy the needs of customers and users. For this reason, an acceptable ontology for integration of human-built products and services can be based similarly on a view that objects have performance, which is in agreement with classical Newtonian physics. We term this as classical integration.

Alternatively, a relational view of products and services can be posited based on the relative measures according to a measurement standard or standards. For example, instead of integration based on a set of objects with various performances for their functions, relational integration depends on the interactions between objects to derive the functions which in turn are measurable. We term this as relational integration. It is only when two

---

* Might not.
† German, bremsen "to brake," and strahlung "radiation."

objects interact that they exhibit the function. In the relational integration case, functions are relative to the objects that interact, rather than as a single object with its individual performance. This view of the world is described in a quantum mechanical sense as relational quantum mechanics (Rovelli 1996, 1997).

And lastly is a worldview that recognizes the objective nature of objects by their measurements (typical of engineering) and the subjective nature of people's interactions with those objects (typical of sociology). The ethos for this alternative perspective of integration is based in part on a philosophical foundation that supposes that integration is describable as many, one, and does not exist—a blending and restatement of the concepts expressed by Stanislav Leśniewski (Henry 1972).* The substance of this formal ontology of integration is based on the genres "entity" and "activity" (Machamer 2004). In Machamer's book, "object" and "process" are differentiated (and carrying with them all the historically significant encumbrances that these two words harbor) as a workable ontology. Objects have mechanisms that in themselves are entities and activities, objects and processes. Objects have properties (intrinsic to their being that object), traits (the combination of the object's properties and the object's environment), and attributes (that which is associated with the object, but neither intrinsic nor situational (e.g., environmental)). In keeping with the formalisms of Leśniewskian mereology (Surma et al. 1992) as reviewed by Woleński (2000–2001), objects exist as objects; the constituents (parts) of objects are objects; and compositions, agglomerations, and combinations of objects are objects. No object has any property that is not a property of an object. Leśniewski presented the idea of a theory of relations for parts and whole (i.e., a mereology) (Simon 1987) as objects. Translating and editing Leśniewski's lecture notes, editors Srzednicki and Stachniak presented Leśniewski's development of the logical theory of relations between objects showing that relations do not depend on recognizing objects as sets of points (Srzednicki and Stachniak 1988), but rather as either distinct entities (objects) or domains that embody objects.

Characterizing Objects for Integration

For this book, for the relations for parts and whole (i.e., object-and-object), we view mechanistically through the interactions of objects with EMMI (i.e., EMMI-to-object). By a mechanistic interpretation, the whole of objects that are integrated is the totality of the objects that comprise a system (an object). The notion that objects may be in a specified conceptual region and that the objects are in some way related presumes a relation that is more than spatial. For if it were not for interaction of EMMI, one object would have no influence on another object. The case for EMMI-on-object (i.e., interactions)

---

* Stanislav Leśniewski considered the worldview of one, many, does not exist; in contrast to Parmenides' worldview of one, does not exist, cannot exist.

has been made in various ways (e.g., rock breaks window*; Olson 2002). The presentation in this book applies the logic of EMMI as the harbinger of interaction that is enacted by or through objects and their mechanisms (processes). That the whole of objects does not satisfy various conditions for a system does not detract from either the concept of the whole or from the parts (which are objects in themselves). As a whole, there are various conditions to retain the properties and traits of the whole (i.e., remain as a stable system).

Nature of Intellectual Objects

That objects are physical is only part of the whole. The objective whole for products and services is also comprised of intellectual parts. We must realize that building a product involves both the end result (i.e., the product or service) and the processes employed to produce the product or service. To that end, objects that represent physical objects are also to be considered as objects—the type referred to as intellectual. Intellectual objects means anything that is embodied in a physical object, such as an idea that is expressed on paper or in some tangible form; trademarks that represent a physical object(s); service marks that represent a service that is enabled by a physical object(s); copyrights (by written or other corporeal manifestations) that confer ownership or legal rights in chattel, real property; inventions, methods, techniques, approaches, specifications, know-how, algorithms, data, and software program techniques that are reduced to tangible form whether embodied in design, drawings, or sketches; discoveries that are created, conceived, or reduced to practice *and* documented; or netlists and source documentation, and tables and figures. The determination of intellectual property as objects is that it is both perceived and performed by the intellect and reduced to a tangible form. Intellectual property as objects only becomes a member of the type of objects when the physical manifestation of the intellectual content is reduced to something physical. Telling someone (object-to-object interaction) something that is wholly cerebral and not reduced to a physical entity retains the intellectual nature but not the aspect of property. Intellectual thought may be recognizable, desirable, credible (or not), but unless that thinking (or communication of that thinking) is conveyed in or reduced to a tangible form, the intellectual content is not

---

* The underlying topology of human near rock; human near window; human juxtaposed, picks up rock; throws rock; rock intersects spatial domain of window; rock breaks window can be described mereologically as object (human) performs acts and activities (process) to pick up rock (object); human (object) throws (process) rock (object) (rock is matter (EMMI) that is expelled from human by human mechanisms); trajectory of rock (object) collides with and imparts energy (EMMI) to glass window (object). Energy absorption mechanism of window (object) is insufficient to maintain stability of glass (process); and glass shatters (process), sending glass flying (objects). The topology of the configuration is altered, one piece of glass becoming many.

property, the object is. Property is tangible in form and characteristics, expresses value, and embodies rights, privileges, obligations, permissions, and penalties. Intellectual property (sometimes referred to as intellectual capital) continues to be recognized as the most important asset of many major corporations.* Intellectual property is the basis for new technology, innovation, and invention. The valuation of intellectual property is widely debated and often contentiously determined. When a buyer and seller of intellectual property agree on value, mergers and acquisitions result,† products are built and sold to customers, and international agreements become the means of trade and commerce.

Objective Measures of Performance

Physical and intellectual objects (i.e., intellectual property objects) have objective measures that relate directly to the performance(s) of products and services. There is a wide disparity in defining objective measures, so the term "objective measure" is limited in this book to that which is quantifiable in terms of performance. Objective measures include any item or combination of items that are categorized as EMMI. The amount of money something costs, the measured speed of an object, the amount of energy released from the Sun, and the quantification of mass launched into orbit are all objective measures. Additionally, the belief that an object will have various functions with performances, while being cause enough to invoke processes (cognition, carrying out procedures, and even corporeal realization of that thinking, e.g., writing down thoughts), is considered to be subjective measures. However, the objective measure of having one written document that describes thinking, determining the mass of the paper on which the thoughts are recorded, counting the number of words and number of letters, and determining the size of the document (length, width, and depth) is embedded in the process frame as an object. Similarly, the user's anticipations of an object are determinable and identifiable as subjective measures. These include any behaviors. In this book, the author defines behaviors by the use of an operational definition (Kerlinger and Lee 2000) that particularizes objects and processes in ways that are measurable. Behaviors are the movements of objects by processes, processes that result in objects, and objects interacting with other objects. This definition of behaviors is generalized from that used in systems engineering. Systems engineering considers behaviors (as captured in behavioral diagrams) to be activities, sequences of activities, and states of "machines" (Object Management Group 2007). In software engineering, behaviors are defined as characteristics of equipment, for example, reliability, loss, run speeds, and dimensions of various objects

---

* As early as 1883, the Paris Convention for the Protection of Industrial Property brought consensus to recognizing intellectual property as an important aspect of business.
† Assuming that the acquisition or takeover is not "hostile" in the classic business sense.

(Rockwell Software 2004). Further, behaviors are derived from objects and processes, functions are provided by joining objects and facilitating use by design, and users use objects to perform various types and kinds of work. "In the behavioral approach to system theory a system is regarded as a subset of a function space, the *behavior*, containing the input/state/output-trajectories ..." (Trumpf 2002). Behaviors that do not have empirical data are subjective by this definition.

Sometimes, the term "objective measures of performance" or "performance measures" is used (United States Department of Defense 2010) instead of objective measures. For example, in predicting the objective measures of performance for road pavements, crack widths, crack depths, and joint displacements are considered (Garcia-Diaz and Riggins 1985). Performance measures are observed and measured according to a reference scale or standard of measurement. Every object has at least one objective measure (and most often several), since there is something physical that is usually measurable (in the classic engineering and physics sense). Measures of performance might be the speed of an object or the amount of money an object costs. Often there is a "target" value (quantity and unit, e.g., kilometers per hour) in which performance is measured along with an acceptable variance about that target value (quantity and unit, e.g., kilometers per hour). These objective measures are used in testing to determine how well the object performs to a target value within the bounds of a specified variance that is deemed to satisfy an objective for stability (or quality). For the objects that undergo testing, a value and use can be ascribed, referenced to the perspective of a user and their specific circumstances.

Value and Use: Objects

Rather than contriving and discussing value in an economic sense based on categorizing various things, it is useful to consider the categories of objects first, and then determine their value. Knowing that integration deals with object and processes is a key determinant of determining what is valuable for integration. Integration is concerned with the value of an object that is at its peak performance and how that differentiates from an object that has lost functionality and has degraded performance. Then, thinking generally about commodities, we acquire a sense as to why something is useful and valuable.* To wit, a logical set of choices for characterizing products and services is in terms of use, value, and price. Use can be defined by a set of design parameters that maximize a preference function (Antonsson 2001), value can be described in a subjective and objective sense, neither of which fully captures the broad conceptions of value (von Böhm-Bawerk 2005). In the

---

* Naturally, we consider useful and not valuable, valuable and not useful, and valuable and useful. The preferential order of value and use indicates a priority according to the general notions of the key stakeholders for a specific project, product, or service.

words of Eugene von Böhm-Bawerk, "economic valuation of a good is nothing but a reflection of a more basic valuation which we accord to the life and welfare purposes which goods serve to attain." We extend that fundamental notion of value to encompass the cause by which objects are used to achieve, attain, or accomplish—that of the performance of an object. An object that performs well and is costly is not as valuable as an object that performs similarly and costs less. This tenet has been espoused by Miles and forms the basis of value engineering (Miles 1961, 1972).

The difference between the two approaches of use and value with regard to thinking of objects and their intended development, integration, and uses is first, fundamentally how one thinks and speaks about objects, and second, how one views the nature of importance of objects. Whether our perceptions of objects are influenced by thinking in terms of uses or value guides how objects are construed in the workplace, how they are managed, and how they are accommodated in the systems of financial accounting (whether by principles or by heuristics). This topic is significantly beyond this introductory text as it impacts on business models, enterprise architecture, project organization and management, and business strategy.

Performance-Based Value

There is a cliché that has dominated the scientific world for centuries, permeated the social world nearly to the point of paranoia, inspired dramatic rhetoric and oratory, and has seemingly captured the minds of "thinking" people—to go counter to objective reality spins a web of doom for scientists, offers displeasing moments for sociologists, and takes the form of dispelling remarks from others. Perhaps there is not as much drama as this would seem, but the reconciliation of objective measures with subjective measures remains to be a much-needed exploration and resolution. Since the concepts of integration span all disciplines and fields and all thinking and doing, the ontology of integration must not only be cognizant of the necessity, but also embody the sufficiency of integrating subjective and objective structures. Whether by subjective or objective measures, the presumed goal of integration is to form something of value.

Value is measurable subjectively and objectively. Objective value is often characterized by measures of amount (by numerical counting). Subjective value is often characterized by esteem, opportunity, or some form of intangibles. There are different types of value spanning use, esteem, cost, exchange, scrap, and various performances as compared with standard references. Value can be thought of both for objects and processes. Objects can be imbued with value by their properties (e.g., resilience of gold due to various types of deteriorations (such as rusting)) and by their performance(s). Faster cars often command higher prices than slower cars. The general notion of value seems to require a precondition for establishing the initial value and a postcondition (or anticipation of a postcondition) that presumes a final or residual value. In the case of gold, the initial value may be identical as the final value, assuming that there is no interaction with a mechanism of supply and demand or other influence-driven change from that an individual would self-impose. For example, there are no buy–sell dynamics that influence an individual to think the value of the gold has changed. Leaving the possibility open that there are some values that are determined by self (without outside influence), the value that is ascribed to products and services can be thought of as both driven by self and by nonself influences. Some of the key determinants of subjective value are (1) historical significance, (2) cultural legacy, (3) family heirloom, (4) esthetic beauty, (5) uniqueness, (6) scarcity, (7) marketing, and (8) sales.* For our purposes, objective value will be determined by performance measures that are determined relative to the amount of EMMI required to achieve that performance. EMMI can be considered for an instant, taken over a stage during which the performance is being created (i.e., development), or considered in a lifecycle sense.

Value of a performance measure is nothing more than measuring the function of a product or service that embodies that performance. Every function is characterized by its performance(s). Each function has at least one performance and most often many performances. For example, the function of 'to write' (as distinguished from the process of "to write") has various performances. How fast does one write? How many letters? What are the sizes of the letters? How often are the letters written? Each question is suggestive of at least one performance that relates the function of 'to write' to an empirical result through a performance measure. Person A is paid $60 per hour and writes 12 words per minute. The value of the performance of person A with regard to words/$ is 12 words/$. Person B is paid $45 per hour and writes 10 words per minute. The value of the performance of person B with regard to words/$ is 13.33 words/$. If both person A and person B have the same defect rate, then the value of person B with regard to words/$ is higher than that of person A. Value V per enactment of a function is defined as the ratio of performance P to investment I. This representation is the fundamental premise of value engineering (Miles 1961, 1972). Value compares what is received in performance of a function with what is (was) invested to achieve that performance. If two products with factually comparable functions and performance are offered for different prices and they are normalized in terms of the geneses of both performance and investment,[†] a higher value is associated with the lower-priced product. Whereas the value of a function

---

* Private communication with Stevenson Higa, purveyor of fine art and crafts, chairman of the board, Images International Corporation, Hawaii (1998).

[†] Normalized means the performances and investments are equivalent by their measures and amounts, for example, same conditions, same discount rates, and same period of time.

may vary with time, additional investments made during the system life-cycle to maintain performance are an acceptable way to determine the life-cycle costs for a given function. The aggregation of all performances that are ascribed to a function must be included in the determination of value for that function, or the recognition that there may be one performance of a function that is either more appropriate to the use of that function, or significantly more important in some aspect over that of the other performances. The system value, $V(t)$, is given by

$$V(t) \sum_{F(t)} F(t) = \sum \frac{P(t)}{I(t)} * \gamma$$

where $F(t)$ is a function or nonlinear summation of subfunctions that are enacted by the product or service, $P(t)$ is the performance measure (units of EMMI appropriate to the use of the function(s) $F(t)$), $I(t)$ is the investment* (e.g., dollars or other equivalent convenience of assets that are required to achieve performance $P(t)$, the time, t, measured relative to the onset of initial investment in the project (or a period, or portion of a lifecycle, or over the lifecycle), and γ is the normalization factor (dimensionless). The summation is simplified for the purpose of this discussion, and generally aggregated over all subfunctions, subperformances, and subinvestments.

The change in performance of a system object due to an interaction of EMMI from another object is equal to the work done. Performance can also be described with reference to the cost/(unit time), as well as to the total time over which the performance occurs. Incorporating and factoring the variable of time express the value equation in terms of the measure of *performance* per rate of investment.

$$V(t) \sum_{F(t)} F(t) = \sum \frac{P(t)}{I(t) / t} * \frac{1}{t} * \gamma$$

By including time explicitly, the value of a function can be measured in terms of performance per investment rate (e.g., labor rate) times the normalization factor γ divided by t.

Value is variously described beginning with the ancient Greek Protagoras[†] (an early humanist, a champion of pragmatism, and a master Sophist) continuing subjectively in sociological contexts as illustrated by expressive cognition (Kuwabara 2011), as encompassed by classical sociological theory and behavioral models (Zafirovski 2005) in game theory as an alternative to

---

* In this case, material wealth; in general, energy, matter, material wealth, or information (EMMI).
[†] Protagoras, 485–421 BC … things are to you such as they appear to you and to me such as they appear to me …

objective probability (Kyburg and Smokler 1964; Dastou 1994), in knowledge management for modeling (Rocha 1999), in public administration (Shingler et al. 2008), in business valuation (Kwon et al. 2002), in engineering analysis for reliability (Bhatt 2000), for product support (United States Department of Defense 2011), for management strategies (Chow and Van der Stede 2006), and in economics for the dual nature of money with objective and subjective qualities (Zyphur et al. 2006). In systems engineering, subjective value and its measure are often included as part of building and integrating products or services (Bernstein 2001), as measures of project success (Parsons 2005), for software testing (Hamlet 2007), and as determinants of systems engineering quality (Valerdi and Davidz 2009). Importantly, the recognition of the systems engineers' need to embrace both the subjective and objective components of developing and integrating products and services was instilled from the mid-1960s. By the late 1990s, that favor was both prevalent and acknowledged as an important distinction for systems engineering (Sproles 2000).

Subjective Value: Processes

Similarly, economic value of a process is derivable from Eugene von Böhm-Bawerk, "economic valuation of a good is nothing but a reflection of a more basic valuation which we accord to the life and welfare purposes which goods serve to attain." The words "to attain" are suggestive of a set of measures that can be developed to determine the value of a process. However, the set of measures for a process that focuses on activities are different from that of measures of performance. Measures of performance relate to functions. Various schemes and measurement scales for measuring processes have been proposed, including "… cost, schedule, risk, and improvements" (Millard 1999). The key determinants of value for a process, first, are subjective. That there is no quantitative reference is implicit in subjectivity and subjective measures. However, no doubt there is value in processes. Process can be patented, bought, sold, and improved to increase production, build products, and deliver services. And processes can be ineffective, inefficient, and cumbersome. People conjure processes, communicate to others about their thinking about processes, and manage processes. People estimate costs of and spend money to deal with processes.

Process is the amalgamation of activities and tools that combine ideas. Two processes are differentiable when they require different skills from the same person, need different equipment for the same job, and use different tools for the same activity. Generalizing from definitions of software processes by Humphrey (1989) and Lonchamp (1993), a process is a partially organized set of activities, tools, and practices carried out by humans who are constrained by, for example, resources, budgets, schedules, scope, and policies.

Processes have inputs, outputs, and losses to achieve those outputs. Processes are measurable with objective measures such as cost, number of people, and the amount of loss to achieve certain results. Processes are comparable to other processes subjectively. Yet it is quite difficult to say that one process is better than another. What can be said is that there are noticeable differences between processes, some taking more labor, some requiring more resources, and some costing more money. However, in isolation, these objective measures have little meaning as their bases are quite different and therefore not comparable. Were there only a few processes that when combined produce a certain result, the combinatorial advantages and disadvantages of these processes might be discernable compared to another set of like-kind processes. In this case, the comparison would be to ascertain if there was a combinatorial advantage determined by the number of people involved with the process, the total costs of the processes, and the amount of time it takes to complete the work prescribed by the processes. For example, two sports team compete in a "game," each team bringing its different processes to test the consequences of their processes on "game day." In a similar fashion, processes of a like kind can be "tested" given that the competing process sets agree to a set of rules and standards by which to measure the outcomes of the "test." Short of a game-play equivalency, there would not seem to be an objective, rational basis on which to measure a process empirically.

Further, processes can be measured and improved relative to themselves (Goldberg et al. 1994). If the same process is measured according to a set of objective measures in a simulated "game-play," then enacted again using the same rules and standards in a subsequent "game-play" situation, then the before and after comparisons of objective measures indicate the degree of controls that are operative on the activities within the process. If there are random sources of perturbations, then the variations in the objective measures can be collected and evaluated for a set of "game-play" "tests."

The basic unit of dimension for a process is an act—a single factor signifying that a process might be evaluable in isolation is termed as an act, a single step in a string of steps that when combined are recognizable as an activity. Activities combine into processes. At the level of an act, the *actor* (in this example, the human) may take form as "walking" between a desk and a lab. That "walking" is part of a series of like-kind acts, concatenate to the activity of "going to the lab." That the combination of "going to the lab," "setting up an experiment," "running the experiment," and "taking data" is considered the process of "running an experiment" signifies the manner in which processes and their subtasks can be granularized (or partitioned). There are many ways to granularize acts and activities, and there is no standard. So the practice of valuing a process is problematic. You might note an advantage to moving your desk into the lab to 'save time' (a function). Changing the activities changes the processes. It is notably difficult to perform the same routine task in the same way each time that task is performed. Consequently, there is variation in the acts, in the activities, and in the process(es). Measuring that variation is difficult, if even noticeable. Defining processes is problematic, implementing processes is problematic, describing processes is problematic, and quite intuitively, integrating processes is problematic (GAO 2009).

From an integration perspective, processes guide the work. The systems engineering process model describes the stages in which the project team focuses on various milestones and deliveries. Moving from one stage to the next stage is process driven (by the work to complete the assigned tasks).

The aim of process integration is to improve the management of capital, assets, and operations, that is, increase the value of the organizational efforts. Convenient measures of process integration are operating efficiency and capital effectiveness. These measures of the target process are always referenced to another "like-kind" process, or self-referenced in the manner of "game-play." Particular attention needs to be given to the reference process. Should the reference process be a previous enactment of itself, clear and open objectivity must accompany both the measurements and the evaluations that compare the target process to the reference process. It is very easy to measure that which you want to see, whether the evidence is there or not. An objective third party is one means to prevent intellectual contamination of the measurement process, analyses, and evaluations.

Further, a process can be compared to an entirely different process that has a different provenance, background, subject, intention, and industry focus. For example, compare an apple-growing process with that of painting orange fences. Aside from both processes having a common abstraction* of fruit, the apple–orange comparison has no objective basis from which to make measurements or comparisons. But wait. The assumptions that go into the determinations of the apple–orange processes do not have to be the same or for that matter even be similar. One can have as a reference an entirely different process as long as there is a common overriding theme that is similar for both the reference(s) and the target process. In the apple–orange process comparison, we find that both processes engage workers. The apple process uses seasonal workers and the orange painting process uses student labor. Management for both processes concerns supporting their workers year round, in spite of their seasonal and part-time employments. Embedded in both the apple–orange processes are the vestiges of policies that provide sustainment, support, and team-building benefits. The comparison of processes should be focused on the "game-play" scenario, which in this case is worker retention. Worker retention regardless of the type of work is important to

---

* Abstraction is an insufficiency of details to describe completely all that is needed so that the EMMI that is necessary to enact a mechanism is available as needed. An abstraction that is too high implies less detail than required. An example of too high an abstraction is captured in the statement "Exit room." The level of abstraction is important to convey meaning. "Exit room" may mean to leave the room in which you are currently located; you must go to another room to exit (indicated by the statement "Exit room"; or whichever notion you think is irrelevant, you must exit the room.

both sets of managers. As a result of the retention policies and team-building activities, both sets of workers deliver more productive hours per dollar than their competitors in their respective markets. The key in evaluating the effectiveness of these retention policies is wrapped in process integration. Some retention efforts are based on maintaining critical skills so that new hires can be trained and the workforce is improved and shaped to provide a stable cadre of skilled workers. Another perspective is to think of the skill base as human capital that can be developed and managed (Todd and Parten 2008). In the case of the apple–orange comparison, the process integration effort was based on the specific needs of each individual worker. Rather than view the process as critical-skills-centric retention or human-capital-centric retention, apple–orange process managers focused on people-centric retention. Process integration depends completely on the focus of integration. That focus is more influential for decision-making and management roles than design and architecture.

The aim of process integration may be to (1) improve (e.g., maximize) production (output) efficiency or effectiveness; (2) increase the independence from changes in the operational environment, improve user satisfaction, immunize the assumptions and decisions from technology and legislative vagaries, and enhance the interoperability with known and unknown future systems; (3) expand the partnering opportunities through network-centric operations; and (4) deliver process visibility through standards, shared data, and interpretable protocols.

Measurements performed for one class of abstraction can be combined and correlated with measurements made for the other classes of the same abstraction. These measurements are correlated through the integrative domain framework as long as the conditions for integration within the limitations and constraints are satisfied. Either or both of the subjective and objective frames can be used as the reference frame in which to associate and connect measures and measurements for both frames. The purpose for a reference (in the general sense a class of abstraction, object, or process) is to bring additional information together in such a way as to highlight the relation between the reference and the additional information. The relation between the reference and the additional information is to provide a basis for comparing, contrasting, and evaluating the additional information relative to a set of information. The reference serves as a perspective from which to view additional information. For example, if *object* is chosen as the reference domain for integration, the objects and processes associated with that class would be expressed within the physical context (objects). The perspective posed by the physical context serves as the lens through which to examine the nature and conditions of the objects and processes. If there are various kinds of associations or patterns between the functions or processes with physical entities, then the perspective of *object* might either suggest that relation or explicitly show it. Similarly, the function frame, or in general, the process domain can be used separately as the reference perspective to observe and interrogate the measurements of objects or behaviors. Sometimes there may be one domain that seems natural for a particular measurement—the one in which relevant patterns appear and all of the variables relate causally to the patterns. It is prudent to not only include the other domains in the measurement framework but also to consider the other classes of integration. In other words, relying on only one interpretation of patterns (regardless of their supportive interpretations, and corroboration of variables by others), the alternative views may be the only way to discover a problem, provide insight, or innovate another solution.

Creating value through processes means providing options to deal with contingent needs, establishing supply chains for goods and services, building value chains to manage stakeholder value, and improving governance (Rouse 2004). Processes manage variation, shape interactions, and result in decisions (Rebovich 2005). Processes form interpersonal relations via their human *carriers*, promote project cohesion, and instill metrics for measuring work (Bausch 1997). Systems engineering processes for the U.S. Department of Defense often exclude consideration of certain categories of lifecycle support costs and system readiness (leaving those for sustainment budgets) during development and integration of new products or services (GAO 2003). Project management often considers their primary roles as minimizing distractive external influences and providing sufficient resources to carry out the work. Overall, there is considerable thought and effort expended by project management to maximize the positive mettle of the project, that is, stir the spirit of teamness.

Conceptualizing, defining, designing, architecting, developing, integrating, operating, and disposing of a product or service span the lifecycle processes of a product or service. Processes are enabled by objects. People (objects) using (processes) materials and other assets (objects) build, integrate, and deliver (processes) products or services (objects) to customers and users (objects). Users (objects) use (processes) products and services (objects) to perform or enable (processes) work (objects). Processes are inexplicably intertwined with objects, but processes and objects are dissimilar, distinguishable, and separate. Processes include the cognitive aspects of preparing for and carrying out of the procedures that result in something done. Like objects, processes have value and are measurable. The value of processes is determinable by the results achieved by those processes.

However, there would seem to be no ready method, no easy way of determining value in a process since there are substantial differences in the makeup, uses, and interpretations of processes. Comparing one process with another process in itself is a reference by which unlike kinds can be contemplated. Comparing a process to itself (either before or after changes are made) is a self-referencing process. Either of these two methods allow a summary finding that while there are two different processes, and if whose objective functions are similar, their results will have an equivalency of meaning (subjectively), but not necessarily objectively.

Management Processes

There is a long-standing tradition of separating processes for managing activities from processes specific to doing work. In systems engineering the separation is seen with the project manager (whose role is to manage the project) and the systems engineer (whose role is to orchestrate product development and integration to deliver requisite functionality, performances, and quality). The project manager guides (works with) the systems engineer within the context of the organizational processes and external limitations of time, budget, and resources. The systems engineer works with the multidisciplinary team to determine requirements, and builds the desired solution within the constraints of time, money, and scope. The project management processes deal with the interactions with managers and various team members to plan, organize, direct (command), communicate, control, and team build. Appendix 2 delineates some of the processes of management. The process for systems engineering (managed by the chief systems engineer or the systems engineering manager) follows the systems engineering process model. Integration is concerned with both kinds of processes—that for managing the project and for developing and integrating the product or service. An adequate general reference to project management is periodically updated and released by the Project Management Institute, Inc., an earlier version of which is referenced (Project Management Institute 1996).

Processes as Intellectual Property

Processes can be turned into intellectual property (intellectual object) through cognitive thinking and procedures that result in the patent process. Process patents are recognized as valuable when reduced to an application and granted. In that way, processes become objects. Intellectual objects can become processes and be representative of the value embodied in the intellectual property. Intellectual objects are recognized as valuable when put into use. Physical objects are valuable due to their intrinsic properties as well as their uses and convertibility into other objects. The concept of converting one object into another object through various mechanisms results in EMMI, a EMMI that change with the conversion of objects. This conversion of objects and the resultant changes in EMMI is both a social phenomenon and a means of interaction and integration. Another way of thinking about objects and processes is to recognize that EMMI derived from objects are inputs to objects that enable procedures and procedures result in EMMI to objects; further, objects are the result of procedures and procedures are the outputs of objects. By these we recognize that objects are objectively determined while processes are subjectively determined. Therefore, the distinction between processes and objects is essentially the distinction between the subjective domain and the objective domain. Both domains can be construed as social, or physical, or other means of categorizing, but significantly, subjective and objective considerations are the superclassifications for integration. We determine that the subjective and objective classifications be the frames in which integration is viewed and enacted. Further, the unique relation between objects and processes in their elemental forms shows that they are not comparable, and they relate through other classifications, and have distinct, determinable values. For interaction and integration, the ontology we seek is objects, EMMI, and processes.

Subjective and Objective Ontology

Objects and processes have been both implied and stated as ontological structures for business systems (Tronstad 1997), architecture (Koopman 1995), socioeconomic issues (Osorio et al. 2010), and a myriad of other applications (Gailey 1985; Breuker et al. 1997). Objects as metaphysical entities have a long history of debate and deliberation, extending back in time for 2000 years.

Objects are vestiges of the functional viewpoint—the view that what the product or service must do is based on physical objects that produce various behaviors (Shishko 1995; Harel and Politi 1998; Defense Acquisition University 2001; Rodriguez et al. 2004; Guenov and Barker 2005; Eriksson et al. 2008; Cechini et al. 2009; Do et al. 2010). That objects and processes (among other are choices for representing many things human-built) are the *raison d'être* for systems engineering is based on the necessity of testing and demonstration to verify that the work is appropriate to satisfy the requirements and to validate the product or service for utility and fitness of use. The functional view is objective; objects are objective. Objective measures can be quantitative or qualitative, but both types deal with the numerical counts of items.

Processes are subjective. Subjective implies things influenced by personal feelings, biases, or intuitive thoughts. Subjective measures are reflected in survey or questionnaires that respondents use to express their opinions or interpretations of events, nonfinancial deciphering of business, operations, or product and service utilities, and use (Chow 2006). Sometimes, subjective measures are used to determine how a constraint or condition impacts on development work or integration (for example). The subjective data provide insights into what moderation and interpretations are necessary to go along with the objective data for planning tasks and allocating resources to complete integration (NASA 1990). Processes are subjective both by their design and their enactment.

While both objective and subjective measures are necessary (neither one being sufficient with the other), the correlation between them is poor (Parsons 2005). Systems engineering and in particular systems engineering integration rely on both objective and subjective measures to bring a product or service to its deliverable configuration and operability. The combining of objective and subjective measures is particularly important for our ontology of objects and processes—objective being objective and processes being subjective. The correlation between objects and processes is through the procedures carried out by the systems engineers and the systems integrators to build products and services. That correlation depends on developing metrics for both subjective measures and for objective measures that are related to the same concomitant object and process (i.e., procedure, activity, or act (in decreasing level of sophistication and complexity)). For example, the process of transporting an object from one location to another may depend on the mass of the object, the number of objects, and the size of the objects. If the procedure is to have people lift the objects and all of these three dependencies have high numerical values, then the process of transportation may be dramatically different than if the numerical value for each is one (assuming the same units of measure applied to both situations). Moving a patient that weighs 150 kg from a gurney to a bed may take four orderlies. Moving 300 patients a day may take 1200 orderlies if all patients weighed 150 kg. These objective measures do not communicate the problem, which is the harm done to the skeletal frames and muscles of the orderlies by repetitively lifting these weights. Were such a scenario to be scripted and followed, the first reports from the orderlies would be for "aching muscles and sore backs." After a period of fatigue, the next reports would show more serious injuries, requiring time off from work for rest and rehabilitation for the orderlies. The subjective measures of soreness and feelings of pain vary widely between people, but nonetheless are indicatively correlated to the objective measures described. Both objective and subjective measures need to be considered when building metrics to monitor an enterprise, a business, a project, procedure, or activity.

Since objects and processes in a project are established and carried out in relation to each other, the socioproduct or socioservice nexus is more than an interaction of the objective and subjective issues. Objective and subjective issues are integrated. As such, the proper discussion of a project is within the framework of objective and subjective issues with appropriate metrics.

The mereology we determine is objects as objects, events as actions, and processes as procedures, activities, and acts. The equivalency of EMMI as input for objects is matched with the results of processes as inputs for other processes, and that of mechanisms for objects is matched with procedures for processes. Processes make objects from other objects; objects make processes from other processes. The relations between objects and processes are the topology that signifies their connection as mereological entities. At this point, a rigorous mathematical development could follow, which is beyond the scope of this introductory presentation.*

---

* Mathematical rigor can blind correct reasoning albeit proffering a deeper analysis. But the question is, a deeper analysis of what? Mathematics is a language that models things that resist naive simplicity. To begin with numbers is to admit an uneasiness of the nature of something. (See Endnote 1 at the end of this chapter.)

Further, we infer that mereological arguments for objects need only be contrasted with a socioeconomic view to determine the subjective and objective components of the two frames. The objective frame deals with the objects and the functions that derive from those objects. The subjective frame encompasses the processes and the behaviors. But the characteristics of the objective and subjective frames is determined by the type of business model used in the enterprise. It is here that we must decide what our centric view will be. Further, the process management and product or service management also complicate the simple parsing of mereological entities and ontological constructs along objective and subjective lines. For development projects (product- or service-centric), the frames are best shown as all that is related to the product or service from the user perspective and all that is related to the management processes that govern the project. The interplay between the product frame and the project frame is the systems engineering work that is accomplished to build and integrate the product or service.

Note that the output of the process frame is an object that is related to the project management and the output of the product frame is a process that is related to the product. The integration approach for the product- or service-centric business models is the intersection of the two frames—subjective and objective. That intersection is referred to as the structure for the product- or service-centric business model integration or the product- or service-centric business model integration framework. Collectively, the business model integration frameworks are referred to as the systems integration frameworks.

Business Models

Business models are descriptive of the management of value for an entity, for example, a product or service, or at the business or enterprise level. The process of "to manage" is taken here to mean "planning," "organizing," "communicating," "directing" (or "commanding"), "controlling," and "team building." The essential characteristics of a business model require the enterprise be describable in terms of its key traits, for example, managing the enterprise, delineating the needs of the enterprise, prioritizing the relative importance of these needs, evaluating the scalability and externalities of the internal operations and external processes, identifying the efficacies of the products and services in the user environments, determining the causal boundaries and boundary conditions, and identifying all interactions both internal and external to the enterprise* with internal-to-external and external-to-internal delineations.

---

* Enterprise or business are used interchangeably, whereas, project can also be applicable and serve equally well given the context of the discussion.

Business models are as much about generating revenue and profit as they are about offering the means of deriving value for stakeholders of the entity through its interactions. One can construe three kinds of business models: event-centric, object-centric, and process centric. Event-centric business models focus on the primary interaction(s) that results in economic benefits to the business entity and in utility (or uses) for the customers or users. An event-centric business model is inclined to focus on providing the customer and user the best value for the sales transaction, specifically the interaction between the customer (object) and the seller (object) for the exchange (or barter) of the product or service for material wealth (money). An object-centric business model is concerned primarily with the product or service (object) and its interactions with the user (object) or the user's environment (object). A process-centric business model deals with organizational dynamics and input–output flows. Their focus is on supply chains, logistics, or other means of support. When architecting the business model at the enterprise level, these three business models are singularly concerned with optimizing their particular view and dynamics of operations. When architecting the project model to deliver a product or service, these three business models deal with requirements in entirely different ways. The typical development environment for products and services adopts the object-centric thinking with the elicitation and determination of requirements. Development and integration are particularly concerned with verifying that the product or service satisfies requirements, while validation is left as a final step before placing the product or service into operation. The emphasis on human systems integration, the bringing together of the user, the product or service, and the user's environment are discussed and included in the object-centric discussion, but not to the level of importance or adequacy that would be the focus from an event-centric (or interaction-focused) conceptualization of the development project (Tvaryanas 2010).

The business model encompasses the concepts of enterprise, business, and project. Unlike the concept of enterprise which deals with the "big picture" (Kasser and Mackley 2008), the system of processes and objects that undertake activities that are goal directed (vision driven) and cognitively related, business is focused on the revenue, profit, and loss aspects of survival. "… enterprise is a mental image of that organization's current and future reason for existing" (Morris and Pinto 2004). Business is embodied in the activities that ensure or pretend to ensure survivability. The business model is concerned with the interactions, events, objects, and organization for long-time survival. Within the construct of the enterprise and business is the singular notion of project. Projects have beginnings and ends, specific objectives, measurable outputs, and some real or outward appearance of planning. Enterprise means to attain effectiveness, business means to survive, and project means to accomplish. The enterprise aims to lay out the nature, vision, and boundaries of work. The business lays out the interactions, organizations, processes, and objects. The focus of the project is to deliver a product or service with requisite functionality, performances, and quality. The project is a socioeconomic entity that devises a way to achieve an objective over a specified period within a budget (e.g., in the case of product development and product integration). Enterprise, business, and projects can be modeled, those models representing the key measures of performance and the metrics for success.

Risk and Loss

Determining value can be thought of in terms of a certain amount of performance for a given lifecycle cost. To a first order, this is sufficient for integration. The subtleties of a second-order analysis include a definitive characterization of risk and loss. Risk and loss are sometimes addressed without an ontological framework to structure relations and to derive meaning. Hence, sometimes simple tables or diagrams are used to reflect the views of key personnel as to what is a risk or what the consequences may be if a problem is realized. Simple by their means and interpretation, the risk management guidelines followed by many people in government and industry often incompletely characterize the insidious nature of what conspires to harm people, products, and services (United States Defense Acquisition University 2003).*

Generally, risk is a structural property of the interactions between objects, whereas specifically, risk is inherent in the interactions involving the enterprise, business, and project. As stated by Kuwabara (2011) in discussing social exchanges, referencing Molm et al. (2000), "risk is a structural property of exchange . . . ." The business model must be true and faithful to all types and significances of interactions, capturing all interactions to expose the structural inherences and opportunities for risk.

Along with risk, enterprises face loss. We often think of opportunity or peril in terms of risk and sometimes associate risk with a loss. That loss could be monetary or social or some form of harm or loss of life. This book characterizes loss in terms of a generalized loss function that attributes EMMI losses to deviations from a target performance value (see Appendix 2). Further, not all losses are attributable to variance about a performance target; some losses result from not having a target value (meaning that the function was not provided and therefore had no performance value). By determining the type and value of losses the business model can be evaluated in terms of its overall effectiveness in providing a reliable flow of products or services, support, and maintenance.

---

* A more comprehensive analysis of risk is taught and practiced in various universities (Kujawski and Miller 2007).

Similarly, for the event-centric business model and the organization-centric business model, the determination of frames and the systems engineering framework justly and logically is constructed as with the product- or service-centric business model. For all three types, various conceptions of wholes and parts have been developed independently and adapted to situations great and greater, including a definitive discussion of parts based on the formal theories of mereology from Leśniewski and the works of Leonard and Goodman (Leonard and Goodman 1940). A key tenet of holism in general systems theory builds on the reciprocity of wholes and parts (Ashby 1962) and the notion that objects are simultaneously wholes and parts (Koestler 1968). This evolution of the relations between objects was extended in sociology (Kuhn 1974) to recognize the observation that knowing something about a system portends a level of information about another part of the system. Next, we determine that while there may be differences between objects, there are representative objects that are descriptive of a type of object, described as mereological pluralism—whereby an object may well have systematic differences with respect to their whole or their parts—the principle of extensionality (Korman 2007). And finally, our objects must exhibit mereological constancy (Simon 1987). Specifically, over time the object remains the object maintaining its essential qualities and its properties. The object's traits may change because of the environmental influences on the object and the attributes may change due to aging (for example). However, the object remains the object in its wholeness. For example, a 1928 car ditched in a canyon, overgrown with vegetation, and long forgotten still retains the intrinsic (e.g., basic, essential, and fundamental) qualities of the original 1928 car (i.e., the car that was), with paint giving way to rust, and seat leather pecked to feather nests. The objects that comprise the car remain (even if some were looted).

Prototype-Based Ontology, Logic, and Mereology

We introduce an ontological structure of integration, the starting point and perspective of which incorporates both objective and subjective components. The objective components of the structure are based on objects, EMMI, functions, and behaviors due to same. Since integration by human actions is sociologically based, a subjective component is essential to capture the notion of the processes used to build something. Humans use objects to build and integrate objects. An object is a structure of anything that receives, transforms, and sends EMMI. An object can also serve as a means to facilitate communication or to make something known. For example, a radio (when operated according to its designed uses) produces sound. The radio is a physical object, with the emanating sound, its EMMI. The ontology that encapsulates integration spans the project that produced the radio, the radio station that transmitted the signals which were decoded by the radio, the user(s) of the radio, the listeners of the sound, and the behaviors of the users and the listeners as a consequence of their interaction with the radio, the sound from the radio, the transmitting station, the recording artist(s), and so forth—the list of stakeholders being quite lengthy (estimated to be in the tens to hundreds). By defining the boundary of the "music" broadly, rather than limiting it to one radio, the stakeholders' number can run into the millions and billions by counting individuals rather than groups of individuals. Objects and their EMMI are essential components of the ontology, but only when included in the context of the (social) processes enacted to realize the delivered product or service, the uses, and disposal—in short, the process associated with the product or service lifecycle.

Objects, EMMI, and processes form the subtypes of a prototype-based ontology (Sowa 2000)—a formal distinguishable from their supertype(s). By comparing members of a subtype, the typical member of the subtype is necessarily different from typical members of other subtypes.

The ontology of integration is concerned with all that is product conceptualization, development and integration, and use and disposal (the complete lifecycle sense). This consideration of integration includes objects, EMMI, and sociological processes and behaviors. The object built is the result of the building process. The object used is the result of what was built. The object use is the result of the building process and the functions by design, by use, and by accident. From the perspective of use (i.e., delivered and operational product or service), the objective of an object is corporeal, real in every sense. An object is recognized as having physical meaning; measurable properties, traits, and attributes; and value for its performance(s). The building of objects is also objective, measurable, and valued for the result, that is, the result of building and delivering the object. Systems engineers (and other builders of products and services) are the creators of systems, systems comprised of objects. The process begins with an object and building up of other objects from elemental objects.

Objects appear to be formed from other objects. Building on our knowledge of the most elementary particles (objects), we observe a few properties that seem to be present in other configurations (also referred to as objects). The concepts of interaction and integration appear operative at the most elemental levels of objects to the most integrated forms of objects (galaxies and the universe[*]). Objects are comprised of (or convertible into) matter[†] with mass or electric charge (that is representable as mass).

---

[*] We refer to the universe as the system—the system has no boundaries, i.e., without limit in all respects. As such, the universe is ontologically one.

[†] Matter may not be the only substance that is comprised of energy or mass (or its derivatives such as force and momentum), but for convenience we loosely interpret all such "things" as matter.

Objects as Models

We investigate objects as models and discuss two perspectives: black box and as related to function.

Objects as Black Boxes

With regard to using objects, we distinguish between natural occurrences based on its rules and that of human endeavors that are guided by rules of humans and rules of nature that build systems, and driven by human intentions. We also recognize that humans are not the only animal, and include all living organisms irrespective of their cognitive prowess, physical abilities, relevant skills, and available tools. The expectation is to provide an all-inclusive discussion about interaction and integration using the analogy of human-built objects.

While the focus is on objects, the model is constructed from two perspectives: (1) an observer who sees perhaps only some of what the object is receiving and sending ("black box"* approach (Ashby 1957)) and (2) a functional representation (Van Wie et al. 2005) of the object's uses. From a limited amount of information, the observer begins to infer something about the contents of the object and how the outputs relate to the inputs. Often the observer will recognize patterns in the inputs and outputs. From the perspective of the object (that may "know" nothing of the patterns that are produced because of its output(s)), its internal workings may be fully or only partly engaged. Different patterns may result (again without the tacit oversight from that of an object). Those objects that do not 'manage'† their inputs or outputs lack the substance to achieve or to remain stable. The perspective of an object is one that should reveal natural groupings based on what objects can and cannot do in their environment of rules and intentions.

The black box approach is restricted by its means to provide thorough explanations, thereby limiting its utility to a very simplistic view. For these purposes, the superficiality does not hamper our discussions, but rather points to a great body of literature on mechanisms. In this book, we establish that objects have mechanisms which mediate for both inputs and outputs (send and receive, respectively). We make no effort to detail or investigate mechanisms; there are far too many observed, proposed, and hypothesized. Every field and discipline has had or is still engaged in debate about mechanisms, to which our discussion on integration would hope to benefit. Consequently, the specifics of the output are likewise not detailed (as these

---

* Black box method of testing is the analysis of input transformation to output based on not knowing the internal workings, logic, or configuration of the object.

† The subfunctions of 'to manage' mean 'to plan,' 'to organize,' 'to direct,' 'to control,' 'to communicate,' and 'to team build,' that is, 'to create consensus,' 'to foster teamness,' 'provide for (or support) stability,' or the logical equivalence.

outputs are grossly driven by the inputs and the object's mechanism). As the trend continues and more researchers settle on a mechanistic view, a more fundamental outlook can be prepared for interaction and integration. For now we consider mechanisms as the intermediary processes by which inputs are transformed into outputs (Gross 2009). The multinuanced and suggestive statement that "mechanisms thus constituted functions" (Gross 2009) is elaborated in this book at some length. Indeed, the association between mechanism and function, and function and process has been suggested widely. Consequently, there are great differences in definitions and uses of the terms "mechanism," "function," and "process." Some writers describe mechanisms as functions, functions as mechanisms, mechanisms as processes, processes as functions, and processes as functions. In systems engineering, the common reference is to this function and that process, while then referring to functions that are decomposed into subfunctions and objects and processes that are linked in flows. By most definitions, the use of the term "function" is really defined as process. Then to carry the contradiction even further, the terms "function" and "process" are used interchangeably. But overall there does not appear to be much confusion and it does not seem to pose a problem for building systems, as systems are integrated and built fairly routinely. Instead, this presentation suggests there is a missed opportunity to improve systems engineering by observing the nature of functions and how they differ from processes and mechanism.*

Objects as Related to Functions

The second perspective of objects bridges the gap between objects with spatially extended boundaries to that of objects having point-like structure and a granular hierarchy where one object is comprised of other objects and drawn or grouped together by similarity, proximity, or functionality (Zadeh 1997). The abstraction of an object from a corporeal shape to a dimensionless entity is at once the same as carrying two thoughts about the same subject. The abstraction of a chair is causally carried along with a cognitive representation of a chair—both of which are operative and logically coexistent. As with physical objects, these cognitive objects have boundaries and boundary conditions. Thus, it follows from the black box approach which suggests that a bounded object can have internal mechanisms, inputs, and outputs, and can behave as though the object (with its extended boundaries) is point-like. Whereas an extended spatial model of objects is important for detailing their inner workings, the point-like model is pertinent to looking

---

* Private communication with Dr. James H. Lake, board certified psychiatrist, clinical assistant professor, Department of Psychiatry and Behavioral Sciences, Stanford University Hospital, Stanford, California, 13 September 2011. Dr. Lake comments, "There still does not seem to be a consensus on core definitions in systems theory and this has resulted in confusion in the field and lack of progress in systems engineering theory and applications."

for patterns. Patterns expose different perspectives—no pattern may suggest an ineffective perspective versus a discernible pattern, that can possibly be optimized for greater clarity with regards to the type of pattern needed to coincide with the circumstances of the integration effort. For example, should a pattern of functionality or behavior be expected to result in a robust, integrated structure? If so, then what patterns would be discernible given a particular order of integrating objects? Further investigation and planning may be necessary to observe the modes of operation for the structure. Then a decision can be made as to the notion of working to quell the periodic response, or ignore, or enhance it. The combining of objects provides the opportunity to investigate the functions and their primary dependencies.

To glean more information from objects and their interactions as they relate to integration, a functional model can be developed with the goal of acknowledging the generally discussed natural relations between the topical notion of function (a relation between objects in time and space, i.e., that which is enabled or required to accomplish something) and the entities that are characterized as objects. The functional model proceeds from a functional decomposition of a top-level abstraction of a function, for example, 'to walk.' Subfunctions are developed following the procedures of functional decomposition. An essential part of functional decomposition for integration purposes is to carefully examine the three types of boundaries (physical, functional, and behavioral). For a functional decomposition, mapping to both the physical and behavioral aspects of the product or service lays out the relations between objects that will become the sequencing for integration. For example, to demonstrate a particular function, these decomposition diagrams show the two objects that when connected demonstrate the function. The performance of that function is enabled, first by the connection, second by the partitioning that differentiates the function from other functions, and third by the coupling and cohesion between the EMMI that is exchanged between the two connected objects.

Summary Overview of Objects

From a systems engineer's perspective, an experiment is more than just a scientific investigation based on a hypothesis, carrying out an experiment, and ending with the confidence that the observations and results are somewhat correlated with the experiment. The wholeness that we investigate may not be pliable and yield to the traditional scientific method. Analytical reductionism from high level to lower levels does not seem to capture all of the system parts (i.e., objects comprised of objects do not decompose into objects that are deemed to be the parts) (Koestler and Symthies 1968; Troncale 1977).

Thinking in systems for a moment suggests how the scientific method might be modified to allow for some ambiguity in experimentation, while retaining within a scientifically posed, process-driven method or methodology. Heeding the perils of rigorously and inflexibly following a particular method of investigation, Feyerabend (1993) poses two notions (in the words of the author): if the results of an experiment are not what they seem to be (by a strictly scientific approach), then rather than suggest a flaw in the experiment, consider first the impact of the underlying perspective of the observer, and second, consider the impacts that may occur outside the physical boundaries defined for the experiment. Reflecting on Feyerabend's reference to these two notions, the author poses integration as *thinking in systems*,* and therefore stated as *thinking in integration*. Thinking in integration means developing an approach for investigating the fabric of systems theory and systems integration that is broadly applicable to all experiments, all examples, and grossly produces similar results as the scientific method, yet offers insights that are not apparent.

The following approach addresses the broader question: how best to interpret the results of an experiment involving objects? The approach begins by

> Initial ideation (concept for investigation)
> Statement of the problem (the causal prediction)
> Statement of the need (what question(s) are answered)
> Stakeholders (perspectives)
> Hypothesis (claims about the nature of the variables and effect(s))
> Method (notions, procedures, models/representations): "white box"† tests
> Approach: thinking in systems
> Design of experiment (hypothesis analysis and testing)
> Principles and theory (framework and schemas for asking questions)

Integration Framework

It is reasonable to surmise there are snippets of theory (e.g., general systems theory and social theories) that apply to integration, there are possible theoretical constructs (reference the theoretical worldviews—classical, relational, and mechanistic—described in Section "Nature of Physical Objects"), and there are frameworks that seem to capture key issues, but there is no fully developed

---

* There is a terrifically insightful and significant book by Donella H. Meadows with the title, Thinking in Systems (Meadows 2008). She was the lead author on the book, Limits to Growth. Additionally, thinking in systems for planning was aptly applied for building complex products and services in the early 1980s (Taylor 1981).
† White box method of testing focuses on the identification and evaluation of the internal logic and procedures that are based on knowledge of the workings and configuration of the internals of the object.

theoretical basis for integration. The key may lie with the framework that captures the essential ingredients for an effective theory. While there are specific frameworks that help describe situations of integration in most disciplines, these are *not* particularly well suited to lay a foundation on which to extend their narrowly focused body of knowledge (1) to posit a compendious improvement for the general practice of integration, (2) to identify improvements in current methods of systems engineering, and (3) to apply a consistency of language and meaning to facilitate both these practices and improvements. A general framework characterized by logic, continuity of method, applicability across disciplines, and scalability within the microscale to the capaciousness macroscale would seem rather questionable. A most propitious outcome would be a theory framework that would focus on the eventual goal of helping to show the way to help explain how integration works. Perhaps such a framework might provide additional insight into quantifiable measures, functional metrics, sensitivities of the variables, and a predictive capability. The end result would be a definitive characterization of integration that could broadly be applied.

While we interpret observations, postulate principles, and derive laws (sometimes) beginning with theory, while for integration, theory seems illusive. We can work through theory frameworks based on principles that are fundamental statements that are both comprehensive in their applicability and generally agree with empirical evidence.* Particular care needs to be taken when developing a framework for integration. The framework we seek must be an integrative framework that combines the methods of human (or natural) activities with that of the outcomes of those activities. Integration is for all objects (both natural and constructed). In the most general sense, the framework needs to be relevant to all integrations, regardless of perspective. The framework must encompass the metaphysical bases that have driven our thinking, helped structure our ideas, and supported our theories over the past several thousand years. Narrowly defined frameworks within a discipline must have substance within the general framework or be corrected. This set of requirements could seem to be a forged gauntlet of expectations that when thrown down would challenge a reconciliation of terminology, a reconsideration of earlier results, but most importantly a rather concentrated awareness of what could be gained by exploring the nuances of the framework's structure.

The benefits of a general theory framework for integration are threefold. First, by limiting the framework to methods of human activities and outcomes of those activities, the explanation of integration can be made clear and the discussion clarified by counterposed arguments, both based on principles and assumptions. Second, judgments may be based on the evidence within the framework as validated by empirical observations. Third, in addition to being predictive of relations and explanations of the identified

---

* Laws are recurring rules or collection of recurring rules that have been demonstrated effective under certain conditions.

variables, the framework must suggest new models of practice (and likely, possible extensions of theory).

Simplicity is a desired outcome so that both circuitous and the sublime notions find their places. The ultimate aim is to formulate a theory framework that is both holistic by design and sufficiently introspective to be useful to disciplines (Swanson 2007).

If the framework was particularly theoretical, it might tend to (1) cover a limited number of categories and variability, (2) not contribute to the practice of integration, and (3) overlook what it could not explain. If the framework was overly practical, observations might be made in the absence of their implications; there might be no discernable context for defining relations between variables in the context of what remains unknown; and there would be no reliable method to extend best practices beyond some limits of ambiguity. Systems engineering suffers from an overly practical implementation without a theory framework that is integral to its proclamations and variety of usages (in spite of "standards").

Integration as Mechanism

Objects interact and systems become systems through integration. Working from a sociological neopositivist position—the aim of which is naively stated to associate causal actions with events—a workable framework can be substantiated on theoretical grounds that integration is mechanistic by its nature. The discursive objects of integration—illation, ontology, epistemology, axiology, methodology, metaphysics, and ideological inferences and deductions—form the essence of integration, its nature, that is, the indispensible qualities exhibited by it abstractions, decompositions, and extractions. The concept of integration is itself an integration of cognitive representations or models of reality. If there is logic to integration, then that logic shall (1) be defined as sufficient to support detailed analysis and interpretation of within a framework of relevant variables; (2) be based on a consistent set of assumptions; (3) stipulate the ontology of formalisms that translate into each other; (4) reside within the traditions of epistemology; (5) agree on a narrative that elicits particular interpretations of phenomenology; (6) be a consistent set of metaphysical facts that relate "phenomena as a whole to other genera of existence" (Lewes 1875); (7) support a set of value structures that are at least partially, piece-wise predictable; and (8) apply methodology to define and transform relations into knowledge (Lazarsfeld 1993). The methods of integration are bound in natural processes and inspired by human methods. We refer to integration as a method.

All things in parts that become whole are boffo examples of the processes that result in integration. Integration must deal with the inconsistencies and cross-purposes of all constituent parts, the end result being the integrated whole. The role of integration is not that of a mediator that negotiates or moderates the exchange between objects. Integration is the result of accepting and using what is offered as parts. Acceptance is enforced by a mechanism or a control on a mechanism, or failure of a mechanism or control on a mechanism. The result of acceptance of an exchange between objects is the distinct possibility that integration with that object may ensue.

Integration transforms an interacting set of objects into a system. Objects that exchange EMMI may be interacting, but unless there is a transformation of one or both of the objects, the result of that transformation will not exhibit system properties and attributes.

Every object interacts through EMMI. *If* every object received exactly the same EMMI from the same source at the same time and transformed that EMMI with the same mechanism into the same output simultaneously, then all objects would be exactly the same and their outputs would also be the same. Any deviation from the exactness in any regard signifies difference in the outputs of the objects, meaning the EMMI received will then be different. The greater the variation in any of the parameters of the objects, the greater the differences in the outputs. If the outputs vary, and the receiving EMMI vary, and mechanisms vary, then the variability in EMMI creates a rich environment of EMMI from which to spur diversity of interactions. Should the objects have even a modicum of differences, the variety in the kinds of objects can become quite large after a long period of interactions and integrations (which can change mechanisms substantially).

Each object has its own course of action, reasonably independent of other objects until it becomes entwined in a ProtaSystem or system or system of systems—three structures that exhibit some or all of the properties of a system. With increased dependencies, an object's mechanism may habituate to the norm of its received EMMI within one of these three structures. This means that the mechanisms are exercised by EMMI and the exercise may encourage a preferential response to an EMMI that results overtime in slight changes to the mechanism. Those changes would seem to be to accommodate a slightly higher efficiency in the transformation of EMMI with a resultant slightly lower loss (waster). If this preference does not occur or does not result in a preference, then the objects are merely interacting and in no way have they experienced integration. For example, in the family environment (an integrated object), habituation is the accommodation of an object's mechanisms to the EMMI received. Children's behaviors are influenced by family dynamics—grandparents, parents, and siblings (or their logical equivalents). The mechanisms of the children and that of the direct (first-order) stakeholders in the family are encouraged, energized, or discouraged through interactions. The results of these interactions are for the family members to become aware of the experiences and behaviors of each other, whether deemed acceptable or not. Awareness is at a minimum a conscious appreciation for the manner in which one might deal with another person's behaviors. Whether one acts on that information is an entirely different issue. The point from a mechanistic perspective is that cognitive thinking has taken place regardless of the actions taken. The mechanisms (in this case procedures and mechanisms) have habitualized to their environment through interactions and have not exercised (or discovered) other mechanisms of dealing with a set of EMMI from different objects. A typical example of this is being shown something for the first time and not realizing how to think about it to gain a greater appreciation for what is shown.

Emergence

"Emergent" refers to the unaccountable effects of combining of objects (Lewes 1875). Rather than pit one mind against another in detecting the cause for effects (whether accountable or not), we broaden the view of emergence to any effect that is the result of combining objects through the processes of EMMI is emergence. This more general definition impacts on not only the way we appreciate effects of interaction (i.e., that every interaction has the potential of changing the attribute(s) of one or more objects) but even the least observant or perceptive observer should not be disheartened by their emergence being merely an as yet unexplained phenomena. Conceivably every unexplained phenomena can have an explanation eventually, and therefore the notion of emergence would only be valid until we all gain additional knowledge. It is more satisfying to observe the effects of interaction, and note those changes in attributes that result and then particularly investigate whether any properties have changed. Emergent properties have lasting effects whereas ephemeral emergent attributes are reversible. Emergence is due to traits of an object or objects.

Dynamics of Integration

The combining of acts into activities into processes, the aggregation and correlation of thinking into knowledge, and the enactment of behaviors based on knowledge and various behaviors are future signs of integration. Integration has immediate and continuing influence on society, how individuals think, how individuals and groups behave, and what we build.

Things combining into new things is the result of integration. Systems integration influences society profoundly, but sometimes in subtle ways. How we think, how we behave, and what we build are the manifestations and results of mechanisms of integration. The outcomes of integrating objects are the structures of society, the phenomena of the physical world, our cognitive nature, the products we build, the services we provide, and the essence of existence. We adapt and add to current structures of integrated wholes with newly constructed or changes in artifacts that alter functionality or services or acts that constitute differences in how artifacts are used. Overall, we are affected by a continual change in our behaviors (both individual and social). Combining the objective domain with the social domain results in a dynamic environment, one in which we often struggle to engineer or model effectively. The interplay between the social behaviors of individuals and groups of individuals with that of the physical environment (human-built and natural) changes the stability of social interactions, the physical interactions, and the natural phenomenology. The impacts are nominally destabilizing. These impacts can be observed immediately or felt gradually over time; some aspects short-lived other consequences long-lasting. The importance of discerning that convergence is possible and then recognizing it likely suggests a meaningful appreciation of the impacts of emergence.

Culture is a consequence of such convergence. New needs appear—needs consistent with the desires, capabilities, technologies, abilities, and standards to which a society is accustomed. Changes in physical structures, culture, and a litany of other factors impose determinants on human behavior (Malinowski 1944). This insight is a fundamental step necessary to address the integration of social mechanisms with product functions—the arena of systems engineering and systems engineering project management.

The notion of objects coming together and remaining or forming in a stable configuration is the essential concept of integration. In macroeconomics, integration can be thought of as the removal of impediments that prevent stable associations (El-Agraa 1989). Therefore, successful integration is based on the removal of a political impediment with the causal variables of degree of cooperation and decision making describing the various behaviors formulated in the context of a political framework. The result is a discussion on integration of country economies as a stable political outcome. The mechanism of this integration is the elimination of impediments to enhance cooperation between economies (Pelkmans 1998). Likewise, in designing or merging business processes found in the organizational science field, integration is defined as a mechanism that merges independent processes that result from functional differentiation (Galbraith 1977). When an organization compartmentalizes its business units, there are often mismatched functions from one unit to another. The uniting of the cross-unit functions is termed as integration. These views of integration are adopted in many fields, including business, economics, information technology, and systems engineering. Integration is variously, but similarly, defined in the context of interfaces and data. For example, for organizational information, integration is defined at the network level, the data level, the application level, and the business process level (Giachetti 2004). It follows from these views of integration that the process of removal necessarily precedes the establishment of a whole that exhibits long-term stability. These views are methodological by description.* Further, integration may be viewed on a normative basis—one that is inspired by axiological reference to a standard.

Integration can be thought of as an entity (object or process) acting on another entity in such a way as to change itself, the other object, or both. Such a description follows from the methodological representations of integration, and is consistent with simple models of physics, representative of a widely accepted worldview, and true to various philosophies and principles. Much of what is found in this writing follows such a Newtonian logic.

Integration is the result of interaction and not merely the acts of interacting per se. Interaction is a necessary process to achieve integration. The characteristics of the integrated whole are distinguishable from those of the interacting entities. It is both the activity of interaction and the results of those activities that we distinguish as integration. To explore the nature of integration, it is convenient to work within a framework that characterizes the objects and the concept of change (where "change is the relation between objects separated by a time interval" (Turchin 1977)), the interactions between objects (described as activities), and the consequences of the properties and attributes of objects, their interactions, and change. The integrated whole can be represented by the processes that created its integrated structure(s) and the observable consequence(s) of the integrated whole.

Processes require cognitive direction, purposeful activities that drive from one condition to another (via inputs and outputs), and models that represent cognitive structures as well as corporeal results of the activities. Processes, when combined, form a goal-driven set of cognitively inspired activities that result in physical representations of those thoughts and acts. Managed processes form the mechanisms of integration for directed human artifacts. One dimension of the integration framework is the set of processes that portray the mechanisms of integration.

The other dimension of the framework of integration is the observable consequences of the integrated whole, that is, the product or service. The integrated whole is then said to have various uses—described as product or service functions that are intended to be put to some purpose. Functions are embodied in the structures of physical objects and services. We view the physical embodiments (i.e., objects) through their properties and attributes, the functions (either intended or circumstantial) through their qualities, traits, and performances, and the processes engendered from the product or service through the behaviors of the users. The observable consequences of the integrated whole form from the physical objects. The functions result from the physical juxtaposition of the objects. The behaviors of the users are influenced in some ways by the objects and their physical presentation. The behaviors are difficult to associate causally solely with the objects, as the environmental, social, cultural, and political environments provide an intricate context.

---

* Metaphysical concepts, epistemological essences, and ontological structure are missing.

Overall, the framework of integration is the coalescence of managed activities (to produce a product or service) and the uses (of that product or service).
The managed activities portray the mechanisms of integration while the
uses capture how the integrated whole is capable of acting.

We observe the results of integration as evidenced by our own thoughts.
Thinking is an integral whole of conscious statements, intellectual and physical contexts, critical comparisons and evaluations, analytical and abstract
processes, and individual acts of recall. Our thoughts sometimes cause us to
engage in various behaviors that are sometimes describable as either acts or
activities (i.e., a sequence of acts that have contextual relevance). For reference, a process is a set of activities that are associated with a particular objective. The outcome of an act, activity, or process is a representation or model
of either the cognitive experience or the behavioral experience. The representation of the cognitive experience is behavior, and the representation of the
behavior(s) is something written, spoken, or built.

Integrative Mechanisms

The mechanisms of integration are a universal "cement." The mechanisms of
integration construct, bind, and instigate (or allow for) change in the natural
and social world. The results of integration are both summative and formative—at once being both. Integrations build on previous integrations while
simultaneously forging new arrangements of coalescent parts. Integration
drives the studies of every discipline and field, permeates our thinking about
theory, and guides our research. It seems that it is part of nature's work to
provide for interaction and integration. The animal kingdom endeavors to
build structures (humans build artifacts) and integrate structures (or artifacts) into systems.

We distinguish between those objects that merely interact, leaving no lasting change on one or the other object, with that of a dependency in which one
object relies on the other for sustainment. It would seem a safe presumption
that the essence of integration lies somewhere in the realm of the discussion
on causality. There is academic support for structural realism (also termed
neorealism) and object. But it is insufficient to begin the discussion of integration with causality. While causality is causal, it is not the root of causality. So,
while David Hume has positioned causality as the "cement of the universe,"
unfortunately, causality is a concept with which we have continued to struggle. However, a somewhat satisfying explanation for integration can be
gleaned from a discussion on mechanisms. The relations between causality
and mechanism are often obscured and tortured by confounding symptoms
of causal factors. We sometimes find it satisfying to think in terms of mechanisms. A mechanistic view invokes a sensibility about the relation between
cause and effect. Mechanisms, depending on the philosophical bent, help
determine your view of reality; issues of stability can be thought of as driven
by mechanisms. A mechanism can be thought of as resulting from a process

or a set of processes, an event or sequence or confluence of events, the juxtaposition of something physical, or enactments of something physical.

For human-built products, the mechanisms of integration are the processes comprised of individual acts or combined acts that constitute activities; the events that result from applying the function of a product or service, or the combining or dissociation of objects. These mechanisms of integration construct, bind, and instigate or allow for change in the natural and our social world. Summarizing the general nature of mechanisms: mechanisms have physical structure or result from physical structure, enactments of structure or changes in structure modify behavior(s), and modifications are often observed and measured. Mechanisms of integration are of three types: those that depend on process (Machamer et al. 2000), those that express themselves through events (Bechtel and Abrahamsen 2005), and those that are inherent in the physical domain (e.g., the mechanics of object motion, or avoidance of the effects of object motion) (Glennan 2002). The results of integration are both summative and formative—at once being both. Integrations build on previous integrations while simultaneously forging new arrangements of coalescent parts. Integration drives the studies of every discipline and field, permeates our thinking about theory, and guides our research. Integration is a unifying process that satisfies Parmenides' reality of one.*

Exploring Integration Concepts

Integration deals with the inconsistencies and cross-purposes of all constituent parts, but neither as a mediator nor as an adjudicator. Integration is not the trade space for suppositions either imposed as limitations on a project or as constraints *a posteriori* in the form of allotments of time, cost, or skills. The role of integration is not that of a mediator that negotiates or moderates the exchange between objects. Integration is the result of accepting and using what has been offered *a priori* from the various outputs of the systems engineering lifecycle process: a formative feasibility study that precedes the project; determination of requirements through stakeholder elicitation and analysis; preparation of the design that provides for the general context of the work throughout the solution system's lifecycle; development of an architecture that determines the qualitative worth of the system; and specification of various design models (i.e., representations) that serve as the implementation

---

* It seems to be the nature of enquiry to examine phenomena for patterns, behaviors, and properties. The importance of patterns, behaviors, and properties has been thought fundamental to explaining nature and human habitudes. Whether it is perception or realness, the quest for a set of common, universal observations has rationalized a 2500 year enquiry to discover the essence of the universe. The first to advocate that nature had oneness of character and a reality as only a whole was Parmenides of Elea, [se.500 BC] (Kirk et al. 2009). Parmenides' view of enquiry (What is it that is? What is it that is not? What is it that cannot be?) reinforces unity as the object of knowledge—the universal object of nature. Since 500 BC his tenor of logic has inspired thinking about time, motion, change, and unity.

guidelines for the system's developers (e.g., in development—logic, data, physical, functions, performance, software, etc., and in operations—maintenance, operational reliability, affordability, etc.). The trade spaces examined in each of these systems engineering development stages are inputs to frame the tasks and complete the planning for integration.

Integration is believed to have a reasonably common and acceptable meaning across disciplines, generally expressed as combining things to make a whole. However, there are few mentions of the workings of integration in the literature. How do you combine things? In what fashion should objects be combined? When is a whole a whole? When is a whole not a whole? The subtleties of integration are largely lost or not discussed in presentations on integration, most narrowly focusing on a one- or two-sentence definition. It is as if the subject of integration was unworthy of elaboration, taken to mean the common view of "you know, just do it." From that point onward, the word integration is used with great candor, often applied with zeal to anything that would seem to benefit from being put together with something else.

The subject of integration is the entry discussion for systems, thinking in systems, and systems engineering. Even as a topic for systems science, integration has been substantially ignored. This book offers a look at the essence of integration for the purpose that integration be considered and discussed as an enigmatic science. My aim is to pose the topic of integration as a topic for mainstream enquiry and inference,* to investigate it from beyond the peripheral duties of a few technical artisans, intimate with the skills and techniques of integration. Even for theorist and craftspeople familiar with practical methods and tools, the tendency has been to rely on the sys-

---

* The settled rationale of Western scientific advancement has been to unify existing paradigms of thought around epistemic, methodological, and empirical qualities. Researchers in every discipline apply various principles to developing theories supported by careful examination of phenomenological variables that are thought to be significant. While the wisdoms and knowledge cast during the past several thousand years are closely coupled with basic principles of earthy truths, the procedures of enquiry are driven by methods of experimentation and reason. The verities of biases and perspectives of observation advance knowledge in one of two ways—either through a rather step-wise continuous extension of previous thought and knowledge or as a result of amalgamating precipitous changes in direction of thinking to disrupt an existing paradigm and thrust enquiry into new areas with perhaps different form. The test for knowledge is in our ability to use what and how we think as influenced by emotion and experience. Indeed, the progress of knowledge is merely the integration of inspiration with evidence that is compelling to a preponderance of learned review. This nexus of the theory of knowledge, the principles and procedures of enquiry, and measurable observations is the nature of enquiry that concerns the development of every field of endeavor. Enquiry is the development of "truths" that represent knowledge. The growth in any discipline is immediately hampered if set within a myopic view of domain-specific principles. Without a broad appreciation of abstractions that do or could transcend traditional disciplinary boundaries, precipitous advances are rare. It is the unity of truths which abstracts domain-specific knowledge to a simple, small set of generalizations that has help change our structures of knowledge through better verification and falsification of existing observations. (See Endnote 2 at the end of this chapter.)

tem principles which are apparent across some disciplines and not continue their quest for puzzling information that might challenge and possibly modify an accumulating body of systems knowledge. Pointing out potential new areas of research, reviewing existing "established" knowledge, and posing elementary questions should be wildly endorsed and properly reviewed (Troncale 2006) rather than bewildering the experts or forcing an embarrassing rationalization that their field of expertise did not surface the fundamental nature and contexts of integration. Systems theorists assume that systems are systems, but it will become clear that systems can have great diversity and that one must be careful in extrapolating from one system to another, as there are various types of systems.

Is placing sand inside a plastic bottle integration? Does that activity imply that the whole is the sand in the plastic bottle? Is the sand–bottle object a system? Then turn the plastic bottle upside down. Is that activity disintegration? Is the upside-down sand–bottle object a system? A system upside down? Does the sand–bottle object signify a whole? Does the whole have at least three partitions: the plastic bottle, the sand, and the air? In other words, does the relation between objects satisfy some hypothetical minimal test for the process of integration?

Remove a card from a "full deck" of cards. Is the deck of cards still a deck of cards? Or has the whole (stipulated as the deck of cards) changed so that it is no longer a deck? Does the activity of adding the "missing" card back into the deck constitute integration? When playing a game "with a deck of cards" does the deck of cards still exist in some distributed fashion across several players? Depending on the rules of the game, the deck may again be reconstituted into a full deck. Recognizing the integral whole of the full deck, if these players were asked about the nature of a deck of cards, they might answer they were playing with a deck of cards (or a subset that when combined with the remaining cards would constitute a full deck). Does this card example suggest some type of relation between a deck of cards (as an integral whole of individual cards) and the distribution of the cards according to some rules such that at any moment in the game there exists a relation between the remaining cards of the deck and the distributed cards? In other words, while the dealer acts in such a manner to provide cards under the condition that the cards are to be returned, the cards are still considered to act as a deck. Whether the cards are together or distributed, they remain a "deck" but can be considered to be different levels of abstraction of "deck." While the events of the game change the level of abstraction of the deck (i.e., missing cards are counted as part of the deck), the cards reconstituted as the deck are indeed defined as the deck. However, the two concepts of "deck" are not the same—one interacting due to their placement in the deck, the other interacting card-to-card, but not all cards with all cards. The physical interactions distinguish the two concepts of deck. The functions of the cards are due to the local touching of a subset of cards, that is, the cards that are dealt in "hands" to players representing unique sets of cards that result in functions that drive user (player) behaviors. The functions result from certain sequences or other relations between cards in the "hands" that are discernable by the players. That which could be combined and that which is combined are not the same if integration is as commonly stated and defined. The parts of the unique sets of cards in a "hand" do not represent the whole of the game-play for all rules and conditions of the game. Integration must be more than the mere combining of things to make a whole.

At the most basic level, we first realize that an object can be put together with another object to make something that is comprised of these two objects, neither of which combines with the other in any fashion except as a juxtaposition in space and as events in time.

The abstraction class is comprised of emotions, feelings, and conceptualizations; the social class is comprised of behaviors, money, power, and influence; and the model class is the logical description of a representation of a proposed structure to achieve the best overall effect. The perspectives of the class measurement framework are shown within the domains—objects and processes. For objects, some of the possible key measurable properties are dimensionality, mass, orientation, environment, temperature, pressure, force(s), and motion. For functions, some of the possible key measurable properties are stakeholder need(s), inputs (i.e., the *release*, (e.g., energy, matter (the condition of having mass), information, or capital wealth (e.g., money)), mechanism to transform input into output, condition (or state of the element), regulation (stability of the element), output (performance), and loss (to achieve output performance)). For processes, some of the possible key measurable properties are potential uses (of physical and function items) managed and unmanaged factors, acts and activities, output (results), constraints (time, money, skills, rules, policies), and loss (to achieve results).

Abstraction Classification of Integration

To begin to describe integration we discern the primary classifications of integration—classifications that are completely descriptive of the various aspects of an integrated system. That is not to say that these classifications are only descriptive of integrated systems, rather to say that integrated systems will be represented by the classifications, as well might other kinds of aggregations or combinations of objects and processes. In the early stages of conceptualizing a system, we observe the various types of thinking (which we refer to as commonly defined cognition), the various social mechanisms enacted by people (which we refer to as procedures), and the representations of objects and processes (which we refer to as models or "representations"). The three classifications of integration do not overlap. Therefore, the actions of abstractions span ideas independently of the social or model classifications. We can think of these classifications in terms of business models that delve into the relations between the socioeconomic values in businesses

(Hedman and Kalling 2003; Shaw 2007); within the dynamics of supply chains (Jitpaiboon 2005); for services firms (Ray 2003); and for complex software development organizations (Bodenstaff 2010).

We refer to the class of integration as an *abstraction* integration (or "cognitive integration")—that is, two things connected conceptually and cognitively considered as entities in object or process thinking. The abstraction conceptualizations could be tempered by temporal or cost constraints, influenced by requirements, or limited by project scope without necessarily considering either the procedures needed to carry out the ideas or the representations of those ideas or procedures. It is necessary to think about the objects that are to be built, the functions that are to be delivered, and the actions of the users without knowing how to set in place or document those procedures. By example, reading is an abstraction of visual processing of data taken from a book. Intentions are abstractions, as are feelings, emotion, theorizing, thinking, interpreting, and questioning. Abstraction integration portrays (and structures) objects and processes within the context of the observer's thought structures, but without needing anything more than other abstractions. Abstractions can be used to correlate objects and processes.

Similarly, the concepts involved when playing a game that requires a full deck of cards are representative of integration by abstraction. Removing and not returning any card to the deck destroys the physical "deck"—the deck no longer being fit for use in the game. A deck is a deck, except when it is not a deck. Yet, for the purposes of the game, the deck is still considered a deck in an abstraction sense. The physical and temporal aspects of the location of the cards are accepted conceptually as indicative of the rules of the game, and therefore an abstraction. In this case, the deck is defined as four individual suits (or types of cards) that each have 13 separate cards with the same denomination, for a total of 52 cards. The cards are integrated by abstraction for game-play as determined by the observer's context for their individual cards, according to the rules of the game.

Social Classification of Integration

A second class of integration is termed as social integration, the integration focused on the procedures (or social mechanisms (Moody and White 2003; Reed 2008)) of carrying out the cognitive issues of process and the physical realization of the objects and procedures as a means of documenting the ideas. Unlike integration by abstraction (which is to achieve meaning by cognitive association) and model integration (which is based on the things that represent the product or service), social integration relates the general sociological issues (including economics, political science, and social behaviors) through activities, events, and physical entities. The ways of acting out abstract cognitive thoughts or the involuntary responses to stimuli are considered behaviors. Social integration captures the dynamics of how the products are used by people and how the context and content of communication are intonated through vocalization and gestures for others to hear and see. How the players play their hand dealt according to the rules of a card game is an example of social integration—the bringing together of the individual behaviors of the players within the context of the rules of the game. The strategy involved in the card play belongs to the abstract class, while the written record of the score and the analyses and evaluations regarding the value of the chips belong to the model integration class.

Model Classification of Integration

A third class of integration (model or representation integration) deals with the intended functionality of combining things into a whole. Unlike integration by abstraction where the user ascribes meaning by the association of constituent things (specifically, the abstract notion of one combination of cards winning over other combinations), model integration illustrates functionality according to a purpose. Text is a representation of ideas; a physical image of a broom is a representation of a broom, rather than the broom itself (the broom can also be a representation of a broom); and a symbol is a representation of something such as an object, or cognitive structures, or procedures. The purpose and the game-play is set out in representations (or models) that portray various design specifications. The specifications include diagrammatic forms that illustrate the systems engineered product, the schematics of a physical structure, the information exchanges across a network, and the illustration of flowing energy or pressure in a physical experiment. Examples of model integration include everything that is objectified, such as conceptualizations written down and the embodiments of procedures that are enacted or documented.

For example, in the earliest days of physics, Isaac Newton (1642–1727) formulated three laws of motions. They can be described as (1) no action, no movement, or no change in constant movement (i.e., an object at rest remains at rest, or if in motion continues in constant motion); (2) quantification of action is the amount of matter (mass) multiplied by the rate of change in rate of the motion (i.e., a force is directly proportional to its acceleration (the proportionality constant given by the object's mass)); and (3) for every action there is an equal and opposite reaction. These ideas concerning motion remain foundational to our formulation(s) of physics. They were and continue to be stated in the context of the model class physical domain.

The formal term in systems engineering is for written requirements, documents reflecting design, architecture, integration plans, test plans, verification plans, and validation plans, to mention a few. For the specification plans that are provided to the developers of the product or service, the types of representations include

- Logic and data representations
- Physical entity representations

- Functional representations
- Performance representations
- Quality representations
- Hardware representations
- Software representations
- Usability and user interface representations
- Maintenance representations
- Support representations
- Logistics representations
- Operational reliability
- Manufacturability and assembly representations
- Affordability representations
- Reliability representations
- Behavioral representations

As with abstract integration, the domains of the model class of integration span objects and process, the attributes of things objective and subjective. A product (e.g., a deck of cards) that is manufactured has an intended use. The cards come packaged in a box or wrapper. One of the functions of the physical wrapper is to contain the physical extent to which the deck can be separated into individual cards. Some functions of a deck of cards include 'to entertain' (play games), 'to act as building tiles to construct structures' (build houses made of cards), 'glide' (when 'pitched' in an atmosphere), 'to shim' (act as a jack for leveling uneven supports), 'pick teeth' (flossing), 'jimmy' (open doors, disengage locks), 'act as a paperweight,' 'act as a combustible to start or sustain a fire,' 'cover an eye,' 'to symbolize an abstract or corporeal concept,' and 'act as an injurious projectile.' All these functions of a manufactured deck of cards are determined by the needs and wants of the user. The process domain characterizes the human activities that result in the representations. These processes span the set of acts that when concatenated into like-kind or singularly related activities result in the formation and enactment of a process.

The combination of the abstraction, social, and model classes constitutes the whole of the integration description. These classes link the common set of limitations, the constraints allocated within the project and temporal constraints that synchronize the interpersonal relations, intellectual discussions, and the various corporeal representations for the product or service.

Newton, as observer and subsequently as communicator, was separated from his experiment with the motions of an apple and therefore used the physical domain in which to describe that motion. Carrying out his methods and approach to the experiments falls into the model class (his organization and enactment of tasks or activities fall into the process domain, his milestones are in the function domain, and his experimental apparatus and "laboratory" fall into the physical domain). His intuition, creativity, and analytical thinking fall into the abstract class (his brain was in the physical domain, his activities of thinking were in the process domain, and the events of his discoveries were in the functional domain). His behaviors fall into the social class (his speech and involuntary mannerisms are in the physical domain, his control of language is in the process domain, and his actions are in the function domain). The synthesis of Newton's processes, actions, and physical entities resulted in the communication of his three laws of motion as an integrated set of abstraction, model, and social dynamics. Thinking of modern quantum physics where we observe the influence of the observer's actions on the outcome of the experiment illustrates the importance of incorporating the domains of process, physical, and functions with the amalgamation of the integration classes: abstraction, model, and social.

Consolidation of Thoughts on Integration

That two things can come together in such a way to form a whole, that is, something *new* that is in itself considered complete, is a matter of curiosity. It is both the nature and context of this integration that we observe daily. But that two components somehow can be made to form a whole and that this whole may not resemble its components is mysterious. How can the whole be perceived to have even some properties different from that of its original components? Integration can result in a whole where the properties of the whole are the same as the properties of the individual constituent parts. An example of this integration is the sand in the plastic bottle. The plastic bottle contains the sand, but the containment is not a property. Containment is a condition of the integration of sand in the plastic bottle. The properties of the sand, that is, attributes or characteristics are different and distinct from the properties of the plastic bottle. Whereas the condition of the sand being contained in the plastic bottle is the set of circumstances that circumscribe the sand, in essence they impose a restriction on the sand due to its situation or environment when integrated with the plastic bottle. That an integration should result in a whole that is completely reflective of its constituents, we refer to as a *spurious* system.

We can think situationally of *systems integration* as a constructed balance between the constituent parts and the whole—a set of circumstances and events that once enacted transform the parts into a whole, but without reciprocity to reconstitute the parts from the whole. To illustrate, consider making soup as a metaphor for systems integration. We consider the three classes of integration: abstract, social, and model. Consider the associative feelings that are engendered by the smell and taste of the soup. These feelings may rekindle memories of past experiences, people, or situations. These are abstract in nature, dissociated with present realities, but reminiscent of another time, different circumstances, or emotions. Drawing from a recipe, memory, or trial and error, the plan is to capture that memory and make the soup. The chef gathers ingredients and enacts processes to prepare, combine, and cook (i.e., model) with the intention to provide sustenance, preserve the memory, and likely offer the soup as a means to brand the restaurant and its epicurean delights—the soup offered as the signature dish. The chef, preparer(s), and the soon-to-be eater(s) conduct themselves in appropriate ways, exhibiting manners and deportments indicative of social custom and habits.* Depending on the remembered ingredients and the preparation and processes, the result might be turned into a soup *de jure* or *de facto*.[†] However, unlike the simple admixture of water and salt, for example, where salt can be dissociated from its water bond and both ingredients restored to their original parts. The ingredients for this particular soup undergo an irreversible blending that is more than an interaction. The new signature soup will be the result of integration. The saline solution is a *spurious* system—an assembly of constituent parts revealing its distinct parts before and after processing (e.g., through evaporation). The soup has chemical compositions that are changed by heat.

Integration is more than merely combining or assembling parts to make a whole. Integration is a coalesce of objects interacting in unpredictable ways. Integration, when achieved, may not be repeatable in the same way. The end results involve more than one science, subject, or skill. Integration requires a wide range of activities to put something together, with all fields and disciplines inextricable intertwined. Soup is not conceived, made, or eaten only as product (model class), only through skilled kitchen management (social class), or only because of feelings (abstract class).

The classes of integration are conglomerated, yet specialized and reflective of the differences between them. Their inherent differences are fundamental, and have correspondingly been inculcated into our thinking through our institutions of learning. This book presents the three classes of integration as codependent specializations that are neither normative nor prescriptive. If integration was so shackled to any particular field or discipline, nothing could be discovered, built, or discussed. All disciplines, all fields, and all work transcend the classes of integration as we think to solve problems and

---

* A necessary condition for a successful project is to have broad and substantial stakeholder agreement and stakeholder support. Principle 1: The Principle of Alignment results in better outcomes.

† De jure is defined as the rightful (that which is intended) embodiment of human work in the form of products and services. De facto is the accidental embodiment of systems that are caused by human or natural workings (that which is unintended). De facto systems can be thought of as emergent systems that have developed as a result of circumstances. An example is the introduction of nonnative plants into an area; the dumping of ship's ballast in foreign waters and establishing nonnative species; and the movement of various types of insects (e.g., killer bees) into new regions.

ask questions, gather resources, develop plans, manage "the work," build the products or services, deliver the goods, assimilate feedback on the customer's experiences with the goods, sustain the goods in use, and finally revamp, replace, or dispose. Soup as integration is a nexus of classes (abstract, social, and model), all three contributing at the proper time in the proper way to support the whole throughout its lifecycle. Systems integration is a method of system science; a method of systems engineering; a method of all disciplines; all fields; and all work.

But it is not science to know how to convert one thing into another or put two things together or give something a name. The subtleties of integration are wrapped in connecting things in ways that satisfy the expectations for the degree of dependency between objects (coupling) and the manner in and degree to which the objects relate to each other (cohesion). Defining things and their limits (referred to as granularity) of influence is a significant challenge for integration, as is interpreting the level of utility (referred to as abstraction of categories or hierarchy of uses of the whole). Both abstraction and granularity are particularly vexing for integration. The techniques and formulations of determining abstractions and partitions that are descriptive of things are not science. Integration as a science is part of a continuum of activities starting with an object that is doing something, why it is doing it, and what the consequences of doing it are.

An object that does not interact with another object cannot be integrated, as there is no exchange of EMMI between the objects. In this case the objects are termed as *static*, having no impact on anything else, except conceivably within or on themselves. Thinking about integration presupposes lifecycle issues. When we speak of "having no impact" the implication is there is no impact during the lifecycle of the activities. An object that is static has no impact on any other object during its lifecycle. In other words, there is no interaction with another object, that is, no consequential impact of one object on another (ever). It is likely that static objects exist in the universe. But then by definition, static objects are not detectable. If any exist, we would never know it. Interaction is the correlative influence of two objects on each other.

Static objects are differentiable from *active* objects. First, active objects are measurable, and second, active objects are linked causally to other objects. The object doing something is characterized by doing something. That something is measurable as the release of (e.g., transmitting, sending, causing something to be taken, or acting as an agent with power to convey) energy, matter (the condition of having mass), information, or capital wealth (e.g., money). Collectively, these items are termed the *release*. If the *release* is in some manner taken into another object, the releasing object is said to have *acted on* the *receiving* object. In this initial discussion, the *acted on* object has participated (by taking in the *release*) but not responded to the *releasing* object. The *acted on* object may experience effects ranging from none, to some sort of internal activity, to destruction of some or all of its functionality due to its reaction to the *release*. If the *receiving* object had no response, then the condition of having no impact is satisfied and the *releasing* object is *static*. It is arguable that something always happens to the *acted on* object, but that reaction may not be observable, that is, below the threshold of detectability or causality. If it is determined by reasoning or experimentation, the *acted on* object should have a reaction, but nothing is observed, the reaction is determined to be *unknown*. If the reaction is observable then we need both a perspective from which to consider the observable in the context of patterns in addition to a framework in which to make measurements according to that context. If the functionality of the *acted on* object is degraded or destroyed, then the *capacity** of the *acted on* object 'to manage' the *release* was exceeded. Now, if one object's *release(s)* invokes the receiving object to *release(s)*, the two objects are said to be *interacting*. If the two objects begin to interact in a cause/effect manner, the two objects are said to be *interacting and exchanging* (termed *exchanging*) releases.

Strategy of Integration

To develop an overall strategy for integration, we need to accommodate the types of systems (NotaSystem, ProtaSystem, System, and System of Systems) and bring together the classes of integration (abstract, social, and model) so that the attributes and parameters within the domains (objects and processes) can be related to provide for the causal factors that result in products and services.

At this point, we should be beyond the initial stages of thinking about integration and are now willing to appreciate the nature of integration. What does it mean to not be integrated? It is not integration to know how to configure or test the interfaces between objects and processes. Are the objects necessary, and insufficient? If there is no interaction (direct or otherwise) between two objects, then neither of these objects can be integrated. Without action there is no mechanism to facilitate integration.

Differentiation is determined as a need for integration (Bernstein 2001). If the desire for the whole is to have increased differential value over that of its constituent parts, then the strategy for integration formally identifies the starting and ending structures of integration while recognizing the goal is to increase value (Chapter 2) of the product or service. The value of systems engineering is predicated on its ability to integrate disparate components and a wide range of fields and disciplines. This effort is exacerbated by contradicting requirements constituted at the beginning of the work and then

---

* The maximum power, ability, and extent of storing, containing, absorbing, or grasping.

changed or modified* over the course and execution of the development work. Systems engineering is held to a higher standard than just satisfying the need for providing value through the constituent parts of the product or service. Systems engineering is presupposed to be a redaction of engineering. This view was held by the formulators of systems engineering at Bell Laboratories (Schlager 1956), reinforced by the widespread use by the U.S. DoD beginning in the 1950s with the U.S. Air Force and into the late 1960s with the first U.S. military standards for systems engineering. Even in the near past (Stem et al. 2006) systems engineering has earned the entrusted means of engineering and providing large, enigmatic systems. The problems faced by systems engineering can involve incongruous technologies, components, and systems, each with various intricacies and confounding multidisciplinary problems. The *mischievous* whole exemplified by hundreds of millions of interacting elements is not amenable to simple reductionist methods. Inductive and illative thinking is mandatory. The strategy of systems integration is a second cousin to these discussions. Systems integrators are faced with two problems: first, unraveling the domain parameters as indicated in the objective and subjective frames (objects and process), and second, once partitioned into tasks, these parameterized pieces of work enabled by processes and focused on objects need to come together in an integrated way to provide a network of reliably interacting elements. The elements are the three parts of the subjective domain (cognitive, procedure, and model) and the three parts of the objective domain (objects, functions, and behaviors). The nature of integration is relegated to instance of interactions for objects and processes, humans and products, or humans and services. Detailing the combined objective and subjective frames spells integration. Integration is deliciously detailed—a method in which an insidious mistake is made more distressing by its own consequence.

Differentiation implies power and change. Strategies of integration rely on power—the ability to do what you have sufficient force to do. The result of power is change or status quo. The detectability of change in the inherent traits, attributes, and peculiarities (i.e., properties) of objects or processes is determinable within the context of the classes and domains of integration. That there are objective causalities, objective and subjective measures, metrics, and measurement frames determines that change or status quo can be detected. That there is a sufficiency of power needs to be ascertained. Andy Sage suggested a three-level perspective on applying systems engineering to engineer systems (Sage and Armstrong 2000). Carrying forward with our interpretation of integration through the mereology of processes and objects is strongly suggestive that the drivers for change are judicious use of power. In the social sciences, the requirement to separate the experimenter from the

---

* Changes and modification come from key stakeholders, including customers, users, and project team. Changes are a natural and expected part of systems engineering. However, changes are neither desirable nor acceptable for integration.

experiment is a most arduous task. The social class is inextricably involved with the model and abstract classes in the measurement framework of all three domains. In political science, there is a somewhat pervasive antagonism toward rationalizing power (Diaz et al. 2004).

Power

While much has been written about power in innumerable contexts, including sociology, psychology, and economics, regarding its expression, its relation to causal factors, its capability and ability, and the impacts on the actors (e.g., people, groups, organizations, and governments), very little has been said about power and integration. Such a discussion helps to illuminate some aspects of power that are discussed elsewhere, but perhaps not in the same way. For objects that interact (and sometimes integrate), the concept of power can be expressed in terms of dominance (through overwhelming magnitudes of EMMI) or an ability to influence (through the proper release of EMMI that has effects on another object's mechanisms). At the fundamental level of interaction, power is the limit imposed by one object on its EMMI. In other words, power is both an object's EMMI and the object's constraints that limit another object's access to EMMI. For humans, power is EMMI and access to EMMI. Objects value EMMI as their means to make things happen. One object, in turn, can change the makeup of the EMMI it receives into other kinds of EMMI. Changing the output of EMMI by the rate of releasing outputs, the magnitude of an individual output, the average "power" of a series of individual outputs, and taking care as to which objects have access to the EMMI may bring more power to any object than would otherwise occur. The rate of doing work is another form of power as mediated by an object's output EMMI. The changes in EMMI may be thought of as delegation (in an organizational sense), as veto (in a decisional sense), as a means of sustainment (in a distributional sense), as a means of communication (in an informational sense), as a means of commerce (in a material wealth sense), as a means of influence (in a political or social sense), as a means of leverage (in a economics sense), and as a means of exchanging EMMI (in an integration sense). Fundamentally, every object has a mechanism to transform EMMI into an output; therefore, every object has some measure of power. Metaphorically, we can refer to objects as the constituent parts of products and services, as people, and as organizations of people. Objects that combine into an integrated whole would seem to have more power to influence other objects than do individual objects.

The partitioning (or allocation) of EMMI is sometimes portrayed as exercise of power. In this regard, power has two components, the resource of EMMI and the access to that resource. For integration of a system of systems, the meaningful expression of architecture of this exchange of EMMI is a managed process. How individual systems are given access to or exchange EMMI is reflective of the participation of the individual systems. The management of EMMI within the system of systems structure is portrayed and enforced by the architecture and protocol that affects each system within the system of systems. Essential to managing EMMI is the recognition that the persistence of power is related to the lifecycle of the meaningfulness of the input EMMI, for example, old information has marginal utility. For EMMI to be recognized as power, the architecture must reinforce both the importance of access to receive EMMI as well as the pertinent influence it represents. Whether power is examined as a political relation (McClurg and Young 2010) or a social relation (Nieminen 2005), the results are similar. The conception of power as a relation (Dahl 1969) is the essential driving influence for designing and architecting a system of systems. Partitioning the sending of and access to EMMI is the task of the system of systems integrator.

However, EMMI enshrined as a resource is not power. Rather, it is the access to that resource which represents power, that is, the use of that resource is always mediated by access. Both the resource and access to the resource are power.* Power could be prioritized according to the amount of the resource and the amount of access to that resource. In the subjective domain, the abstract class entangles the thinking prioritization and the means of accessing the resource, the social class of integration encompasses the procedures that define access, and the model class orchestrates the kinds of representations that will be corporeally referenced. In biology, the symbiotic relation between different organisms and plants is often thought to be an essential component in nature's balance. Which class of integration could be ignored? Ignoring the abstract class would eliminate the systems theories that consider the metainteractions occurring across the planet. Ignoring the model class excludes lifecycle analysis from which to determine long-term impacts. Omitting the social class isolates the impact of humans on the environment. At the class level, one scenario suggests that intellectualism drives the need to mitigate "wasteful" social behaviors, which in turn results in "improved" models of behavior. Within the integration measurement framework, the object and process domains provide each class with substantial preferences, always any one over any of the others. The benefit of analyzing the structures of integration reveals the causal relations between factors (influences, circumstances, and elements), mechanisms, and enablers. However, the difficulty with viewing the results of integration is that much of the causal, interpretive facets, and mechanisms are concealed. Processes, behaviors, and events that occur between structuring the integration prod-

---

* Along with the World Wide Web evolved a number of business models that advantaged access over distribution. The concept of power did not change. An example of broadening the power by empowering more people (through greater access and independence is peer product (Koszarek 2008).

ucts (classes and domains) and the final resultant integrated whole are often invisible. This observation suggests a strategy for integration.

Making appropriate investments in key resource areas such as information systems, tools to improve efficiency, training and education for the project team, the customers, and the users* provide substantial paybacks in both systems engineering and integration. For the specifics of the project, these concepts extend to increase the likelihood that integration will provide the highest amount of strategic value through the product or service. Another investment that pays off is for systems that can be modeled. Model-based systems engineering offers the allure of describing both the objective and subjective domains for the objective causalities.

Model-Based Systems Integration

Systems engineering has evolved a more formal approach to integration, albeit still methodological. The U.S. Department of Commerce, National Institute of Standards and Technology (NIST) derive a model of integration for systems engineering starting from a defined systems perspective (Barkmeyer et al. 2003) that is premised on the interplay of models. At the most abstract level, integration is a process (in contrast with this book's careful characterization of integration as a method in which processes play a dominant role) for bringing parts together to form a whole—in this case a system. What is meant, termed as "technical integration," is to ensure interoperability by turning an integratable component into an interoperable component. Technical integration is a procedural list of tasks that identify interface requirements (inputs and outputs) and data entities that when transferred ensure interoperability between the identified components. The systems perspective is comprised of models that describe the organizational structure of resources (the system structure model), the rules for conducting business operations (the policy models), and the logical, physical structures of communications (the network models). These three models form a web of planned interfaces that presumably span the scope of interactions that will result in integration. For technical integration to succeed, the behaviors of the components must be completely represented by the models. The difficulty of building all of the known and emergent behaviors into the models is compounded by the continuing iterations of discovery of what a component does and is supposed to do, and then coupled with the addition of new requirements that naturally result from the interactions with stakeholders during design, architecting, development, and testing, it can completely undermine the efficacy of the models. On large-scale system integration or system of systems integration efforts, technical integration is challenged and has not yet been shown to be effective. As stated in the NIST report, there must be a consistency between the functional and behavioral characteristics

---

* Users who become mixed up turn into Suers.

of components and their roles in the business process activities. Change a function, add a function, change a business process, and the characteristics of the components change. Changes in characteristics at the component level change the interfaces or data or both. Moreover, the NIST report further stipulates there must be technical and stakeholder agreement on the (1) form, fit, and function of information; (2) interpretation of data to be consistent with the business processes; (3) rules signifying the business process enactments to be represented fully by the components and their behaviors; and (4) combined effects on integration of increased time, costs, and reduced performances. Some of the notable recent failures in integration (reference the U.S. Army's Future Combat System, and other GAO referenced failures) suggest that technical integration may not encompass a sufficiency of integration theory to offer a favorable guide for systems engineering. Consistent with the view that system integration is operative at both the system level as well as through its elements, the *Systems Engineering Handbook* published by the International Council on Systems Engineering (INCOSE) states that integration is performed on the system, its elements, and external systems (INCOSE 2008), and the Institute for Telecommunications, U.S. Department of Commerce states that integration is defined as the progressive linking of system components to merge functional characteristics into an interoperable system.

Most Effective Strategy for Integration

After investing in systems engineering and systems integration and improvements in the effectiveness of skills and efficiencies for work, arguably the most effective strategy for planning to integrate a human-built system is first to represent the totality of the system's uses through its system-level functionalities, that is, in terms of a simulation model of what the system will do and how the system will operate when completed. The fundamental difference between the strategies of object-to-object versus object-to-system model integration is the inherent inaccuracies of piecewise continuous structures. The effectiveness of object-to-object integration assumes that the objects can in fact be integrated and further that the interfaces and data exchanges can be identified and characterized *before* the onset of the integration work. Were this assumption of piecewise continuity untrue, the individual tasks involved in the integration effort would need to deal with an unknown set of interfaces, undetermined data types and characteristics, and perhaps different protocols for the data transfers. If the interfaces are known, the data types determined, and the protocols specified, then the effectiveness of object-to-object integration simply rests with no changes being made during development. However, very few (if any) development projects are completely, precisely, and accurately defined before planning for integration (typically begun in the first few months of the development project); very few (if any) development projects experience no changes once integration has begun; and very few (if any) development projects find no errors with the objects or in their EMMI before integration is complete. Even if all the issues, rework, and problems were both tractable and recognized so they could be included in the early integration planning, not all of the daily changes can be accomplished within the period allocated for the specific integration tasks. The allocation of project resources and the requisite skills of the individuals working the tasks constrain the completion of work that is anticipated but as yet unplanned. Further, *all* interfaces and *all* data exchanges for *all* objects need to be defined for the entire system before the assumption of piecewise continuous, successful integration of parts is an effective strategy, deterministically. The issues with piecewise continuous in conjunction with integration are twofold: first, the individual tasks associated with integration of an object-to-object strategy are nondeterministic, and second, the number and duration of iterations to complete the integration is also nondeterministic. Piecewise continuous integration (object-to-object) is inefficient, portending unexpected delays in integration and perhaps unplanned, additional expenditures.

A more effective strategy for integration is to represent the totality of the system's objects, identifying the expected (1) system-level functionalities, performances, losses to achieve those performances, and boundary and boundary conditions; (2) physical entities and their mechanisms, EMMI, boundaries, and boundary conditions; and (3) the expected behaviors from users of the new system as well as their behaviors due to their anticipation of tasks for the new system, their boundaries, and boundary conditions. In essence, it is a simulation model of what the system will do and how the system will operate when completed. A simulation model (Hoover and Perry 1989) uses the variables that comprise relations between the system functions in logic that addresses the impacts of context and environment through system behaviors. System properties and attributes are discussed in terms of objects and EMMI. The importance of using a simulation model to facilitate planning for integration is to predict how each object will interoperate with the system (as a whole). Joining each object through EMMI with the system-level perspective reveals the service each object provides to support system functionalities. The paradigm of integration is not achieved by an object-to-object but rather an object-to-system model. Change the model of the system and simultaneously change the actions of all objects. To integrate is to unite an object with the system model—the result of revealing, identifying, specifying, describing, and detailing the functions enabled by the system's interface with that of the object. Therefore, integration is not object to object, not interface to interface, and not data for data.

Unlike systems engineering that is intensely iterative, integration of human-built systems is system focused to achieve end-to-end performances. Integration is neither systems engineering nor profoundly repetitious. Repetition in integration is expensive, time consuming, and tactically inefficient. Moreover, iterative integration is strategically ineffective.

Human-built systems integration achieves an architecture that provides services to objects with EMMI enabling the mechanisms of objects.

Consequently, system design and architecture are profoundly important to integration. The system architecture is a comprehensive statement of the system's physical configuration and connectivity in terms of its infrastructure of objects (and their mechanisms), support for EMMI, provisos for functions, preference for various performances, and origins and impetuses for losses. The social situation of human-built systems requires integration to accommodate environments and behaviors beyond that of any single discipline.

> Systems integration is the unification of the objects and their interactions of energy, matter, material wealth, and information to provide system-level functionalities and performances.

The systems integrator is concerned with connecting objects through EMMI to provide the requisite emergent properties and attributes.

Axioms of Integration

First axiom—inaction: An object that does not interact with another object cannot be integrated. An object that interacts with another object can be subject to integration, under certain conditions. However, conditions must be right for integration. We consider four types of conditions: conditions that relate to boundaries of functions, objects, and behaviors due to objects (or in anticipation of objects); conditions within each of the interacting objects that cause or enable the continuance of mechanisms whether in their operative or initial phase of action; conditions that maintain the isotropic properties of the objects; and conditions that satisfy the constraints imposed by the *releases*. That is also to say that for integration to occur the conditions must support it. In nature, the conditions appear to be predicated on low-energy interactions, sometimes initiated with one-time actions of substantial energy.

Second axiom—action: An object that interacts with another object *releases* (gives up or loses) EMMI. Friction (due to mechanical drag, electromagnetic fields, gravitational fields, and molecular bonding) counters forces. There is no free lunch. You must lose something to do something. Perpetual motion is impossible. Integration of objects must consider the losses due to the actions of integration and the results of integration. Integration results in a loss caused by the collective releases and acceptances of energy, matter, material wealth, or information.

Third axiom—mechanisms: An object is enabled through the properties and traits of its structure. The intrinsic makeup and circumstances of structures is influenceable by forces (EMMI) to change or be restored. The result is to transform EMMI that is incident onto an object into an output across the physical boundary of that object.

Fourth axiom—interaction: An object that *releases* energy, matter, material wealth, or information in time or space as a consequence of another object is limited. An object interacting with another object is subject to time limitations (e.g., delay time for mechanism to operate and release energy, matter, information, or material wealth, spatial distance between the objects, and types of *releases*). Interacting objects are subject to conditions and limitations.

Fifth axiom—inactivity: A process that does not interact provides no access to "power." Power is encapsulated in resources, human skills, rules, budgets, schedules, or EMMI. Integration requires access to power to be activated and operative.

Sixth axiom—degrees of freedom: A process that interacts with another process *constrains* the other processes or objects. Interaction is a means of giving up, expending, or losing EMMI.

The axioms of integration are indicative of a physical reality that is posed through a small number of simple observations and interpretive logic. Posing a question about the outcome(s) of interaction between objects (or alternatively stated as that which is induced to occur by *releases* from objects) highlights the measures and the measurement of change as important issues.

The functions reside within the physical world: functions are enacted at the boundaries of physical entities and express themselves in the functional world, while processes are carried out by physical entities, as observed in the physical, functional, and behavioral worlds.* In sociology, the concept of functionalism, that is, the functions of individual objects (e.g., people and the boundaries between the physical entities that impact people's behaviors and the people) combine to make the whole of society that is expressive through a set of "norms," "customs," "traditions," and "institutions" (Giddens 1986). Functions are typified by the events which result from enactment of the mechanisms which produce output performance. The framework that captures functions was expanded to include the physical environment (Gailey 1985). The discussion on functions, processes, and physical entities requires the construction of a framework that maps the reality of functions through each of the three classes of integration—abstraction, model, and social.

To investigate the essential objects of integration, a framework needs to reflect causality in a system. A framework needs to be based on a set of principles and laws that have standing both theoretically and empirically. Integration can be thought of as having two provenances: (1) principles which are intradisciplinary and (2) holism. Integration relies on the principles of causality (Simon 1966; Pearl 2001), the principle of perturbation (Langford 2006; Groah 2007), and the principle of action (Langford 2009). Integration extends this paradigm by attempting to encompass metalessons from all other disciplines.

To arrive at a framework that shows improved utility (over that of other frameworks) across many experiments (with some modicum of variability) and accounts for unobserved phenomenon (that which we surmise exists, but as yet have not confirmed), principles representing the best practices of

---

* The term "world" is used generally to categorize loosely. The intention is to "bag" them up so as to not lose something.

human integration and a set of generally applicable principles and fundaments need to be considered as the baseline of experiments. The principles and fundamentals help formulate a theoretical and practical basis for integration. The test for selecting principles and fundaments is to discern their relevance, applicability, and reflectivity of the best practices used by researchers and practitioners in the various disciplines and fields involved with systems integration (either by its execution or from is application for analysis). The framework must be constructed to maintain a balance between theory and practice that are the distilled and abstracted lessons* and essences of empirical observations and anecdotal evidence. This balance is maintained by adhering to a governing set of rules:

- For every theoretical construct (theory, model, framework, or frame), there shall be corresponding best practices that typifies the application.
- For every best practice there shall be a corresponding theory that embodies its use and relates to the framework through its context.

A theoretical construct for systems integration spans all the apparent relations and underlying principles that relate to the empirical observations and conjectural notions the key variables that have been shown to represent the substantial essences of integration. Some degree of verification of each relation and principle is required. The form and method of verification are not stipulated, but must pass scrutiny by reasonable inquiry, analysis, and evaluation (often considered part of peer review). As the relations between key variables begin to form, the categorization may be accomplished in any number of ways—the result being a consistency of defined terms that can be seen replicated from experiment to experiment. Here, the notion of experiment is broadly considered as any grouping of tasks that are goal directed. A project is intended to be an experiment for this purpose. The critical determinant of a theoretical construct is to first ensure the causal objects have been identified; second, define the objects in a manner consistent with their relations and categorization; third, parse the categorizations into like-kind frames (frames reflect the commonality typical in a discipline or field that have proven efficacious in analyzing, evaluating, and predicting in a verifiable way); and fourth, determine the relations between the frames (if any) that when juxtaposed provide a causal mapping between the relations (and thereby the frames). This portrayal is consistent with Michael Stankosky's view that if a subject (in his case knowledge management (Stankosky 2000)) "... is to be applicable, universal, and relevant across all enterprises, and rightly claim its place among academic disciplines in this knowledge age, knowledge management requires theoretical support" (DiGiacomo 2003).

---

* Lessons learned; lessons spurned.

To be effective as theoretical constructs, some means of determining what is most widely used and found to be effective within a discipline or field is described as best practices (or more succinctly stated, those practices that have achieved a modicum of success, but without knowing why they were successful). Best practices are the heuristics and methods used widely in the discipline or field that have been proven (empirically or anecdotally) to provide some degree of acceptability by their practitioners. Often, a hallmark of acceptability is the assumed repeatability of the practice. Other times, the practice is presumed to result in losses that are acceptable to the stakeholders. Some of the best practices in acquisition for products and services provided through systems engineering are modularity, use of commercially supported procedures and methods, performance-based standards, technology updates and insertions that support lifecycle cost affordability, and the funding pilot programs to demonstrate concept feasibility (United States Department of Defense 2010). In some instances, commercial best practices from one industry are incompatible with commercial best practices from other industries and often both incompatible with military best practices (Pennock et al. 2007). Even like-kind industries may have incompatibilities. Often, only a subset of best practices is widely appropriate, suggesting that a more general practice is apropos. For systems integration, we must take great care when considering the broad implications of best practices.

On a procedural level, the same U.S. Department of Defense acquisition guidebook suggests incorporating best commercial practices (evaluation of commercial off-the-shelf equipment, lifecycle planning, fostering strong relationships with vendors and subcontractors, and protection of intellectual property rights), collaborative team environments, modeling and simulation, and electronic business solutions. Best practices now extend beyond a few routines that were found to be useful on some projects, to comprehensive studies that are reflected in the U.S. Department of Defense directives that are mandated by law for all systems engineering efforts: using DoD standard data and following data administrative policies, and assessing information operations risks.

Best practices should be based on principles and standards. The project manager, charged with using systems engineering to deliver a product or service, must grapple with either using standardized agreements, parts, processes, and methods or building and tailoring their own set of principles and project standards. Locally contrived standards are a natural consequence of new technologies and new approaches to design and architecture. Usually the projects are a mix of work that include compliance with standards in many, but not all instances of work. Recognized international and national standards usually represent voluntary acceptance and wide commercial market adoption, but the genesis of new standards are often stimulated by project managers that move outside of the standards to innovate. The project managers and systems engineers look to best practices based on systems engineering worldwide. Systems engineering knows no geopolitical or disciplinary borders.

The sources for best practices come from any place where successful methods and techniques are discovered and used (INCOSE 2007).

Best practices are determined to be what is necessary to be successful. Moreover, best practices are generally based on principles. For example, the systems theory principle that systems can be conceptualized as self-reliant entities that are simultaneously wholes and parts (Koestler 1968) is applied to an issue for managing the project's use of random access memory (RAM) inside computers. By treating RAM as an integral part of the systems engineering process (reviewed at milestones and discussed in status meetings), the reliability of a subsystem can be calculated based on architecting varying degrees of parallel and series processing to manage the reliability of various components in a computer or network. These allocations and reliability calculations assist the systems engineer to determine the real implications of various allotments of memory. Further, the allocation and testing of RAM can be coordinated across the development team and across the phases of development. Since the context for RAM is often performance for products and services, integration is particularly impacted should the allocation be insufficient to buffer the required amount of data (i.e., may reduce performance) or is in excess of the needs for optimum data transfer (i.e., increases costs and does not impact performance). In either case, the reliability of the transfer process may be affetced because of throttling or overflow conditions (Government 2005). For integration, the architecture is the essential guidepost—the roadmap for what is connected to what and how those connections facilitate or squash various system behaviors.

The imposition of following best practices as a dominant view in systems engineer should make systems thinker wary. The reason for standards and best practices is to create an environment of greater predictability in product and service developments that, while new in many aspects, push individual intellectualism to the edges of knowledge. Such projects are the most daunting. New technologies may be immature at the beginning of work and never rise to a level of predictability that endows it to be included in the development effort. Integration of the outcomes of imperfect intellectualism or snarling, untameable technology is difficult and doomed. That systems engineering is sometimes effective in delivering needed performance is often shadowed by costly overrun budgets and schedule slippages (Table 3.1).

Similar to a generalized framework for functions and physical entities, processes are formulated as activities (and primal acts). Processes result in decisions about what to do (in contrast to functions that are the behaviors that the user wants to perform as a consequence of the product). Processes describe the intentions of the architecture. As a process, integration is the combining of a systematic series of actions that take place in a definite manner, directed to bring about a particular interaction between objects and sets of objects.

TABLE 3.1

Systems Engineering Is Sometimes Effective in Delivering Performance

Programs	Program Costs (Total)	Development Costs	Development Schedule	Comments
Crusader artillery vehicle	$4.3 billion (estimated in 2001)	+ 55% (cost excess over initial estimates, as determined in 2001)	+ 26% (schedule slippage in 2001) with the expectation of 14 years of development	Program began in 1999 with expected production in 2008. Program was cancelled in February 2004. It was determined there was no need for the program. Emphasis was for applied technology development.
Comanche helicopter	In excess of $14.6 billion	+ 127% (cost excess over initial estimates, as determined in 2001)	+ 119% (schedule slippage in 2001)	Program began in 1994 with 2 years of systems engineering to determine if requirements were feasible given cost and schedule limits. The result was a determination that an additional $0.5 billion was needed to develop immature technology. A key driver was to reduce operating and support costs. The competing requirements "resulted in an inflexible system solution." Emphasis was for applied technology development.
Caterpillar 797 mining truck		+ 5% at completion	0% at completion Emphasis was for on-time delivery	Program began in 1997 and completed in 18 months. The need was identified in 1996 and the project met the expectations of the customers and users.
NASA FUSE		+ 20% at completion	0% at completion Emphasis was for on-time delivery	By making trade-offs of resolution, bandwidth, and time on-orbit, the program met the critical NASA need.

Source: Data from GAO 2001. GAO-01-288, *Best Practices: Better Matching of Needs and Resources Will Lead to Better Weapon System Outcomes.* Washington, D.C., U.S. Government Accountability Office, 80pp.

Endnotes

1. Numbers for the sake of numbers can invariably and conveniently be expressed in some form of statistics. That averages and distributions represent our world is presumptuous. This may seem polemic, but it is not intended to disparage mathematics. In fact, many of my best friends are numbers. They are who they are and represent what they say. However, not for expediency, not for deference to the trappings of intellectualism, and not for pretence of elegance should thinking be so restricted at the onset of a quest for knowledge. When the hardships of cognitive toil have fallen to gumption, when all that could be wrong is found wrong, and when there is no inconsistency of a worldview, then the journey must begin again with the elegance of mathematics. If one's view was that events are causal and objects and processes are necessary and sufficient, mechanisms would be deterministic. According to this view, however contrived or misguided, there is no chaos, no aggregate, nor a strictly summative notion. There is no like-kind object, and no two objects would necessarily be identical. To this end, the universe is a library filled with information, awaiting sublime, correct reasoning. To this end, there are systems and there is integration.

2. The trade practiced by researchers is to apply notional contributions to theory that reflect established and accepted principles of formal and systematic investigation or examination. The result is an effort to persuade others and bring forth new knowledge that is enacted through the processes and functions of enquiry—the acts and mechanisms of asking and answering questions. Yet what results from enquiry is a discord between theory and fact—an ever-present misalignment of theory, principles, and observations. Often this disagreement is based on interpretations of patterns, behaviors, and properties. At issue is our explanation of phenomena that is derived by postulating principles, and inferring laws. We seek truths through enquiry. But an occasional result of enquiry is a theory that explains certain phenomena and but challenges some of the means of actualizing knowledge. The impact of enquiry can be said to relegate the dogma of present theory to the history of future inferences.

References

Aerts, D. 1983. The description of one and many physical systems. *Proceedings of the 25th Cours de Perfectionnement de l'Association Vaudoise des Chercheurs en Physique.* Les Foadements de la Mécanique Quantique, Montana.

Antonsson, E. K. 2001. *Imprecision in Engineering Design.* Pasadena, CA: California Institute of Technology.

Ashby, F. G. 1962. Principles of the self-organizing system. *Principles of Self-Organization: Transactions of the University of Illinois Symposium.* London: Pergamon Press.

Ashby, W. R. 1957. *Introduction to Cybernetics.* London: Chapman & Hall Ltd.

Barkmeyer, E. J., Feeney, A. B., Denno, P., Flater, D. W., Libes, D. E., Steves, M. P., and Wallace, E. K. 2003. *Concepts for Automating Systems Integration*. Gaithersburg, MD: National Institute of Standards and Technology, 90pp.

Bausch, K. C. 1997. The Habermas/Luhmann debate and subsequent Habermasian perspectives on systems theory. *Systems Research Behavior Science* **14**: 315–330.

Bechtel, W. and Abrahamsen, A. 2005. Explanation: A mechanist alternative. *Studies in the History and Philosophy of the Biological and Biomedical Sciences* **36**: 421–441.

Bernstein, J. I. 2001. *Multidisciplinary Design Problem Solving on Product Development Teams*. PhD thesis, Technology, Management, and Policy Program. Boston: Massachussetts Institute of Technology, 259pp.

Bhatt, S. 2000. *The Application of Power Quality Monitoring Data for Reliability Centered Maintenance*. Palo Alto, CA: Electric Power Research Institute (EPRI), 19pp.

Bodenstaff, L. 2010. *Managing Dependency Relations in Inter-Organizational Models*. PhD thesis, Centre for Telematics and Information Technology. Enschede, The Netherlands: University of Twente, 307pp.

Boldyreff, A. W. 1954. *Systems Engineering*. Santa Monica, CA: Rand Corporation, 16pp.

Booher, H. R. 2003. Introduction: Human systems integration. In H. R. Booher (Ed.), *Handbook of Human Systems Integration*. New York: John Wiley and Sons, chapter 1, pp. 1–30.

Boulding, K. 1956. General systems theory—The skeleton of science. *Management Science* **April**: 661–671.

Breuker, J., Valente, A., and Winkels, R. 1997. *Legal Ontologies: A Functional View*. New York: ACM.

Broersen, J. 2003. *Modal Action Logics for Reasoning about Reactive Systems*. PhD thesis, Dutch Research School for Information and Knowledge Systems. Amsterdam: de Vrije Universiteit, 256pp.

Buckley, W. F. 1967. *Sociology and Modern Systems Theory*. Englewood Cliffs, NJ: Prentice Hall.

Buede, D. M. 2009. *The Engineering Design of Systems: Models and Methods*. Hoboken: John Wiley & Sons, Inc.

Cechini, F., Woolley, R., Ghotb, H., Krueger, M. E., Tighe, W., Lewis J., Jacoby, C., and Rantowich, N. 2009. *Systems Engineering Guidebook for Intelligent Transportation Systems*. California Division of the United States Department of Transportation Federal Highway Administration and the California Department of Transportation.

Chow, C. W. and Van der Stede, W. A. 2006. The use and usefulness of nonfinancial performance measures. *Management Accounting Quarterly* **7**(3): 1–8.

Dahl, P. A. 1969. *The Concept of Power*. New York: Free Press.

Dastou, L. 1994. How probabilities came to be objective and subjective. *Historia Mathematica* **21**: 330–344.

Defense Acquisition University 2001. *Systems Engineering Fundamentals*. Fort Belvoir, VA: Systems Management College, 222pp.

Diaz, C., Sassaman, L., and Dewitte, E. 2004. Comparison between two practical mix designs. *Proceedings of the 9th European Symposium on Research in Computer Security*. LNCS.

DiGiacomo, J. 2003. *Implementing Knowledge Management as a Strategic Initiative*. MS thesis, Graduate School of Business and Public Policy. Monterey: United States Navy Postgraduate School.

Do, Q., Campbell, P., Cook, S. C., and Solomon, I. S. D. 2010. Tailored systems engineering processes and artifacts for small-scale projects. *4th Asia-Pacific Conference on Systems Engineering (APCOSE 2010)*. Keelung, Taiwan.

Ducamp, C. and Lagarrigue, A. 2007. MV² tool presentation: A management tool for the validation and verification of requirements by Airbus. *INCOSE 2007—17th Annual International Symposium Proceedings*. San Diego: International Council on Systems Engineering, 13pp.

Dueker, K. J. and Vrana, R. 1995. Systems integration: A reason and a means for data sharing. In H. J. Onsrud and G. Rushton (Eds.), *Geographical Information Systems*. Brunswick, NJ: Center for Urban Policy Research (Rutgers), 149–155pp.

El-Agraa, A. M. 1989. *The Theory and Measurement of International Economic Integration*. London: Macmillan.

Eriksson, M., Borg, K., and Borstler, J. 2008. Use cases for systems engineering—An approach and empirical evaluation. *Systems Engineering* **11**(1): 39–60.

Fairbanks, A. 1898. *Parmenides: Fragments and Commentary*. London: K. Paul, Trench, Trubner, pp. 86–135.

Feyerabend, P. 1993. *Against Method*. 3rd Edition. New York: Verso.

Forrester, J. W. 1958. Industrial dynamics—A major breakthrough for decision makers. *Harvard Business Review* **36**(4): 37–66.

Gabora, L. and Diederik, A. 2002. Contextualizing concepts using a mathematical generalization of the quantum formalism. *Journal of Experimental and Theoretical Artificial Intelligence* **14**: 327–358.

Gailey, C. W. 1985. State of the state in anthropology. *Dialectrical Anthropology (Dordrecht, Springer)* **9**: 65–89.

Galbraith, J. R. 1977. *Organizational Design*. Reading, MA: Addison-Wesley.

GAO 2001. GAO-01-288, *Best Practices: Better Matching of Needs and Resources Will Lead to Better Weapon System Outcomes*. Washington, D.C., U.S. Government Accountability Office, 80pp.

GAO 2003. *Best Practices: Setting Requirements Differently Could Reduce Weapon Systems Total Ownership Costs*. Washington, DC: U.S. Government Accountability Office, 77pp.

GAO 2009. *Defense Acquisitions: DoD Must Balance Its Needs with Available Resources and Follow an Incremental Approach to Acquiring Weapon Systems*. Testimony before the Committee on Armed Services, United States Senate, Washington, DC: U.S. Government Accountability Office, 25pp.

Garcia-Diaz, A. and Riggins, M. 1985. Serviceability and distress methodology for predicting pavement performance. *Transportation Research Record* **997**: 56–61.

Gershenson, J. K., Prasad, G. J., and Zhang, Y. 1999. Modular product design: A lifecycle view. *Journal of Integrated Design & Process Science* **3**(4): 13–26.

Giachetti, R. E. 2004. A framework to review the information integration of the enterprise. *International Journal of Product Research* **42**(6): 1147–1166.

Giachetti, R. E. and Rojas, J. A. 2007. Simulating coordination of human-robot teams for military operations. *Proceedings of the 2007 Industrial Engineering Research Conference*, Nashville, TN.

Giddens, A. 1986. *The Constitution of Society: Outline of the Theory of Structuration*. Berkeley: University of California Press.

Glennan, S. S. 2002. Rethinking mechanistic explanation. *Philosophy of Science* **69**(September): S343–S353.

Gold-Bernstein, B. and Marca, D. 1998. *Designing Enterprise Client/Server Systems.* Upper Saddle River, NJ: Prentice Hall.

Goldberg, B. E., Everhart, K., Stevens, R., Babbitt III, N., Clemens, P., and Stout, L. 1994. *System Engineering "Toolbox" for Design-Oriented Engineers.* Marshall Space Flight Center, Alabama: National Aeronautics and Space Administration (NASA), 306pp.

Government, U.S. 2005. *DoD Guide for Achieving Reliability, Availability, and Maintainability.* Department of Defense. Washington, DC.

Groah, J., Smoller, J., and Temple, B. 2007. *Shock Wave Interactions in General Relativity: A Locally Inertial Glimm Scheme for Spherically Symmetric Spacetimes*, New York: Springer.

Gross, N. 2009. A pragmatist theory of social mechanisms. *American Sociological Review* **74**: 358–379.

Guenov, M. D. and Barker, S. G. 2005. Application of axiomatic design and design structure to the decomposition of engineering systems. *Systems Engineering* **8**(1): 29–40.

Hamlet, D. 2007. Software component composition: A subdomain-based testing-theory foundation. *Software Testing, Verification, and Reliability* **17**: 243–269.

Harel, D. and Politi, M. 1998. *Modeling Reactive Systems with Statecharts: The STATEMATE Approach.* New York: McGraw Hill.

Haskins, C. 2007. A systems engineering framework for eco-industrial park formation. *Systems Engineering* **10**(1): 83–97.

Hassellbring, W. and Reichert, M. (Eds.) 2004. *Proceedings of the GI-/GMDS Workshop on EAI (Enterprise Application Integration).* Oldenburg, Germany: EAI Industry Consortium.

Hedman, J. and Kalling, T. 2003. The business model concept: Theoretical underpinnings and empirical illustrations. *European Journal of Information Systems* **12**: 49–59.

Henry, D. P. 1972. *Medieval Logic and Metaphysics.* London: Hutchinson & Co. Ltd.

Herald, T. E., Verma, D., and Lechler, T. 2007. A model proposal to forecast system baseline evolution due to obsolescence through system operation. *Conference on Systems Engineering Research CSER.* Hoboken, NJ: Stevens Institute of Technology.

Holland, J. H., Holyoak, K. J., Nisbett, R. E., and Thagard, P. R. 1986. *Induction: Processes of Inference, Learning, and Discovery.* Cambridge, MA: The MIT Press.

Hoover, S. V. and Perry, R.F. 1989. *Simulation: A Problem-Solving Approach.* Reading, MA: Addison-Wesley.

Humphrey, W. S. 1989. *Managing the Software Process.* Reading, MA: Addison-Wesley.

INCOSE 2007. *Systems Engineering Vision 2020.* INCOSE, San Diego, California, International Council on Systems Engineering, 32pp.

INCOSE 2008. In C. Haskins (Ed.), *INCOSE Systems Engineering Handbook, Version 3.1. Systems Engineering Handbook: A Guide for System Life Cycle Processes and Activities.* San Diego, California, 304pp.

INCOSE 2010. In C. Haskins (Ed.), *Systems Engineering Handbook.* San Diego: International Council on Systems Engineering (INCOSE), 382pp.

Jain, R., Chandrasekaran, A., and Ozgur, E. 2010. A systems integration framework for process analysis and improvement. *Systems Engineering* **13**(3): 274–289.

Jain, V. 1981. Structural analyses of general systems theory. *Behavioral Science* **26**: 51–62.

Jitpaiboon, T. 2005. *The Roles of Information Systems Integration (ISI) in the Supply Chain Integration Context—Firm Perspective.* PhD thesis, College of Business Administration. Toledo: The University of Toledo, 279pp.

Kasser, J. E. and Mackley, T. 2008. Applying systems thinking and aligning it to systems engineering. *18th INCOSE International Symposium*. Utrecht, Holland: International Council on Systems Engineering.

Katifori, E., Szollosi, G. J., and Magnasco, M. O. 2010. Damage and fluctuations induce loops in optimal transport. *Physical Review Letters* **104**(29 January): 048704–048708.

Kerlinger, F. N. and Lee, H. B. 2000. *Foundations of Behavioral Research*. 4th Edition. Belmont, CA: Cengage Learning.

Kirk, G. S., Raven, J. E., and Schofield, M. 2009. *The Prescocratic Philosophers*. Cambridge: Cambridge University Press.

Kleindorfer, P. R. and Van Wassenhove, L. N. 2004. Managing risk in global supply chains. In Gatigon, H. and Kimberly, J. (Eds.), *The Alliance on Globalization*. Cambridge: Cambridge University Press, Chapter 12.

Klir, G. 2001. *Facets of Systems Science*. New York: Kluwer Academic-Plenum Publishers.

Koestler, A. 1968. *The Ghost in the Machine*. New York: Macmillan.

Koestler, A. and Symthies, J. R. (Eds.) 1968. *Beyond Reductionism. The Alpbach Symposium*. Tyrol, Austria, Hutchinson of London, 438pp.

Koopman, P. J. 1995. A taxonomy of decomposition strategies based on structure, behavior and goals. *Design Engineering* **83**(Design Engineering Technical Conferences Vol. 2): 611–618.

Korman, D. Z. 2007. *The Naive Conception of Material Objects: A Defense*. PhD thesis, Austin: University of Texas

Koszarek III, W. E. 2008. *Peer Production in the U.S. Navy: Enlisting Coase's Penguin*. MS thesis, Systems Engineering Department. Monterey, CA: United States Naval Postgraduate School, p. 79.

Kuhn, A. 1974. *The Logic of Social Systems*. San Francisco: Jossey-Bass.

Kujawski, E. and Miller, G. A. 2007. Quantitative risk-based analysis for military counterterrorism systems. *Systems Engineering* **10**(4): 273–289.

Kuwabara, K. 2011. Cohesion, cooperation, and the value of doing things together: How economic exchange creates relational bonds. *American Sociological Review* **76**(4): 560–580.

Kwon, D., Watts-Sussman, S., and Collopy, F. 2002. Value frame, paradox and change: The constructive nature of information technology business value. *Sprouts: Working Papers on Information Environments, Systems and Organizations* **2**(4 Article 11): 196–220.

Kyburg, H. E. and Smokler, H. E. (Eds.) 1964. *Studies in Subjective Probability*. New York: John Wiley & Sons.

Langford, G. 2006. Reducing risk of new business start-ups using rapid systems engineering. *Asia-Pacific Systems Engineering Conference*. Singapore: National University of Singapore Temasek Defence Systems Institute.

Langford, G. 2009. Foundations of value based gap analysis: Commercial and military developments. *19th Annual International Symposium of the International Council on Systems Engineering*. Singapore: National University of Singapore Temasek Defence Systems Institute.

Langford, G. O. 1971. *Experimentally Obtained Metastable Atom Excitation Functions for Helium, Methane, and Ammonia*. MS thesis, Department of Physics. Hayward, CA: California State College.

Lazarsfeld, P. F. 1993. *On Social Research and Its Language*. Chicago: University of Chicago Press.

Leonard, H. S. and Goodman, N. 1940. The calculus of individuals and its users. *Journal of Symbolic Logic* **5**: 45–55.

Lewes, G. H. 1875. In R. James (Ed.), *Problems of Life and Mind*. Boston: Osgood and Company.

Lewis, P. J. 1994. *Information Systems Development: Systems Thinking in the Field of Information Systems*. London: Pitman.

Lonchamp, J. 1993. A structured conceptual and terminological framework for software process engineering. *Proceedings of the 2nd International Conference on the Software Process (ICSP 2)*. Berlin, Germany: IEEE Computer Society Press.

Lorenz, E. N. 1963. Deterministic non-periodic flow. *Journal of Atmospheric Sciences* **20**: 130–141.

Machamer, P. 2004. Activities and causation: The metaphysics and epistemology of mechanism. *International Studies in the Philosophy of Science* **18**(1): 27–39.

Machamer, P., Darden, L., and Carver, C. 2000. Thinking about mechanisms. *Philosophy of Science* **67**(March): 1–25.

Malinowski, R. 1944. *A Scientific Theory of Culture*. Chapel Hill: The University of North Carolina Press.

McClurg, S. D. and Young, J. K. 2010. *A Relational Political Science. Political Networks Paper Archive*. Carbondale: Southern Illinois University, 11pp.

Meadows, D. 2008. In D. H. Meadows (Ed.), *Thinking in Systems: A Primer*. Sustainability Institute, White River Junction, Vermont: Chelsea Green Publishing.

Miles, L. D. 1961. *Techniques of Value Analysis and Engineering*. New York: McGraw-Hill Book Company, Inc.

Miles, L. D. 1972. *Techniques of Value Analysis and Engineering*. New York: McGraw-Hill, Inc.

Millard, R. L. 1999. *Value Stream Analysis and Mapping for Product Development*. MS thesis, Aeronautics and Astronautics. Boston: Massachusetts Institute of Technology, 139pp.

Miller, J. G. 1978. *Living Systems*. New York: McGraw-Hill Inc.

Molm, L. D., Takahashi, N., and Peterson, G. 2000. Risk and trust in social exchange: An experiment test of a classical proposition. *American Journal of Sociology* **105**: 1396–1427.

Moody, J. and White, D. R. 2003. Structural cohesion and embeddedness: A hierarchical concept of social groups. *American Sociological Review* **68**(1): 103–127.

Morris, P. W. G. and Pinto, J. 2004. *The Wiley Guide to Managing Projects*. Hoboken: John Wiley & Sons, Inc., p. 213.

NASA 1990. *Manager's Handbook for Software Development, Revision 1*. Greenbelt, MD: National Aeronautics and Space Administration.

Nieminen, A. 2005. *Towards a European Society: Integration and Regulation of Capitalism*. Helsinki: University of Helsinki, 465pp.

Object Management Group 2007. *Unified Modeling Language: UML Superstructure Specification, v2.1.1*, 732pp.

Olson, E. T. 2002. The ontology of material objects. *Philosophical Books* **43**(4): 292–299.

Osorio, C. A., Dori, D., and Susman, J. 2010. COIM: An object-process based method for analyzing architectures of complex, interconnected, large-scale socio-technical systems. *Systems Engineering*. Wiley Online Library (wileyonlinelibrary.com) 1–19.

Parsons, V. S. 2005. Project performance: How to assess the early stages. *26th ASEM National Conference. Organizational Transformation: Opportunities and Challenges*, Virginia Beach, American Society for Engineering Management.

Pearl, J. 2001. *Causality: Models, Reasoning, and Inference*. Cambridge: Cambridge University Press.

Pelkmans, J. 1998. *European Integration: Methods and Economic Analysis*. London: Longman.

Pennock, M. J., Rouse, W. R., and Kollar, D. L. 2007. Transforming the acquisition enterprise: A framework for analysis and a case study of ship acquisition. *Systems Engineering* **10**(2): 99–117.

Project Management Institute 1996. *A Guide to the Project Management Body of Knowledge, 1–930699–45-X*. Pennsylvania: Project Management Institute, Inc.

Project Management Institute 2000. *A Guide to the Project Management Body of Knowledge*. Newtown Square, Pennsylvania: Project Management Institute, Inc.

Quine, W. V. O. 1960. *Words and Object*. Cambridge, MA: The MIT Press, Massachusetts Institute of Technology.

Rahming, K. B. 2009. *Applying Risk Management to Reduce the Time in Lay-Up While Increasing the Cost Effectiveness of a Nimitz (CVN 68) Class Aircraft Carrier in Dry Dock during the Execution Phase of a Refueling and Complex Overhaul*. MS thesis, Systems Engineering Department. Monterey, CA: United States Naval Postgraduate School, 137pp.

Rapoport, A. 1986. *General Systems Theory: Essential Concepts and Applications*. Tunbridge Wells, Kent: Abacus Press.

Ray, A. S. 2003. *Adapting the Building System Integration Method to Portray Architectural Organizations*. MS thesis, Architecture. Texas A&M University, 134pp.

Rebovich, G. 2005. Systems thinking for the enterprise: New and emerging perspectives. *Enterprise Systems Engineering Theory and Practice*. Bedford: Mitre Corporation, 2, 85pp.

Reed, I. 2008. Justifying sociological knowledge: From realism to interpretation. *Sociological Theory* **26**(2): 101–129.

Rocha, L. M. 1999. Evidence sets: Modeling subjective categories. *International Journal of General Systems* **27**(6): 457–494.

Rockwell Software 2004. Arena packaging: Template user's guide. *Template User's Guide*, Wexford, PA: Rockwell Automation, 112pp.

Rodriguez, W. D., Massenburg, W. B., Lengerich, A. W., Stone, D. T. K., and Catto, W. 2004. *Naval Systems Engineering Guide*. Department of the United States Navy. Washington DC: Defense Acquisition University, 295pp.

Rouse, W. B. 2004. Value-centered R&D organizations: Ten principles for characterizing, assessing, and managing value. *Systems Engineering* **7**(2): 167–185.

Rovelli, C. 1996. Relational quantum mechanics. *International Journal of Theoretical Physics* **35**: 1637–1678.

Rovelli, C. 1997. *Half Way through the Woods*. Pittsburgh: University of Pittsburgh Press.

Sage, A. P. and Armstrong, J. E. 2000. *Introduction to Systems Engineering*. New York: John Wiley & Sons, Inc.

Schiemenz, B. 2002. Managing complexity by recursion. *European Meeting on Cybernetics and Systems Research*. Vienna, Austria.

Schilling, M. A. and Parparone, C. 2005. Modularity: An application of general systems theory to military force development. *Defense Acquisition Review Journal* **12**(3): 278–293.

Schlager, J. 1956. Systems engineering: Key to modern development. *IRE Transactions* **EM-3**(3): 64–66.

Senge, P. M. 2006. *The Fifth Discipline: The Art & Practice of the Learning Organization*. New York: Doubleday Currency.

Shaw, D. R. 2007. Manchester United Football Club: Developing a network orchestration model. *European Journal of Information Systems* 16(5): 628–642.

Shingler, J., Van Loon, M. E., Alter, T. R., and Bridger, J. C. 2008. The importance of subjective data in public agency performance evaluation. *Public Administration Review* 68(6): 1101–1111.

Shishko, R. 1995. *NASA Systems Engineering Handbook. National Aeronautics and Space Administration Jet Propulsion Laboratory.* Pasadena: California Institute of Technology, 186pp.

Simon, H. A. 1962. The architecture of complexity. *Proceedings of the American Philosophical Society* 106: 467–482.

Simon, H. A. 1973. *The Organization of Complex Systems.* New York: Braziller.

Simon, H. A. and Rescher, N. 1966. Cause and counterfactual. *Philosophy of Science* 33: 323–340.

Simon, P. 1987. *Parts: A Study in Ontology.* Oxford: Clarendon Press.

Sousa, G. W. L. 2004. *Impact of Alternative Flow Control Policies on Value Stream Delivery Robustness under Demand Instability: A System Dynamics Modeling and Simulation Approach.* PhD thesis, Industrial and Systems Engineering. Blacksburg, VA: Virginia Polytechnic Institute and State University, 268pp.

Sowa, J. F. 2000. *Knowledge Representation: Logical, Philosophical, and Computational Foundations.* Pacific Grove, CA: Books/Cole.

Sproles, N. 2000. Complex systems, soft science, and test & evaluation or 'Real Mean Don't collect soft data'. *SETE2000 Conference*, Australia.

Srzednicki, J. T. and Stachniak, Z. (Eds.) 1988. *S. Leśniewski's Lecture Notes in Logic Nijhoff International Philosophy Series.* Dordrecht: Kluwer.

Stankosky, M. A. 2000. A theoretical framework. *Knowledge Management World* (special millennium issue), p.15.

Stem, D. E., Boito, M., and Younossi, O. 2006. *Systems Engineering and Program Management: Trends and Costs for Aircraft and Guided Weapons Programs.* Santa Monica: Rand Corporation, 199pp.

Stone, R. B. S. 1997. *Towards a Theory of Modular Design.* PhD thesis, College of Engineering. Austin, TX: The University of Texas, 289pp.

Stryker, A. C. and Jacques, D. R. 2009. Modularity versus functionality—A survey and application. *7th Annual Conference on Systems Engineering Research 2009.* Loughborough University, UK: CSER 2009, University of Southern California.

Surma, S. J., Srzednicki, J. T., Barnett, D. I., and Rickey, F. (Eds.) 1992. *Leśniewski, Collected Works. Volumes I and II.* Dordrecht: Kluwer.

Swanson, R. A. 2007. Theory framework for applied disciplines: Boundaries, contributing, core, useful, novel, and irrelevant components. *Human Resource Development Review* 6(3): 321–339.

Taylor, T. C. 1981. Perspectives on some problems of concept selection, management and complexity in military systems development. *Naval War College Review* 34: 55–65.

Todd, H. M. and Parten, D. S. 2008. *A Systems Engineering Approach to Address Human Capital Management Issues in the Shipbuilding Industry.* MS thesis, Department of Systems Engineering. Monterey, CA: United States Naval Postgraduate School, 247pp.

Troncale, L. 1977. Linkage propositions between fify principal systems concepts. *North Atlantic Treaty Organization Conference Series: International Conference on Applied General Systems Research*, New York: Plenum.

Troncale, L. 2006. Towards a science of systems. *Systems Research and Behavioral Science* **23**: 301–321.

Tronstad, Y. D. 1997. A business systems engineering model architecture. *Proceedings of the Seventh Annual International Symposium of the International Council on Systems Engineering*, Los Angeles.

Trumpf, J. 2002. *On the Geometry and Parametrization of Almost Invariant Subspaces and Observer Theory*. PhD thesis, Mathematics Department. Wurzburg: Universität Wurzburg, 206pp.

Turchin, V. F. 1977. *The Phenomenon of Science—A Cybernetic Approach to Human Evolution*. New York: Columbia University Press, p.128.

Tvaryanas, A. P. 2010. *A Discourse on Human Systems Integration*. PhD thesis, MOVES Institute. Monterey, CA: United States Naval Postgraduate School, 642pp.

United States Defense Acquisition University 2003. *Risk Management Guide for DoD Acquisition*. Fort Belvoir, Virginia: Defense Acquisition University Press.

United States Department of Defense 2010. *Defense Acquisition Guidebook, Version 5 May*. Department of Defense. Washington DC: Defense Acquisition University, 310pp.

United States Department of Defense 2011. *Defense Acquisition Guidebook, 29 July 2011*. Department of Defense. Washington DC: Defense Acquisition University, 381pp.

Valerdi, R. and Davidz, H. L. 2009. Empirical research in systems engineering: Challenges and opportunities of a new frontier. *Systems Engineering* **12**(2): 169–181.

Van Wie, M., Bryant, C. R., Bohm, M. R., McAdams, D. A., and Stone, R. B. 2005. A model of function-based representations. *Artificial Intelligence for Engineering Design, Analysis and Manufacturing* **19**(2): 89–111.

von Bertalanffy, L. 1928. *Kritische Theorie der Formbildung*. Berlin: Borntraeger.

von Bertalanffy, L. 1968. *General System Theory: Foundations, Development, Applications*. New York: George Braziller.

von Böhm-Bawerk, E. 2005. *Basic Principles of Economic Value*. Grove City, PA: Libertarian Press, Inc.

Wiener, N. 1948. *Cybernetics: Or the Control and Communication in the Animal and the Machine*. Cambridge, MA: MIT Press.

Woleński, J. (2000–2001). In S. J., Surma, J. T., Srzednicki, D. I., Barnett, and V. F., Rickey (Eds.), *S. Leśniewski, Collected Works*, Vols. I–II, Nijhoff International Philosophy Series, Vol. 44/I–II Dordrecht: PWN-Polish Scientific/Kluwer Academic Publishers, 1992, pp. xvi + 382 (Vol. I), pp. 383–794 (Vol. II) ISBN 0–7923–1512–X. *Modern Logic* **8**(3 & 4): 195–201.

Zadeh, L. A. 1997. Toward a theory of fuzzy information granulation and its centrality in human reasoning and fuzzy logic. *Fuzzy Sets and Systems* **90**(2): 111–127.

Zafirovski, M. 2005. Social exchange theory under scrutiny: A positive critique of its economic-behaviorist formulations. *Electronic Journal of Sociology* **2**: 1–40.

Zyphur, M. J., Islam, G., and Franklin, M. S. 2006. *On Money and Valuation*. Saussure, Applied Semiotics/Sémiotique appliquée.

4

Systems

The fundamentals of interaction are predicated on the action of one object influencing another object. The objects may be same or different. For our purposes, an object is something that exists (i.e., has meaning in the physical world). This is not to say that *things* that are not physical have no meaning. It rather implies that there are things that exist that are not objects, and that are considered to exist yet differently than as objects. Our mereology distinguishes between objects and processes, where processes are comprised of things not physical. Objects have properties and characteristics, and can be considered either a fundamental constituent (at subatomic dimensions) or an aggregation of sorts of other objects (parts). We use the word *object* to signify an item of unit importance that is appropriate and consistent with the level of analysis necessary and sufficient to relate objects and processes to integration.

Systems are made up of objects. For convenience, levels of detail are often related by the single descriptive word *object*. Development efforts generate words such as components and subcomponents, assemblies and subassemblies, and modules and submodules to coincide with a particular phase of work, some measure of a task, or in a gross sense, a category of like-kind entities. In a hierarchical fashion, greater detail is relegated to lower-level positioning in that hierarchical description. We have defined a framework in which to organize the variables of groupings of objects. Some of those groupings may be systems or system of systems, some may not. Those that are not systems fall into two categories: not a system (NotaSystem) and a prototypical system (ProtaSystem). NotaSystem has only a few properties of a system and ProtaSystem has many. There are no characteristics of NotaSystem that differ from two objects interacting occasionally. A single object receives no EMMI, is therefore not under the influence of a force that activates its mechanism, and consequently has no output. A single (by definition) noninteracting object is NotaSystem (by definition). However, two objects can be distinguished by their properties, traits, and attributes as NotaSystem, a ProtaSystem, a System, and a System of Systems.* That objects have other characteristics has been covered in the previous two chapters. Those characteristics include properties (intrinsic to the object), traits (properties posed in

---

* Recall that we do not distinguish between objects by their design or amount of detail. An object is an object.

a particular context, such as environment), and attributes (those items that are neither intrinsic properties nor traits, but associated with the object).

From Chapter 1, we summarize the principles of integration, and then discuss the implications of those principles with reference to systems.

Principle 1: The Principle of Alignment: Alignment of strategies for the business enterprise, the key stakeholders, and the project results in better outcomes for product or service development.

Principle 2: The Principle of Partitioning: Partitioning of objects can create tractable problems to solve if and only if boundary contiguity is achieved.

Principle 3: The Principle of Induction: Inductive reasoning should guide integration management and recursive thinking.

Principle 4: The Principle of Limitation: Integration is only as good as architecture captures stakeholder requirements.

Principle 5: The Principle of Forethought: Integration is a primary, key activity, not an afterthought considered as a result of development.

Principle 6: The Principle of Planning: Integration planning is predicated on pattern scheduling (lowest impact on budget), network scheduling (determinable impact on budget), and ad hoc scheduling (undetermined impact on budget).

Principle 7: The Principle of Loss: When two objects are integrated, both objects give up some measure of autonomous behavior.

From Chapter 2, we defined the mereology of systems as objects and processes. The framework for integration is shown in Figure 4.1.

From Principle 2: The Principle of Partitioning, we note that objects can be grouped and partitioned, functions may or may not track with groupings of

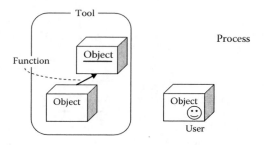

- Direction of arrow shows movement of EMMI from end to head
- Dashed line illustrates the perceptions from the user's perspective

FIGURE 4.1
The mereology of systems.

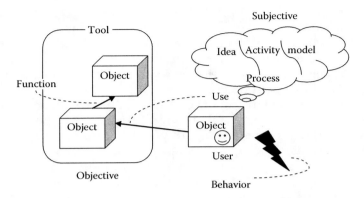

- Direction of arrow shows movement of EMMI from end to head
- Dashed line illustrates the perceptions from the user's perspective

FIGURE 4.2
The ontology of systems.

objects, and behaviors may track with objects and have somewhat some degree of coupling and cohesion with functions. The ontology of systems is depicted in Figure 4.2.

One object releases EMMI that impinges on another object. Should the objects exchange EMMI, an interaction is said to have occurred. The release of EMMI from object "A" is "a"; the release of EMMI from "B" is "b". Release "a" causes object A mechanism "A_m" and control "A_c" to produce output "A_d". "A_d" causes mechanism "B_m" (of object "B") with control "B_c" to produce output "B_d".

Exchange interaction "A with B" results in no change in "A" or "B". Sustained exchanges require drawdown of resources from both "A" and "B" as both experience losses attributable to their releases. A and B can be assembled as A and B. For example, under optimal conditions, two distinctly different types of musical instruments can be distinguished by the human ear. When the two instruments are played together they present as sounds A_s and B_s. The playing duet is an aggregation of A and B. The musicians are A_p and B_p; the music is A_x and B_x. The conduction of the two musicians follows the cognitive structures that drive orchestration processes and are written down. A and B are juxtaposed to gain the optimal experience at one location in the music hall. The plan (design) is laid out, the players are informed and working as a team, and the physical layout of the musicians is followed. The totality of the layout, music, orchestration, and conduction is the architecture of the concert (a process). The concert is carried out; the audience distinctly hears both instruments and their harmonization. The concert is over, everyone leaves. NotaSystem? ProtaSystem? System? System of Systems?

The service provided by the concert is entertainment. For integration classes consider the classes: abstraction, social, and model. The abstraction class of integration deals with the emotions and feelings of the players and conductor and audience. The social class of integration deals with the behavioral mechanisms of people. The model class of integration deals with the physical representations of the concert.

Consider the boundaries of the objects and processes, that is, the greatest extent in which an object, combinations of objects, functions, and behavior due to the presence of object (or anticipation of the presence of objects) has on the stakeholders during the lifecycle of these same objects and processes *and* all other processes and objects with whose lifecycle they may interact. Each of these concepts relates to stakeholders; therefore, the number of stakeholders for the simplest of systems may in general be quite extensive. For human-built systems, changing or establishing the boundaries may change the *dimensions* of the physical, functional, behavioral, or processes. A pertinent issue for human-built systems is the impact of not including stakeholders whose lifecycle processes, physical entities functions, and behaviors interact in a causal manner with that of the system under evaluation. In the case of developmental products and services, missing stakeholders may represent requirements that are omitted from consideration. If those requirements are identified, the system will need to incorporate them, ignore them, or plan their inclusion during an upgrade phase. Of particular interest are the integration aspects of the *concert* (generally referring to all objects and processes). Performing a stakeholder analysis is key at this point. The implication of the analysis so far is that there will be a number of hidden requirements that need to be brought out with objects and procedures during the integration process. Flexibility and scalability must be built into the *concert*.

Systemness

A system requires sustained interactions that arise from a sufficient density of the appropriate types of objects with the appropriate types of mechanisms, fed by the appropriate types of EMMI. The bringing together of objects and EMMI in sufficient densities and interactions is a minimum step in integration. When integration occurs, a system is said to have come into operation, that operation is stable, and has a lifetime. Before integration occurs, however, parts of the system may exist in various forms (termed as a ProtaSystem). A ProtaSystem is both a primitive system with inconsistent or partially enacted sustainment mechanisms, with the first manifestations of systemic behaviors. These behaviors may include (1) equifinality (various sorts of degeneracy that are inherent in the existing structure of objects);

insufficient density of objects of the kind and location needed to sustain interactions; inadequate EMMI needed for sustainment, including squelching of the mechanism due to saturation, below threshold inputs, and insufficient suffusion; and incompatibility of EMMI and mechanisms (Troncale 2011); (2) isomorphicity—the similarity of mixes and kinds of objects observed in systems (von Bertalanffy 1968); and (3) inadequate or inappropriate emergence(s); and unsustainable losses of EMMI from any cause. Integration is the process of setting up or by chance satisfying the conditions that lead to an integrated set of objects (i.e., system).

The integration framework is shown in Figure 2.8. The framework shown is an expanded 3×3 matrix. The relations between the subjective domain and the objective domain are spelled out as the items in each frame intersect. Objects are designated as products and services; processes are designated as abstractions (cognitions), mechanisms (procedures), and models (representations).

Objects achieve their usefulness from a systems perspective when connected and interacting. The object's mechanisms and properties provide capability when not connected, but no uses are possible without interaction with another object.* Objects behave differently when connected compared to when not connected.

Emergence

Consider a simple pendulum suspended from a wooden beam above the floor, which leaves freedom of motion in circular motions and only restricted in vertical motions by the height of the beam above. Let the pendulum swing in the Earth's gravitational field. The pendulum is comprised of a string (typical of that used to fly kites—lightweight, pliable, highly susceptible to transverse forces (gets pushed around a lot)), but quite strong in the longitudinal direction; tightening is braid when pulled; a mass, m, that is one hundred times more massive than the string to which it is secured by a tied knot; a point of suspension, comprised of a loop of high-grade brass which is screwed into the beam and securely tightened to the point that only a mechanical steel tool is capable of loosening and removing the brass loop; the beam is massively supported by a structure of walls and joists, all of which have a mass in excess of one million times that of mass, m; the structure is securely embedded in a foundation that rests on terra firma.

---

* A person (object) picking up a hammer (object) is interacting with the hammer (from the perspective of the person). Without the interaction between the two objects, no functions exist, no use is attempted, no performance is accomplished, and no result is achieved.

There are several aspects of the pendulum that are germane, including the function of friction that occurs at the connection of the string to the brass ring; the maximum storage of energy in the string-mass object at the highest point in its swing; the vector force of gravity (presumed to be radially toward the Earth's center); the frictional force of air molecules on the moving mass (from the perspective of the moving mass); and the initial input of EMMI that displaces the pendulum and starts the pendulum's oscillations.

The initial position of the mass m is either displaced from the vertical that is established along the line between the pivot and the center of Earth's mass or along that centerline to the Earth's center of mass. In either case, energy will be expended. If the mass is not displaced from the centerline, then additional displacement would be required to raise the mass along the circumference of a circle with radius of length of string, l. If displaced, the mass wall "fall" and be put into motion along the arc of the circle that is scribed by the string from the pivot point at the location of the brass ring.

The motion of mass will continue along the arc (precessing due to the Earth's motion beneath the swinging mass, as permitted by the manner in which the string was tied to the pivot point). The interaction between the pivot material and the string's material is expected to create friction with the resultant loss of energy of the string-mass object due to its rubbing. There may also be additional losses due to the string and mass colliding with atmospheric molecules, or perhaps eddies and currents of low and high densities of air molecules (referred to as turbules). All effects of friction impart losses to the string-mass object. The net result of these effects will be to dampen the swinging motion of the mass, resulting in a decay in the height that the mass will return to on each swing. Over time, this dampening motion will result in the mass being returned to its nondisplacement position along the line drawn from the pivot point on the beam to the Earth's center of mass. Missing from the objects needed for sustainment is a source of EMMI needed to overcome the losses, that is, there is no restoring force to make up for the losses that are expected from the pendulum experiment. An object will provide the initial EMMI to begin the oscillatory motion of the mass relative to vector representing the centerline of gravity. It is expected that the simple pendulum will become an object that will swing for multiple minutes before losing sufficient energy so as no longer reach one-third of the height displacement of the first swing position of the mass. The string places a constraint on the mass and the mass constrains the string. The mass is restricted to the swing and will not move lower than the length of the string and the string is pulled taut, acting more like a rod than a limp twine. The performance of the string-mass object can be stated as maximum displacement of mass, m, or the height of mass m above some measurable distance from the floor or from the beam. Identifying a standard of measurement will provide a consistent error in distance regardless of the location of the string-mass object during it oscillations. A measure of effectiveness might be the number of oscillations per unit time or the rate of change in the height of mass m above the fixed reference point. The effectiveness reflects the ability of the string-mass object to minimize its losses and maximize its performance within the context of the connections and environment.

Interactions between the ground and the building structure, the structure and the beam, the beam and the brass ring, the brass ring and the string, and the string and the brass ring reveal emergence, that is, the trait of tension on the string. Emergence is necessary for integration, but by itself is insufficient.

There are three types of emergence: intrinsic, temporary, and reversible. The intrinsic emergence results with a change in the properties of one of the objects participating in an exchange of EMMI, that is, an interaction. The property changes of an object that are irreversible become permanent emergent properties. Changes in attributes (e.g., change in the markings on the mass or change in the checking of the beam, or change of clothes for one of the experiments) are reversible. Reversible changes showcase temporary emergence. Temporary emergence can be sustained for as long as the appropriate EMMI is available and the context and circumstances support the emergent attribute(s). Irreversible changes in properties due to interactions are typical of emergent properties of an object(s). It is possible to have reversible changes in properties given the appropriate EMMI and conditions and circumstances (termed as reversible emergence). Such reversible changes in properties represent stable changes in objects that are stable in two EMMI conditions—one before the (reversible) change and one after the change has been reversed. Both situations are stable, the one that represents the change is termed as metastable. The (reversible) temporary emergence is also termed as metastable. Constraints placed on objects result in emergence (all three types).

Identifying the function(s) of an object begin by noting how the object is and can be connected to other objects. Once connected, determine what the various combinations of connection and objects can do (in the physical, functional, and behavioral sense). Since the function derives its existence from the interface between objects (with the object's mechanisms working in concert) to support the needs to maintain the interface, a requirement for interface stability is necessary.

Interface

Interfaces (in their simplest form) are comprised of two objects connected together by EMMI. Interaction requires both objects to exchange EMMI. That means the connection derives from EMMI. For example, consider two similar pieces of wood (similar in terms of properties, traits, and attributes). One piece of wood placed on top of the other piece of wood provides a connection via physical contact. The two pieces of wood will remain connected (in contact) as long as their forces of friction generated by their resistance to movement are stronger than the forces imposed by external events to move one piece relative to the other piece. The function of the two pieces of *connected* wood might be to make a cavern in which to build a fire and create a thermal cavity for generating a higher temperature (e.g., building on the concept of cavity radiation). There is no force expected that would move one piece relative to the other, so with gravitational and lateral stabilities the fire can reach higher temperature with a stable cavity in place, that is, the dimensions of the cavity remain constant. In this case, the function of the connection between the two pieces of wood is to create a physical cavity.

That objects can be configured to provide functions is an inherent trait of interacting objects. The interface between two objects can be viewed as a boundary, that which separates one object from another object. That there are three types of boundaries (physical, functional, and behavioral) points integration efforts in three, sometimes different, directions. Whether the physical boundary of the product or service is the most important aspect is normally ill defined in the set of initial requirements. In general, products and services are considered to be tools for carrying out particular functions. Since functions are distinguished only by the type of function, the performance(s) of the types of functions, and the quality related to the performances (sometimes, quality is ascribed to the function directly—functions with high quality equates to a product or service of high quality). Therefore, most new products or services angle to take advantage of advancements in technology, so that new products or services offer different functions (or combinations of functions), improvements in performance or quality (or both), or show increased value for the same sets of functions (performances and quality) by delivering at a lower price. From an integration perspective, functions are functions, only the objects and processes may change. As such, boundaries become more important for proving functionality and significantly important when integrated to achieve a systemic relation between constituent objects. The interface is the result of objects and EMMI. Managing an interface means managing the symptom, not the cause. The distinction is mighty when integration efforts are confronted with functions that fall short of their performance requirements. Passing unit tests and interface checking should not suggest nor necessarily imply that the risk has been reduced in a development effort. And neither does passing these early development tests mean that earned value is as calculated.

If an earned value calculation indicates 25% of the work is completed and is on budget and on schedule, *and* the work completed for early testing indicates that all is progressing "nicely," the underlying premises may be faulty. If integration budgets swell to 50%, then the risk apportionment should be normalized to the allocation of risk across like-kind projects. Currently, earned value is based only on the project that is ongoing. Rather, it would be interesting to determine if any advantages accrue to an index baseline of "standards of reference" to gain better insight into the problems faced by integration activities.

Functional Analysis

Fundamentally, functions serve as the means to constrain an object of which it is a part through connection. Constrained objects, in turn, constrain processes. And, if processes are constrained by budgets, schedules, skills, scope, policies, or rules, then objects and functions are constrained. Functional analysis attempts to partition and provide more detail to delineate functions so that they can be mapped into objects that can be built and integrated. Systems engineering perform decomposition, analysis, and synthesis in a highly iterative fashion to weed out overlapping and underlapping functions, provide unencumbered interfaces, and provide a workable structure of events to facilitate clear and unambiguous integration.

The value of functions is to describe either the intended or incidental uses of connected objects. Objects can be of two types—physical and intellectual. Function can be explained in terms of performance of objects and by the individual losses of the objects attributable to achieving those performances. Functional analysis explains the capacities of an object (Cummins 1975) through these performances and losses. Functional analysis views an object by its possible uses given those performances and losses, in other words, by their mechanism(s). Functional analysis is widely applicable and found in use from the physical world of engineering to biology (e.g., cardiovascular system, neurological system, endocrine system), architecture (e.g., placement and access to services; views of nature and lighting of streets and buildings), sociology (e.g., the set of rules governing safe travel by vehicles on the roads; the criminal justice system of incarceration versus supervised rehabilitation in half-way houses, the communication of an order to evacuate in the face of a pending disaster by civic leaders), and economics (e.g., the introduction of wireless smart-card technology to aid in commerce crossing international borders, and identification of card holders via biometrics).

Functional analysis is often thought about and described as a hierarchical decomposition of terms from top-level abstractions to lower-level details. From the perspective of objects, no such reductionist portrayal is accurate or representative of how functions are constructed or conceived by users (objects). One can think of a network of objects or more appropriately, an interconnected set of objects whose EMMI interoperate as per design or matter of expediency.

Cummins (1975) makes the case that functions are inherently related to behaviors of objects. In that interpretation of relations between properties and traits of physical objects, functions and their induced behaviors would seem essential complements to the roles of the user. That view is suggestive of the functions as being both part of the decision-making processes of the users and the corporate enactments with which the objects are employed to gain access to the desired performances. These behaviors are both formative when anticipating the use of an object and summative when actually using the objects. For integration, the functional boundaries with the users are not tested to any extent as part of the systems engineering activities. Some degree of consideration is given to the human systems integration concept. However, full implementation in an operational environment has been traditionally generally left to the user. That is changing. As an example, the role of the acquirer (i.e., the customer) is moving from that of the buyer who takes delivery to that of a participant in the decision-making processes of the project. The customer has begun to identify their ownership of "trade-spaces." Taking on this role of an active participant will precipitate changes in the integration strategy, the extent of testing and integration, and the purview of development. Traditional development may extend in the "after-sale" environment to bring about integration of processes. The result might be to better target interoperability requirements, more focused integration needs, and alignment of product and service architectures with that of the operational environment, typically a system of systems environment.

Systems and Integration

There are five necessary and sufficient conditions for integrating to achieve a proto-system (whether natural or human-built). They are the requisite number of objects, the kind of objects, the density of objects in a region of interaction, the adequacy of the EMMI in terms of rate (average EMMI for a period of time), and the specific characteristics of the EMMI. The factors that drive objects and that are found causal can be mixed to achieve various levels of systemness. The factors that are significant are object, boundary, function, property, trait, attribute, output, self-reliance, control, and performance. The duration of lifetime or stability of systemness is determined by the boundary conditions and variances about the performances of the functions. Within these categories of factors we find emergence (trait) and trust (self-reliance). Table 4.1 compares the types of systems with these factors. The systems listed are NotaSystem, ProtaSystem, System, and System of Systems. These system types are summarized by a one- or two-word characterization of the type.

Systems have properties that are different from those of their constituent objects. Systems show emergence related to the properties of their objects and in aggregate reveal themselves through system properties. This type of emergence is different from that of interacting objects, which is reversible.

TABLE 4.1

Factors That Determine Systemness

Factor	No Interaction	NotaSystem Mutual	ProtaSystem Consolidate (Synthesis)	System Integrated	System of Systems United
Object	One	Two (minimum)	Two (minimum)	Two objects, each with three mechanisms, minimum	Two objects, each with minimum of six mechanisms
Boundary	One	At least one object's boundary is extended physically	Objects' boundaries are extended physically	Objects' boundaries are extended physically, functionally, and behaviorally	Objects' boundaries are extended physically, functionally, and behaviorally
Function	No functions	No new functions, each object retains its functionalities	Function(s) added due to changes in mechanisms, resulting in new objects	New functions added due to changes in mechanisms, many new objects results	New function added due to changes in objects
Property	Intrinsic	Intrinsic	A few objects with irreversible properties change to new objects	Irreversible properties at the system level	Reversible properties at the system level
Trait	Property in context	Property in context, no change in mechanisms	Object's properties in contexts with each object's properties	Objects' properties in context with another objects' properties	Objects' properties in context with another objects' properties
Attribute	Characteristic of like-kind objects	Reversible changes in attributes	Some irreversible changes in attributes	Many irreversible changes in attributes	Reversible changes in some attributes
Reciprocity	None	Minimum interaction (requires two objects)	Minimum required to sustain some interactions	Satisfies the law of system reciprocity for some EMMI	Satisfies the law of system reciprocity

continued

TABLE 4.1 (continued)

Factors That Determine Systemness

Factor	No Interaction	NotaSystem Mutual	ProtaSystem Consolidate (Synthesis)	System Integrated	System of Systems United
Output	Single object	As independent objects	Some objects' outputs may be synergistic	Object's outputs affect system-level outputs	Objects' outputs are autonomous, yet affect system-level outputs
Self-reliance	Self-reliant (trust in their own experience)	Objects maintain self-reliance (trust in their own experience)	Most objects maintain self-reliance (some rely on others and must trust)	Objects are dependent, simultaneously wholes and parts (all must trust in each other)	Objects maintain self-reliance, but are simultaneously wholes and parts (objects independently verify)
Control	Self-control	Objects maintain self-control	Few objects show dependencies, most not	Control unifies or guides action(s) of objects	Control unifies or guides action(s) of group of objects, with autonomy
Performance	Object has independent performance(s)	Objects have independent performance(s)	Objects have few dependencies with most objects still autonomous	All objects are dependent and contribute to system-level performance(s)	All objects are independent, nonaligned, and contribute to system-level performance(s)

Systemic emergent properties are irreversible. From a causality perspective, systems are predicated on sustained interactions between objects. The antecedent events that postured some objects for salient contributions are no longer operative in a system. This formalism results from the irreversibility of the systemic emergent properties of some of the objects. Other objects that are not so postured retain their reversibility and exhibit emergent attributes. When these objects are combined, the aggregate behavior can be representative of the emergent properties of the objects, or the emergent attributes of objects, or a combination of both types of objects. Antecedent events, although causal in terms of the formation of a system, serve to provide the objects and EMMI, but are precursors rather than directly involved in the integration process. Therefore, systems are of two types, those formed from only proximate events and those formed by proximate and objective events. Systems that are summative are built up of individual objects and smaller clusters of objects (i.e., proximate and objective events), while normative systems are based on proximate and objective events that come together as agglomerates to form a system of systems.

All the existing definitions of systems capture a portion or all of the superficial notions that elements interact (e.g., work together) within a boundary to perform functions jointly that are unachievable as individual elements. Such definitions generally lead the discussion on how systems thinkers think of their tasks. Specifically, these definitions are how systems engineers go about their building of systems. Somewhat intuitively, thinking in systems' context fosters a general feeling about what is important. Systems engineers "feel" that importance—considered to be the art of systems engineering. But when it comes to designing and architecting a system, that art needs to be integrated into a product or service. Even the seasoned professionals are overwhelmed with thousands of items to track and 100,000 objects to develop. The intricacies of interacting objects overwhelm the marked abilities of the seasoned group of tenacious systems engineers. In any reasonably sized system, there are literally millions of EMMIs (objects interacting in ways that are much less predictable and significantly more volatile than can be tamed). It falls on these systems engineers and systems engineering integrators to apply their skills learned from wrangling over decisions based on trade studies, test results, stakeholder needs and wants, political expediencies, and most aptly their best-informed guesses. But this is not the way to bring order and stability to planning, budgeting, and scheduling.

It should now be intuitively clear that a precise definition of system is naïve. Even trying to encapsulate a clear, comprehensive, and simple conception of systems is challenging.

Following the definition of a system according to Palmer (2009), a system can be conceptualized in terms of the behavior of its objects (descriptive of the essence of their system); the context of the minimum energy structures (reflective of the design and architecture); the perspective of the definer (providing a referenced view); and the methods that epitomize its functioning (the socioeconomic realities of projects). Working these concepts into a definition that can be used in a precising manner:

> a system is a bounded, stable group of objects exhibiting intrinsic emergent properties that through the interactions of energy, matter, material wealth, and information provide functions different from their archetypes.

Said more abstractly and succinctly (but with loss of precision):

> a system is a bounded, stable group of objects exhibiting intrinsic emergent behaviors based on interactions of energy, matter, material wealth, and information.

And finally, paired down to its barest abstraction (with loss of precision and accuracy):

> a system is a group of stable objects showing intrinsic emergence based on interactions.

The systems engineering integrator concept of a system is a provisional goal that results in a valuable product or service. The systems engineer's concept of a system is a provisional goal that results in the design, building, and integration of objects through their interactions to deliver the functions, performance, and quality needed by their customers. Systems engineers are concerned with both systems and system of systems by design and architecture. Systems engineering integrators are concerned with both systems and system of systems within the context of the product or service operating in its operational environment. Systems engineering is a collaborative, interdisciplinary approach to managing and carrying out the transformation of requirements and resources into a system through design, building, and integration of objects. Systems engineering integration is a collaborative, value-enhancing approach to demonstrating functionalities and performances of products and services.

But it may be fruitless to focus more effort to move beyond these general statements of systems engineering and systems engineering integrators as there is a public recognition of several myths in systems engineering (Kasser 2010). As Joe Kasser points out, there are (1) a plethora of standards; (2) many process models that can be used; (3) more dependences on people for success than on any one systems engineering method; and (4) needs for better tools, techniques, and procedures. There should be no illusion that there is not just one "brand" of systems engineering widely accepted as standard practice. In fact, in spite of their differences, several versions of systems engineering have proven reasonably effective in building products and services (Honour 2011). Several examples of systems engineering guides are the *Naval Systems Engineering Guide* (Rodriguez et al. 2004), two guides from NASA (Shishko

1995; National Aeronautics and Space Administration 2007), and the INCOSE handbook (SE Handbook Working Group 2010). The principal difference in success with various systems engineering approaches appears to depend on the ability of systems engineers to use systems thinking, know how to adapt the systems engineering processes to the project, and possess excellent oral and written communications skills (de Souza 2008). In addition, the systems engineer must create innovative solutions to mischievous problems, exercise sound engineering judgment, and apply effective management and leadership skills.

At the heart of systems thinking and systems integration is the set of issues that are deemed important to include in research (Ferris et al. 2003). The framework for systems integration reflects the engineering domain specialties, and the disciplines of sociology and management.

> Systems integration is the unification of the objects and their interactions of energy, matter, material wealth, and information to provide system-level functionalities and performances.

System of Systems and Integration

The general type of engineered and a systems-engineered system is one in which people are involved in the system's use during operations and sustainment. People are systems in themselves. People exhibit individual and group object emergent properties, having boundaries and boundary conditions, and are integrated into a metastable state that has a lifecycle. The occurrence of people's behaviors in conjunction with that of the engineered system is often referred to the system. If what is built as a product or service is a system, then that artifactual object when combined with the human system(s) becomes a system of systems (Ackoff 1971; Osmundson et al. 2007; Lane and Boehm 2008). The casual nature of building a system of systems is similar to that of building a system if and only if the goal of the system of systems is thought of as an all-encompassing system. In other words, even though each individual object (system) is completely bounded and integrated and exhibits system emergent properties and attributes, the system of systems needs to be thought of as a system with antecedent parts (each of which are systems). Each part needs to be viewed as a part of a larger system (i.e., a system of systems) and not as individual systems with some connectivity and interactions with EMMI. The *United States Department of Defense Acquisition Guidebook* (United States Department of Defense 2010) defines a system of systems as a "set or arrangement of systems that results from independent systems integrated into a larger system that delivers unique capabilities." Similarly, the U.S. Department of Defense has adopted a guide for systems of systems (Director of Systems and Software Engineering 2006). These guides further illustrate the intent of integration of systems into a system of systems. But the salient issue, the number one concern, for integrating individual systems into a system of systems is that of sharing EMMI in such a way so as to provide the system with various functionalities and performances, robustness and resilience, and predictable and acceptable losses to achieve and participate in the system of systems.

Applying the framework of objective causality to an example of sharing the functions of "to manage" in a system of systems, consider an incorporated (doing business under the fictitious name of Doctors, Ltd., a legal entity) of a group of doctors entering into a contract to use the services and facilities of a local hospital (also incorporated as a legal entity, doing business under the fictitious name of Hospital, Ltd.). Doctors, Ltd. is a system comprised of all six types of objects—object that interact with EMMI by send, by receive, by send and receive, by receive and send, as an extension, and as a source. Each medical professional in Doctors, Ltd. is a person (i.e., a system). Doctors, Ltd. has an office with equipment, patients with whom they consult, suppliers who transact business with the doctors (both individually as well as a group). Currently, Doctors, Ltd. uses Hospital, Ltd. services and facilities on an ad hoc basis. Patient records are kept by both Hospital, Ltd. and Doctors, Ltd. Both send invoices to the patients and often the patient is billed for duplicate charges, as the exact nature of the activities is not clearly delineated. The responsibility for billing is muddled. The informal arrangements are not resulting in the desired behaviors and effects. Both Hospital, Ltd. and Doctors, Ltd. desire to remedy this confusing situation as the patients have complained and some have gone to other doctor groups having affiliations with other hospitals and have not returned to either Hospital, Ltd. or Doctors, Ltd. The interactions between Hospital, Ltd. and Doctors, Ltd. have not resulted in a system of systems with acceptable performance of their macrofunctions. Losses in goodwill and future revenues are significant and increasing. Indeed, Hospital, Ltd. and Doctors, Ltd. are interacting as two systems, but they are not a system of systems. The metasystem functions of unified billing, governance, and accountability do not exist. There are no metafunctions. Hospital, Ltd. and Doctors, Ltd. agree to enter into an agreement to provide metafunctions, improve their macrofunctions, and realign their microfunctions to enable improved interoperability. In addition to resolving billing issues, the Hospital needs to increase its medical staff of doctors without incurring additional overhead charges. Doctor, Ltd. agrees to take adjunct positions on the hospital staff. Yet, both Hospital, Ltd. and Doctors, Ltd. desire to remain somewhat independent, even while participating in their system of systems. The first step in their efforts to build metafunctions is to identify the kinds of functions that would benefit both parties to the agreement as well as the patients. Without presenting the details of the systems engineering aspects in developing a system of systems, the focus will be on the particulars of integration. Applying the framework of conditional causality, the processes that will result in changes to physical entities (perhaps beginning with new procedures to deal with the problems identified with patient billing), the additions of metafunctions that serve some of the needs of all parties within the system of systems, and the desired behaviors of the stakeholders involved in all causal effects that deal with or are a consequence of interaction with the system of systems, will be defined and agreed to before beginning the planning for integration. For planning purposes, the sharing of management functions is deemed to be the best means to manage the functions of the system of systems. There are many means and options to enact and manage the system of systems functionalities, including, by tacit agreement, through procedures, and with various types of physical entities (e.g., computers and software). The management functions of 'to plan,' 'to organize,' 'to direct,' 'to control,' 'to communicate,' and 'to provide teamness' are laid out for Hospital, Ltd. and Doctors, Ltd. Table 4.2 illustrates a few of the considerations for objects, events, and EMMI that exemplify the partitioning of power.

Table 4.2 illustrates the comparison for objects, events, and EMMI for Doctors, Ltd. (DL) in a systemic relationship with Hospital, Ltd., illustrating the reciprocal relations between objects based on a sharing of power. The architecture that supports and enables the interactions is one comprised of people who perform the management functions at all three levels

TABLE 4.2

Objects, Events, EMMI Mapped to "To Manage" Processes

To Manage Processes	Doctors, Ltd. (DL)			Hospital, Ltd. (HL)		
	Objects	Events	EMMI	Objects	Events	EMMI
Plan work	DL doctors	Request facilities	Information	Administrative staff	Reserve facilities	Information
Organize work	DL doctors	Take adjunct positions	Material wealth and matter	Patients	Schedule HL patients with DL	Information
Direct work	DL doctors	Schedule DL and HL patients with DL doctors	Information	Administrative staff	Request scheduling HL patients with DL doctors	Information
Control work	DL doctors	Comply with HL rules	Information	Administrative staff	Promulgate policy and rules	Information
Communicate	DL doctors	Convey status	Energy and information	Administrative staff	Convey status	Energy and information
Foster teamness	DL doctors	Show reason	Matter and information	HL staff	Show reason	Matter and information

of abstractions for the system of systems. For planning purposes, both Hospital, Ltd. and Doctors, Ltd. manage their own personnel; for organizing and directing work, the subservience of doctors from Doctors, Ltd. tracks whether the patients are from Hospital, Ltd. or Doctors, Ltd. Controlling work is principally with Hospital, Ltd, as the established work relation is arms-length, at-will, for hire status. Teamwork is expected at all levels of interaction and is encouraged by accommodating reasonable requests and maintaining frequent and congenial communications.

Whether natural or human-built, the process of integrating systems (the joining of objects to achieve an effect) is describable with the framework of objective causality.

Organizational Models

Organization and architecture (i.e., structure) are the two key structural factors that make up the fundamental functionalities at the metalevel of a system of systems. Once the structure is formulated and constructed, other issues need to be considered and dealt with, such as how each *sub*system interoperates with the metasystem model, how effective are the functions that form the metafunctions enabled by the *sub*systems, what overall *sub*performances cause substantial losses for the *meta*performance, and how the metafunctions are constrained and limited by the operations, supporting *sub*functions, and various human behaviors (if people are involved in operations). While there are any number of ways to structure a system of systems, presumably large variability in organizational performance is normally not characterized by its structure. It is usually assumed that to a large degree organizational structure is a mentioned factor, but not necessarily significant. The voluminous literature on measures of effectiveness for organizations is replete with discussions about horizontal and vertical integrations, matrix and line organizations, and lifecycle models (Quinn and Cameron 1983) that for the most part describe their physical, functional, and behavioral aspects. These publications fit within lines of inquiry that follow strategies of staging where an organization fits within a continuum of changes (lifecycle); surmising what is right (or wrong) with existing organizational notions (diagnosing and evaluating); and analyzing the organizational dynamics based on presumptions of static structures (work groups and management), external pressures (business market competitiveness), or human dynamics (covers mergers and acquisitions). Within these broad strategies, various attributes are investigated, including influences, preferences, and power. This portrayal is not meant to be either a survey or an exhaustive description, but rather to suggest another strategy which applies directly to the integration of a system of systems. That is, organizations seem to be distinguishable by a structuring of power (including distributing power, storing power, and using power). By power, it is meant to control the access to something that is useful to others. Much has been said about gaining power, applying power, retaining power, and using power, all of which apply quite aptly to people and organizations, organizational effectiveness, and countering the effects of changing organizations. Consider an organization that is rich in material wealth. They judiciously use their wealth to work with others through loans of various types of resources and joint ventures that utilize this wealth. The company develops partnerships, creates a strong interest for others to do business with them, and promotes a goodwill that earns respect from their peers and solicitors. Contrast this company with another company that has resources that are equal in value and importance to potential solicitors and partners, but has not engaged in partnerships or allowed access. The company that extends access to its material wealth exerts power, whereas the company that hoards has expressed no interest in allowing access, nor engages in any way to use those resources. These two companies stand contraposed to a company that has material wealth and uses it to engage a select few or feigns a threatening demeanor to those in disfavor. Providing access to power (which is not a new theme for consideration) lies at the heart of the design and architecture of a system of systems, and therefore, is essential to the integration strategy and outcome. Recognizing there are different kinds of power (economic power is recognizable in political arenas, but may not be equivalent to social power or knowledge power), access to power is the common link for planning integration. Architecturally, the access may be person to person or by agreement through some automated trusted agent.

Power structures (the essential elements of an organization that protect, store, support, and manage* desirable resources) form the key elements that make the valuable resources a sought-after commodity. These essential elements may be formal as is the case with set policies and procedures or they may be informal (Land 1985), on a person-to-person or person-to-machine basis. For an integrated system of systems, those valuable resources are EMMI, the users of the valuable resources are *sub*systems (objects), and the power structures regulate access. This discussion concerning access to power and power structures presumes that the systems engineering has provided an acceptable level of reliability, maintainability, stability, boundedness, and predictability. Mahoney and Weitzel (Mahoney and Weitzel 1969) determined that reliability was one of the most critical requirements of organizational effectiveness for both general business and research and development organizations. Reliability of access to valuable resources is particularly important for system of systems.

---

* "To manage" is to "plan," "organize," "direct," "control," "communicate," and "team-build" those essential elements.

From a systems engineering integration perspective, the need for an effective system of systems presumes that requirements are determinable, objectifiable, and buildable. Requirements determinability is a matter of identifying the needs of the designers, customers, and users* of system of systems, eliciting data and information from knowledgeable stakeholders, then applying a process to deliver the physical, functional, and behavioral properties and attributes desired in the system of systems.

In the most general sense, there are multiple combinations of measures of effectiveness which are relatable to organizations and organizational processes. Without the benefit of significant research to guide this discussion, there seems to be four kinds of organizational power structures. These kinds appear to be consistent across organizations, whether they are inculcated into groups of people, companies, or governments of countries. These organizational power structures are termed as the *item*, the *syndicate*, the *aggregate*, and the *agglomerate*.[†] No particular preference is given to any one or another of these power structures. No one is better or worse, only tailored to match circumstances. They are merely structures that need to be accommodated with design and architecture when building a system of systems. We can think of the architecture of a system of systems as that of a system, with the architecture supporting the governance of the system of systems, the processes that provide for the metafunctions, and the power structure of its organization. In many ways, the architecture of the organization is akin to that of the architecture of the product (Yassine and Wissmann 2007), with the notable exception being the power structure. The reason why organization power structures are distinct and not included in products derives from the nature of the *sub*systems and their independence. Within a product that is a system, autonomous operating capability with connectivity to other systems operations is inconsistent with the product being a system. Semiautonomous operation that has connectivity to a system is termed as a system of systems.

The item is a unitary power structure with a single entity that determines access to valuable resources. An example of an *item* power structure is a monolithic organization that has a consistent set of policies that governs uses of power and access to valuable resources. Decisions made by central governance are policy, with no exception other than as is agreed to by the central governance. An example is that of the small business owner, whose authority over access to valuable resources is inviolable. There may be one or many objects at the level of governance of an *item* power structure.

The syndicate is typified by many objects which have decision-making authority, act as a group once consensus is reached, but have a myriad of

---

* More generally, the set of stakeholders (significantly beyond the designer, customer, and user) for a typical system number in the multiple 100s and grow geometrically with the number of subsystems within a system.

† The vaguest of association with dictionary definitions or those defined with more precision is the intent for their usage in this presentation.

informal habits that can dominate with local decision making for exceptions. Yet, in spite of these exceptions, the group's consensus is both tolerant and forgiving of local decisions. An example of a syndicate power structure is the World Wide Web. The governance is fundamentally through protocols and policy implemented through standards which represent central governance. The users of the web participate and by their adherence and compliance with the standards and protocols they participate with the common goal of accessing valuable resources. Power is locally managed by independent *sub*systems (web portals, as objects) which have local authority to grant and provide access to valuable resources.

The aggregate is exampled by a football team or a group of doctors who are the managers of a private hospital. Each member of the football team (object) can arrange for a few complimentary seats (or if that is in keeping with the team's policies, then a few complimentary season tickets in the most sought-after seats). Individual members of the governing board of doctors may arrange for a private room, when otherwise there would be no preferential treatment. The staffs of both organizations accede to such requests in part because it is considered acceptable behavior for key personnel or there may be an implied coerciveness about the demand (request) that is made. The Internet does not exhibit the power structure of an aggregate organization because its central governance regulates the broad policy for accessing all resources as a necessary condition (but it is not sufficient in the main to access all valuable resources). Power structures that have a distributed quality are not *items*, and those whose distributed nature does not represent the interests of the requestors for access are not *aggregates*. And those power structures that have a distributed quality that does not represent the interests of the requestors for access and that does not have access to all valuable resources is not an *agglomerate*.

The agglomerate is a combination of the syndicate's and the aggregate's organizational power structures. The agglomerate has centralized control and governance for those issues that have significance in binding the *sub*systems together, similar to a syndicate's power structure. But the *sub*systems act as autonomously as possible within the limits imposed by policy, similar to an aggregate power structure. An example of an agglomerate is a multinational corporation that is contracted with suppliers competing to provide parts on a competitive basis. The suppliers are integrated into an inventory management system (perhaps just-in-time delivery-style supply chain) to facilitate a significant reduction in the inventory that must be stocked to support production. The suppliers retain their autonomy and independence, while participating in a metafunction at the system of systems level. The approach taken by the *sub*systems is to gain special favors that are made available by decision makers in the multinational corporation.

Countries exhibit organizational behaviors that are consistent with a system of systems power structures. Singapore presents an *item*-like power

structure—highly focused, centralized, and monolithic from their governance to the ideals and responsibilities of their citizens. The People's Republic
of China operates as an aggregate power structure—centralized, yet responsive to local needs. The United States and the European Union (Nieminen
2005) are examples of an agglomerated organizational power structure—
operating with highly independent states (in the United States) or countries
(in the European Union) where both *sub*systems have high degrees of autonomy and authority. The differences between the United States and the
European Union, albeit substantial in most dimensions, are materially alike
when it comes to their architectures which enforce and sustain access to
power. The rules can be quite different, the circumstances distinct, and the
particulars novel, but the physical entities provide substantially the same
functionalities at the metasystem's level, with a resultant access to valuable
resources that is substantially similar.

The two organizations, Hospital, Ltd. and Doctors, Ltd., are representative
of two different categories of system of systems. Doctors, Ltd. is a syndicate
of doctors with a limited set of partners acting in concert to set policy and
providing centralized management. The essential characteristic for a system
of systems is that there are many individual systems with common skills
and generally the functional goals. Another example of a syndicate is the
group of people who use the Internet. The Internet users have common tools,
similar skills to exploit Internet and web content, and all share in the same
benefit of gathering, exchanging, or posting information. Pursuit to their
particular rules, the syndicate may share as partners in some benefit (e.g.,
profits, in the case of Doctors, Ltd.).

Hospital, Ltd. is a privately held entity with shareholders, who elect a
Board of Directors to manage the affairs of the business. For our purposes,
we refer to this category of organization as an amalgamation—a blend of
different interests and skills that have particular influence on the outcome
of the business operations. The difference between the categories of amalgamations and syndicates is the predominant theme of stakeholders deriving
different benefits from their association with each. Stakeholders in Hospital,
Ltd. may be participating purely for the cachet of being involved with the
medical profession (and a hospital in particular) and the political benefits
derived from the same in local government. While this difference between
system of systems (as syndicates and as amalgamations) may seem particularly inconsequential, the differences imply quite different architectures
and therefore very different planning for integration. The simplistic difference of quite different goals is handled easily through architecture by
emphasizing that all parties can accomplish their goals (once it is recognized how to phrase and present the relation between the goals and the
benefits for the business that are derived from the architecture). The more
troublesome factors that impact on architecture are the degree of autonomy
expected by the two organizations based on their mindsets and previous
business structures. Changing a business architecture is not only quite a

difficult set of tasks, but also most disruptive to operations and worker's habits. Each organization has a "style" and culture of work that is instrumental in bringing out the subtleties of the architecture. Even recognizing the nuances is often insufficient to keep from disrupting workflow and efficiency.

The fourth category of a system of systems organizational power structure is the agglomerate—a decentralized scheme of controlling access to valuable resources, with central control over policies that impact on various groups. As with the syndicate and the amalgamation, the agglomerate has groups organized for a particular purpose or convenience, but each with a different mix of incentives. Those incentives are sometimes locally contrived and enacted without the limits set by the central controlling entity. As with the syndicate, the amalgamation is comprised of like-minded individuals who are focused on nearly the same goal—supporting the organization with various specialties that add to the total skills of the organization. And consistent with an agglomerate, individuals in the amalgamation have very different agendas for their association with the organization, differences with regard to accessing valuable resources, and perhaps local control over such access. An amalgamation is comprised of a group of stakeholders or *sub*systems with (perhaps significantly) different views about the operations, value of the valuable resources, and access to those valuable resources. The *sub*systems and individuals in the organization want to be in the organization and participate in the same operations as are offered by the system of systems. They have a preference to continue working within the current group or *sub*system, but their allegiance can be compromised (as referenced to either the policies of the central or local governance). Members and *sub*systems of the syndicate have greater solidarity than do members or *sub*systems of an amalgamation.

The group of doctors at Doctors, Ltd. self-organized into a syndicate, each doctor agreeing to give up some degree of autonomy and self-reliance to join with other doctors. The syndication of professional skills as a legal entity (Doctors, Ltd.) has value to Hospital, Ltd. as an organized group of medical practitioners with whom they have an established rapport, credibility, and trust. Rather than dealing with one doctor at a time, the system of systems (Doctor, Ltd.) has its own metafunctions that illustrate coupling, cohesiveness, and connectivity through EMMI sufficient for metastability. The essential emergent property of Doctor, Ltd. is suggested through the filling in by one colleague for another colleague in case of emergencies. This "covering" holes in a schedule are typical systemic behavior that is motivated by factors such as desire to help, economic incentives, sacrificing for the good of the group, reinforcement of teamness, wanting to maintain the relationship with the patient through continuity of experience with the same medical practice, and so forth. Such behavior is less common in private, individual medical practice because of the limited amount of time in a doctor's schedule to cover for another doctor.

Conclusion

The discovery of the many aspects of integration is akin to the search for an ontology that is coherent within a discipline or field of research and required to be interoperable across disciplinary boundaries. A most elementary test of an ontology is consistency. That consistency should be tested at the concept level, shown by practice, locally explainable by frames, comprehensible by an integrative framework of those frames, a posited theory that explains and predicts through empirical measurements based on multiple perspectives, and subjected to expert peer review (Smith 2008).

References

Ackoff, R. L. 1971. Towards a system of systems concepts. *Management Science* **17**(11): 661–671.

Cummins, R. 1975. Functional analysis. *Journal of Philosophy* **72**: 741–765.

de Souza, R. A. 2008. *Maturity Curve of Systems Engineering*. MS thesis, Systems Engineering. Monterey: The Naval Postgraduate School, 114pp.

Director of Systems and Software Engineering 2006. *System of Systems Systems Engineering Guide: Consideration for Systems Engineering in a System of Systems Environment*. Washington DC: Office of the Under Secretary of Defense (Acquisition Technology and Logistics), 92pp.

Ferris, T. L. J., Cook, S. C., and Honor, E. C. 2003. A structure for systems engineering research. *Proceedings of SETE 2003*. Canberra, Australia: Rudges Capital Hill.

Honour, E. C. 2011. Improved correlation for systems engineering return on investment. *Conference on Systems Engineering Research (CSER)*. Redondo Beach: University of Southern California.

Kasser, J. E. 2010. Seven systems engineering myths and the corresponding realities. *Systems Engineering Test and Evaluation Conference*, Adelaide.

Land, F. 1985. Is an information theory enough? *The Computer Journal* **28**(3): 211–215.

Lane, J. A. and Boehm, B. 2008. System of systems lead system integrators: Where do they spend their time and what makes them more or less efficient? *Systems Engineering* **11**(1): 81–91.

Mahoney, T. A. and Weitzel, W. 1969. Managerial models of organizational effectiveness. *Administrative Science Quarterly* **14**(3): 357–365.

National Aeronautics and Space Administration 2007. *Systems Engineering Handbook*. Washington DC: NASA, 360pp.

Nieminen, A. 2005. *Towards a European Society: Integration and Regulation of Capitalism*. Helsinki: University of Helsinki, 465pp.

Osmundson, J. S., Langford, G. O., and Huynh, T. V. 2007. System of systems management issues. *Asia-Pacific Systems Engineering Conference 2007*. Singapore: National University of Singapore Temasek Defence Systems Institute, 9pp.

Palmer, K. D. 2009. *Emergent Design: Explorations in Systems Phenomenology in Relation to Ontology, Hermeneutics and the Meta-Dialectics of Design*. PhD thesis, Division of Information Technology, Engineering, and the Environment. Mawson Lakes, University of South Australia, 679pp.

Quinn, R. E. and Cameron, K. 1983. Organizational life cycles and shifting criteria of effectiveness: Some preliminary evidence. *Management Science* **29**(1): 33–51.

Rodriguez, W. D., Massenburg, W. B., Lengerich, A. W., Stone, D. T. K., and Catto, W. 2004. *Naval Systems Engineering Guide*. Department of the United States Navy. Washington DC: Defense Acquisition University, 295pp.

SE Handbook Working Group 2010. *Systems Engineering Handbook: A Guide for System Life Cycle Processes and Activities*. San Diego: International Council on Systems Engineering (INCOSE).

Shishko, R. 1995. *NASA Systems Engineering Handbook*. National Aeronautics and Space Administration Jet Propulsion Laboratory. Pasadena: California Institute of Technology, 186pp.

Smith, B. 2008. Ontology (Science). *Nature Proceedings*. hdl:10101/npre.2008.2027.2.

Troncale, L. 2011. Would a rigorous knowledge base in systems pathology add significantly to the SE portfolio? *Conference on Systems Engineering Research (CSER)*, Redondo Beach: University of Southern California.

United States Department of Defense 2010. *Defense Acquisition Guidebook*, Version 5 May. Department of Defense. Washington DC: Defense Acquisition University, 310pp.

von Bertalanffy, L. 1968. *General System Theory: Foundations, Development, Applications*. New York: George Braziller.

Yassine, A. A. and Wissmann, L. A. 2007. The implications of product architecture on the firm. *Systems Journal* **10**(2): 118–137.

5

Integration in Systems Engineering Context

Introduction to Systems Engineering

Systems engineering was originally envisioned to deal with the complexity of products from inception through their delivery (Schlager 1956). Ten years later, with the promise of being able to develop new, flexible means of controlling equipment through software programming structures, systems engineering changed its repertoire and emphasis from discovering and integrating satisfactory components to a broader systems perspective which spanned the lifecycle of the product or service. "The integration phase is usually a long one, since it extends throughout the entire period of design and construction of the experimental system equipment" (Schlager 1956). The backbone of systems engineering is integration, integration in both the general sense of bringing ideas, people, and objects together in the form of a project to deliver a product or service, and in the specific instance of discovering and integrating satisfactory components.

Much has changed since Kenneth Schlager wrote these early thoughts about a new field of engineering called systems engineering. According to a RAND Corporation report (Stem et al. 2006 citing Przemieniecki 1993), systems engineering came into its own right as a discipline along with the development of the U.S. missile program in the 1950s. Prior to that, systems engineering was a fledgling, deracinated from the rank-and-file engineers who knew their problem set was broader than their training. A progeny of two needs, one commercial and the other military, systems engineering was challenged to address two seemingly different types of problems—those that were defined in terms of requirements (for customers who had specific needs) and those that were driven by the economics of services (those who wanted to lower costs and improve the user's experience). Customers with specific needs often drive development with hard requirements for technical performance—more than currently available in the existing product, sometimes pushing the edge of what is physically possible. Customers driven by the economics of better service to their customers often consider schedule before cost. Commercial firms must be attentive to market demands that are often driven by product fashion and perceptions of quality of service. The result is a presumed difference in management methods and systems engineering approaches to solve problems.

While these problems appear different due to the driving influences of performance, schedule, or costs, they are most alike in many respects. Performance-driven requirements rely on a collective of parties in opposition to achieve an "at will" consensus. In other words, a buyer (one who puts forward a set of requirements) and a seller (one who proposes to satisfy those requirements) come to an agreement on a proposed schedule and budget to deliver various product or service performance(s). Systems engineering provides the thinking and the approach to establishing performance, cost, and schedule trade-offs, in deference to the needs of the buyer and the seller. With the approach carefully planned, the systems engineer originates tasks that become the mainstay of the work for budgeting, assigning skilled workers, and monitoring progress. The premise of the deal (e.g., contract) is that two parties (at "arm's length": without conflicts of interest) agree on the deliverables, progress milestones, payment schedule, acceptance criteria, and methods.

In contrast, consider the firm that has an existing installed base of products or services or an infrastructure in use by its users (or customers). If the seller determines that an economic advantage is possible through innovation or changes in the functions (for example) for its offerings, systems engineering first sets out the objectives (i.e., requirements). The technical approach, management supervision, progress milestones, and acceptable product or service performance(s) are proposed by the systems engineers and, if acceptable, agreed to by the decision makers overseeing the activities supporting the installed customer base. In both cases, the systems engineering planning guides the development work to produce the desired product or service performances. The results of systems engineering fall into two domains. The subjective domain (i.e., the cognitive structures that provide planning, methods, and approaches; procedures that follow a model of steps that signify the phase of work and the expectations for each phase; and models and representations of the results of planning and procedures, such as requirement documents, trade-off analyses, and build-to specifications). The purpose of the subjective domain is to engage engineers and subject matter experts to work with the systems engineering plans to build and test physical entities with the appropriate functional traits so that the users can exhibit the sets of behaviors that effectively exploit both the physical entities and their resultant functions. Systems engineering takes subjective information and turns it into objective properties, traits, and attributes. The process of transforming knowledge into objects that work as a part or as a whole is integration. Systems engineers deliver their most beneficial performance on problems whose boundaries (physical, functional, and behavioral) reach well beyond what is often presented in a set of requirements. Even a widely defined scope of work may impact on systems of import unbeknownst to the buyer and the seller. A review by a qualified systems engineer is most appropriate for any project, including building a building (e.g., impacts environment, transportation throughput, electrical utilities, and potable water distribution); building infrastructure (e.g., impacts on environment and movement of people); and building means of transportation (e.g., impacts on environment, movement of people, means of commerce, density of people).

The problem that systems engineers faced then was in determining which new technologies could be used to improve existing services while simultaneously lowering the cost of delivering those enhanced services. With a head of steam that sometimes drove adoption of new, immature technology, systems engineers took on the role of being the objective, rational gatekeepers to protect the investment in infrastructure and systems. That role required skills that spanned various types of research, mandated familiarity with fundamental development that extracted techniques and ideas from research and prepared them for pilot modeling, prescribed an appreciation for the infrastructure that was already in place and the user behaviors that were already adapted to certain types of products and services, and the sensitivities of economic implications from operations. The early systems engineers focused on establishing performance and cost objectives for technologies they determine would improve existing infrastructure and services. Today's systems engineers are sometimes handed technology and saddled with making it work within a schedule and budget.

With an appetite for systems and system of systems that have more functionality with greater depth and sophistication, systems engineering is challenged. However, it is not that the new technology is inappropriate for consideration or maturation. "The development and acquisition of new systems usually requires the use of new technologies in order to meet requirements unachievable with the current state-of-the-art" (Ender et al. 2009). Rather, it is the complexity and reliability that baffle systems engineering and confound integration. It is inconceivable that any other method, approach, or discipline besides that embodied in systems engineering could deal with the vagaries imposed on building complex systems and achieving sufficient levels of system reliability.

Of course there is a method to this decision process, which may be dependent on the structure of the acquisition system. For commercial ventures, new technologies undergo extensive testing before subjecting customers and users to unreliable products or services. Marketplace feedback can be swift and thunderous. Given alternatives, customers will begin to work with other vendors and suppliers. The less reliable, less functional, and less performance systems are replaced by others. For entrepreneurial ventures, new technologies must be proven along with an objectively supportable view that there will be marketplace acceptance before the entrepreneurial venture may be considered for institutional funding from sophisticated investors. The funding and support go to those whose ideas will dramatically change existing markets or establish new markets that are both strong revenue gainers and offer high profit margins. The "play" for the investors is to sell their investment positions for high profits. For governments, specifically, the U.S. government DoD, acquisitions of large, complex systems often signify a "grab for gusto" mentality. Pushing for the incorporation of new, immature technology into system or system of systems designs (supposedly as production or manufactured items) is nonsensical. Production and manufacturing are designed to take a product or service and make more of them, not fix problems that are within the domain of product or service development. The province of production and manufacturing is to take the integrated product or service and to replicate the interactions in finished goods. Immature technology might offer the promise of new performance and exciting functionality; however, that promise is only realized through research and development. A production product or service will not (and does not) function or perform any better than the initial protoproduct or protoservice. Without extensive rework, or redesign for production or manufacturing, a development product or service is just that—development. Limiting production to a few number of "demonstration" products or services is often a means to begin improving reliability. Working together, engineering developers and systems engineers mature the engineering embodiments of technology, all the time improving the system reliability. For example, the Hubble Space Telescope case study (Chapter 1) emphasized the lesson learned from having to delay the launch to orbit. That time spent was integration time to mature and achieve the requisite interactions on a sustained, reliable basis. Had the Hubble Space Telescope been pushed into operational status on orbit according to its intended launch schedule, NASA would have an unreliable, crippled system that would most likely have had to be returned to the development environment for retrofits, improvements, and upgrades (Mattice 2005). In other words, the integral set of ideas embodied in the system design, concept of operations, and architecture is no better than the reliability and maintainability of the constituent components. Integration is the aspect that wraps all the systems engineering, engineering, and management efforts into the product or service and the product or service into the operational environment.

Nature of Systems Engineering

The cognitive functioning of the systems engineer is unique and as distinguishable from all other disciplines or fields as buyers are to sellers. As the topic of the product or service may be the same for all parties, the methods, approach, and tools are different. Systems engineering is distinguishable from all others by its emphases on three aspects of providing solutions to problems: (1) the product or service as an enabler of the desired user behaviors, (2) satisfying the stakeholders' needs can be done in one of many ways, (3) and the desired consequences of the product or service for people, infrastructure, and environment need to be incorporated by design and architecture. These three considerations are of equal importance to the systems engineer.

The product or service is the focus for the development project and with which responsibility rests for assuring no adverse consequence accrues to users and other future stakeholders. The product or service perspective concentrates on the lifecycle of product lines or service(s). Recognizing a substantial investment has been made in infrastructure, the systems engineers are keenly aware of the mandates from people and respect for the environment that embodies the mantra "do so, but do no more harm than it takes to do so." This real cost of doing something should not be greater than the benefit of what is done. The accomplishment should be greater than the effort it takes to achieve the accomplishment. The loss to achieve a level of performance should be kept to a minimum. The systems engineer thinks in lifecycle issues, the net action accomplished for the total investment.

Giving preference to the perspective of stakeholders and the product or service results in potential carelessness with regard to lifecycle stakeholders, collective investments, and the environment. The needs of each of these concerns must be considered in the design and architecture to place the product or service on a presumed no-net harm path, and the actual work and materials used becomes the means of carrying out the plan of no-net harm done. When any one of the focuses (product or service, stakeholders, infrastructure, or environment) is given preference over the others, a lifecycle issue needs to be analyzed and resolved before development begins. What distinguishes systems engineering is that of thinking in systems, enabling by engineering, and integration by preference. Integration is broadly considered by systems engineers as the most important aspect of systems engineering. It is through integration that all thoughts come together and result in ideas that are different from each of the thoughts, that products and services emerge from objects and labor, and that all disciplines can combine to tackle complexity.

Systems engineering remains rooted in the classical formulation of reductionist theory—that which supposes a hierarchical decomposition of the highest, most general-level conceptualization downward through successively greater detail. It is not that the hierarchical schema is inherent to systems engineering or that it is even necessary. It is a comfortable approach to the thinking of the practitioners—one that provides a good-enough reconnoitering of the relations between objects, their functions, and the behaviors of the eventual users of the product or service.

The descriptive formulation of systems engineering casts a summative perspective of what systems engineering is. That pervasive is often fostered by back office conversations centering on wasted efforts to define problems (when it is clear what needs to be solved), iterative thinking about what individual items mean (when it is clear what needs to be done), and paper-work that documents processes and functions (when it is clear that when an expert is hired, all that matters is the result). The author believes it is good practice to approach systems engineering and systems integration as key to the science of thinking, which does not depend on luck for their results. To this end, the systems engineering methodology is prescriptive about what must be done and how to do it.

The development of systems engineering process models (the structure of determining which stage of work should be done and for how long) standardized a meaningful way to consider project status, technical progress, and the impacts of constraints on allocations of resources. There has been neither any empirical study on the efficacy of systems engineering process models nor an enduring debate as to the appropriateness of one model versus another given circumstances, constraints, and the kinship of project variables with technology. The lack of foundation for systems engineering (other than "it worked better than what was tried before") is troubling.

Issues with Systems Engineering

As systems engineering is currently instantiated, sometimes it is limited in terms of its capacity to consistently (1) determine the correct problem (e.g., through gap analysis); (2) identify critical stakeholders before architecting the system; (3) determine the breadth and depth of requirements; (4) integrate cross-disciplinary knowledge, and (5) account for lifecycle needs, to name a few. A better set of data and information about the premises and limitations of the current practice of systems engineering would aid in resolving the issues. The common practice of systems engineering currently deals with many problems, five of which are highlighted below:

- If systems engineering processes are employed but result in a product that does not achieve one or more of its goals—performance, budget, or on-time completions—then the stakeholders have not received the solution that was intended.

- Since stakeholders drive requirements, it is prudent to identify which stakeholders most determine the driving requirements. For example, the adversary is often one of the most influential stakeholders, and is often characterized as a "threat;" the full consideration of all aspects of the adversary is rarely included. While the interaction between the adversary's radar and the surface of an aircraft will be a paramount issue, the time line for an adversary's decision process is not included in the aircraft's skin design. This time line can be extremely important when the focus is on designing an aircraft skin that has a variable signature rather than a fixed signature. Stakeholder analysis is not taught, defined, or detailed in any forum. This often results in a set of requirements that are either missed or known but not acted on.

- There are the inadequacies in the process of determining the minimum set of requirements needed to assure that the constraints of budget and schedule are met. These inadequacies include poor stakeholder analysis, subsequent steps involved in the logic, prototyping, modeling, simulation, and analysis that is often flawed. Many requirements are discovered late in the development cycle, resulting in significant losses caused by extensive modifications or scrapping of the project. Integration will only be as good as what is provided. If the objects do not encompass the requisite functionality, then integration will do nothing to change that. The responsibility for providing the essential subfunctions is solely that of development. Integration merely combines objects in such a way to highlight both the functionality and the performance(s) that result from integration with other objects. Systems engineering is the domain of producing a product or service, development provides the requisite objects, and integration brings those objects together in such a way and manner to realize the essences of the system design, the stakeholder preferences through the architecture, and the requirements to satisfy the concept of operations.

- There are inherent difficulties when integrating cross-disciplinary knowledge. Consider, for example, human systems integration and the question of potential or actual losses incurred at the system level for two design strategies for movement of information: just-in-time versus on-demand. At issue is the timeliness and latency of flow of information, that is, the losses incurred due to poor integration and delaying a decision maker. The integration inefficiencies stem from the concurrency of data within the system (e.g., correlation of one data item with another data item representative of the best correlation of these two data items given that there may be a more relevant data item in the system that is unavailable due to lack of access to it or because it is unknown at the time of correlation). Integration brings out the direct measures of coupling and cohesion between objects and their relations through EMMI.

- Lifecycle issues are regarded as important when scoping work; however, initial budgetary constraints often restrict the efforts devoted to lifecycle engineering and planning that can be considered or designed into the system. The result is to sacrifice the long-term interests of minimizing the total lifecycle cost for the expediency of a lower up-front investment.

While these problems are addressable in multiple ways (such as training, following best practices (as defined in this book), and education), the issues and compounding factors may reflect a more general problem that potentially overwhelms the ability of the systems engineer to deal with the many conflicting interests. In keeping with the reasoning of the U.S. Government Accountability Office (GAO 2009a), established dogma has failed to achieve desired results because many government acquisitions "lack early and disciplined systems engineering analysis" and allow "new requirements to be added well into the acquisition cycle." These two issues are indicative of a mind-set by the acquirers that there are more important factors than simply following what is known to work reasonably well. Systems engineering is coopted by the dictates of pushing forward with little regard for neither engineering logic nor systems engineering reason. Whereas it is the purview of systems engineering to (1) translate needs into requirements, (2) build a solution that is responsive to those requirements, and (3) deliver a product or service that solves the problem, systems engineering cannot solve problems associated with human frailties. Most specifically, it is the determinable trait of inflexibility in those that make demands on systems engineering and quite explicitly engineering systems integration to provide more system functionality, performance, and quality for less (time and money). When viable system solutions exists that meet most of the stakeholder needs, or all of the requirements, the quest for new technologies in a "would be production or manufacturing environment" is wishful thinking. Wishful is making a decision when you know better; wanting is making a decision when you do not know any better. Need is quite a different matter. Such wishful thinking should not be regarded as statistical reasoning and incorporated into a risk analysis and shown as a risk item. Regardless of the acquisition mentality that precedes buying a product or service, if the intended outcome is a production or manufacturing environment, pushing immature technologies into a development environment should most likely not be thought of as risky from the outset, but rather as problematic from the outset, bordering deterministically, as a failure. However, it is pure folly to insert those same immature technologies into objects that are integrated when neither the existence of functionality nor the reliability of performance has been reasonably demonstrated during development. Integration cannot shed new insights that were not previously known by the developers. Neither does the integration process even pretend to show developers their possible options to rectify problems. Integration is not the "work around" or "rescuer" of the development problems that will somewhat mask or ameliorate the conditions under which those problems persist. Pushing ahead and integrating with the hope that the new technology will emerge in an acceptable implementation is wasteful of resources (time, talent, money, facilities, and equipment).

But there are mitigating factors that underlie the wishful thinking. Sometimes, there is thought to be no choice. For the entrepreneur, there is no choice. To do what others can do is failure before start-up. No investor will plop down their money on last year's story or yesterday's technology. If there is no sizzle, there is no sale. For the military planners in the United States, there is perceived to be no choice. To secure project or program funding, the new product or service must have significantly more performance than is currently available, satisfy needs that look into the future, and technical dominance that is the envy of the best dreamers. Consequently, military acquisition professionals and planners expect to have overruns in both schedule and budget. They also expect to fall short of the lofty set of requirements, but still take delivery of a product or service that will engender pride and *esprit de corps*. The landscape of the future military means offering pretty cool things to attract talent and interest. Unlike the start-up entrepreneur whose goals might include fame and fortune, the military planners must envision tools to prevail in conflict. Those tools mean products, services, and competent military personnel. For the commercial developers, the "litmus test" is different. The determination of the correct set of requirements is market acceptance leading to revenues and profitability. Each product or service does not need to be a major success, nor does each product or service require a built-in or implied synergy with other product and service offerings by the same company. Comparing the commercial products and services with military products and services is more than just competitive issues in operational or marketplace environments. The results of a bad day are quite different—there is no comparison between a few sales lost to competition and lives lost. Systems engineering and systems integration are active enablers in all three situations: the entrepreneur pushes technology into prototype form, and then develops full-up working systems that are then worked over by manufacturing engineers before going into production or manufacturing. A problem in the marketplace with a one-product company is disastrous for the company and they most likely will go out of business. Commercial companies fare better with multiple products and services than entrepreneurs. Commercial products and services can be ventured into markets to try out various functions, to determine buying habits, and establish pricing. But unlike entrepreneurial and commercial ventures, military planners can ill afford to be wrong in functionality or performance, but they can spend more than expected and take longer time than projected (as long as the product or service is available when needed). Following recommendations by the U.S. Government Accountability Office, the U.S. military needs to incorporate systems engineering into the early planning stages of new products and services (GAO 2001) and adopt an incremental, capabilities-based view of adding capability and achieving product and service objectives for functionality and performance (GAO 2001, 2005, 2008). In the words of General George S. Patton, Jr., "Never tell people how to do things. Tell them what to do and they will surprise you with their ingenuity," as cited in the United States Naval Institute Proceedings (Christie 2006).

Supporting evidence that systems engineering is quite effective in delivering requisite performance, within schedule and budget limitations, has been studied meticulously and analyzed by the U.S. Government Accountability Office (GAO 2001). Their summary view is shown in Table 3.1.

Limits of Systems Engineering

The inability to show improvement over time in developing systems is sometimes perceived as a suggestion (1) that systems engineering lacks a sufficient theoretical foundation on which its practice is built, (2) that an improved model of the underpinnings of systems engineering from systems theory may help ameliorate the above deficiencies; and (3) that systems engineering as it is currently described and enacted does not reflect an accurate determination of the boundaries and implications of what is possible.

Broadly considered, concepts (associated with a set of properties (Poh 1993) and context (Aerts and Gabora 2005)) and categories (Draucker et al. 2007) form the basic components of any structure or classification that describes work. While some variability in how systems engineering is applied and how the techniques are presented in books could be an issue in themselves, there is a generally agreed set of terms in common usage with 60 years of development history. Systems are built with lessons and practices handed down from project to project. For example, the systems engineering body of knowledge is scheduled for release in 2012 as a second draft for review and broad-based comments. A body of knowledge should point out the boundaries of usefulness, the conditions under which techniques are applicable, and when practice does not reconcile with the knowledge. The discipline of systems engineering continues in its tradition of refining its practices. Then why should one question whether, in theory, systems engineering is the panacea for projects with greater complexities? Is change in what we instruct and practice necessary? These questions are generally becoming less and less the norm as systems engineering matures. Returning to the aim of this book, it is not that systems engineering is flawed and unacceptable, but rather that the flaws need to be addressed so systems engineering can scale in robustness to take on the complex problems that inevitably confront decision makers. Arguably, the key to success in any endeavor is being able to put ideas, people, processes, and things together. That requires interaction and integration.

The difficult problems faced yesterday have no comparison and little in common with the problems that must be solved in the future. Systems engineering as currently practiced has great utility and is useful for solving many problems, but it does not scale to the solutions needed tomorrow. For example, environmental issues continue to defy solutions that have known side effects. The populations of planet Earth have the desire for lower-cost products and services (that challenges our quest for new and appropriate materials); food and shelter (continue to haunt governments' mandate to provide for the needs of their people); and new energy sources (that do not contribute deleteriously to the *evergreening** of our planet).

Ask "Why?"

To test the limitations of systems engineering, consider the consequences of a rapidly changing environment. As such, there would not be any reasonable moment of stability; instead the only predictable events would be that what is happening now would soon be changed. The result would be an expectation of instability predicated on the expectation of frequent unpredictable events. If one were to harbor an attitude in such a dynamic and changing environment, that either change was harmful or living in the past was in some way better than the uncertain future. Stakeholders might be rendered incapable of focusing on any one problem, the result of which would carry over into the solution domain. The consequence might be their shaping of the problem space as an expression for their need of stability. Without applying a diversity in thinking through some structured methodology (e.g., systems engineering†), a haphazard response with hodgepodge solutions to ill-defined problems is bound to occur. Cognitive structures and thought processes that are fully responsive to defining, scoping, and characterizing the problem should be the preferred treatment. But what is the test for such thinking. Best beware if the logic presented does not survive the "why" test. You can ask the question "Why?" of the materials presented and you can also ask as the presenter "Why?". When presented with a statement, simply ask the question, "Why?". Listen to the answer. Then ask the question "Why?" again, and then again until the responder indicates they do not know the answer to your last "Why?". The importance of the question is not to expose anyone's ignorance other than that of yourself. The question "Why?" by itself is liberating. The answers of the question "Why?" should be thought provoking. And the implication of the question "Why?" will free your mind (and those inquisitive souls still listening) to be questioned, share knowledge, investigate, analyze, and remain flexible (Semler 2004). A person thinking in systems who engineers a solution for a problem that satisfies the needs of a stakeholder(s) will have structured their answers in logical fashion, be cognizant of the trade spaces and the trade-offs, have thought through a system design that captures the essence of a set of solutions, have architected the needs, preferences, and influences into the enactments of the systems design, and can relate all the previous items to the end result product or service in a descriptive fashion of the behaviors that should be expected when the solution is implemented. Compare that response (from a systems engineer) with

---

* Evergreening is a made-up term that refers to the systems engineering, building, use, and sustainment of holistic living conditions.

† Admitting bias to systems engineering thinking is respectable, even without writing it.

that from others—the difference is profound and enlightening. Without some depth of knowledge about conceptualizing, designing, architecting, developing, and integrating (i.e., systems engineering) products and services, the actions to be responsive fully to situational changes and complex problems are deficient. Any deficiencies should be immediately suggestive of a potential problem that results from the realities of the work that could be accomplish in keeping with the evolving expectations of stakeholders (along with their appetites for new and novel solutions).

Principle of Constraints

When a set of solutions that appropriately satisfy the needs of key stakeholder(s) is identified, a problem can be solved potentially. And when key stakeholders have need(s), a problem exists. Finding the problem rather than merely identifying the symptoms of the problem is the first task of systems engineering. "When a problem is framed appropriately, we have a clear purpose: we agree about what we're doing, why we are doing it, and how we will know when we're successful. We also have explored the context of the problem, and have identified a perspective or point of view about what needs to be examined" (Spetzler 2003). A principle of constraints could be stated as: change without full measure of consideration and response eventually results in failure of action influenced by the environment. Or, alternatively, enlightened diversity in thinking spawns invention and innovation consistent with ever-changing circumstances. But this is not to say that only systems engineering relies on principles, but rather that principles should guide the analysis and evaluation of information. Since systems engineering began its evolutionary development, it has merely been slow adapting to emerging and transforming surroundings and conditions. In this manner, systems engineering is similar to all disciplines and fields, and changes occur only after it is reliably shown that changes are merited. We observe a trend in ever-increasing sophistication and complexity in solutions desired by stakeholders. And similarly, stakeholders are also lagging behind in their needs as all parties struggle to identify the problem that appropriately can be solved with incremental improvements in technology within the limitations of funding and best estimates of when the new product or service is needed. Needs notwithstanding, stakeholder demands are the driving force for systems engineering. By their nature, systems engineers will push their skills to the limit to solve problems.

Clarion Call for Changes in Systems Engineering

If it were not for a preponderance of agreement among practitioners, educators, and policy makers, systems engineering might have fared better. Keeping up with increasing complexity; attempting to adapt to an exponentially progressive advance in intricate, interwoven, and conflicting set of stakeholder requirements; and more emphasis on training in the practice of systems engineering rather than educating for insights and research opportunities have stifled creativity and innovation in the very people who have the predilection for such work. The illusion that tends to silence the cynics, contrarians, and curmudgeons is that the demands of stakeholders would seem *not* to be inherently evocative of structural difficulties in the practice of systems engineering. The "If it isn't broke, don't fix it" mentality, or "It's good enough"* wisdom reinforces a pervasive elixir that systems engineering is sufficient in the main as currently devised. And therefore, the effective instructional strategy should be to continue with current materials and methods of presentation and practice. The warning signs of an inconsistency between stakeholder purpose and the means and methods of systems engineering have been visible and chronicled by the U.S. Government Accountability Office for the last decade (GAO 2001, 2008, 2009b).

Holism

Systems engineering is used throughout this book to highlight, expose, and example the application of the structures and concepts of integration. Integration is a part of systems engineering as much as it is a part of all disciplines, fields, and thinking. By applying the principles of holism[†] to promote metathinking (thinking in systems through interaction and integration), isomorphisms[‡] build on the correspondences and similarities in form and relations across disciplines (von Bertalanffy 1968, Finkelstein 1993), reductionism[§] is used to analyze and separate constituent elements, and perturbation helps identify nonlinearities in performance and results to quantify losses (Taguchi 1986, Groah 2007). The principle of perturbation (that nonlinearities cause measureable loss) suggests that the nonlinearities of governance and work activities are indeed the realities of integration. And further, that integration is inherently nonlinear helps answer the question as to why the whole is equal to more than the mere summation of the parts. Systems are inherently holistic, interacting in nonlinear fashion.

---

* The most annoying response!
† Holism is defined as the fundamental principle of a whole made up of parts, interconnected parts that cannot exist independently without the whole. Systems are holistic, and since the universe is made up of parts (interacting and integrated), the universe is holistic by its nature and construct. This definition is by no means meant to trivialize a subject that in itself dominates the great thinkers and many lifetimes of scholarly works.
‡ Similarities in concepts or structures, objects or behaviors, all things and factors considered.
§ Systems engineers need to be wary of a strictly reductionist schema for systems engineering. The holistic perspective is that of the system, not reductionistic schemas. "Western man needs to balance his intense devotion to analytical reductionism with anasynthetic words which link his success at reduction to needed successes in holism" (Troncale 1977 citing Koestler and Symthies 1968).

Synthesis

Holism considers the elements of a system as connected causally (ubiquitous, universal dependencies formulated both temporally and spatially). The mechanism that forms ". . . a whole from open, interacting parts such that the whole may exhibit desired, or requisite, emergent properties, capabilities and behaviors" is synthesis (Hitchins 2007). However, integration is more than synthesis. Synthesis is that intermediate step which encompasses emergence. However, emergence by itself does not result in a system, but rather a ProtaSystem. Synthesis is founded on the notion of action at a distance—the impact of forces acting on things possibly displaced in time and place from the original action. Synthesis joins and merges the results of interactions between system elements to sustain the emergent properties that distinguish ProtaSystems *de jure* or *de facto*. Using reductionist methods, systems engineers reduce abstract issues into a formal hierarchy of attributes, traits, and properties (i.e., physical, functional, and behavioral aspects that can be mapped into tasks that seemingly represent smaller, more tractable packages of events or activities). In an iterative fashion a "high-level" task is decomposed in a set of subtasks, then into supporting tasks, and so forth. The tasks conform to processes that have been predetermined to satisfy skills, budget, and schedule constraints.

From this descriptive definition, classical systems engineering formulates and combines objects at various stages in the systems engineering process. To this end, classical systems engineering integration is thought to imply a sense of participation or membership, one that supposes participation results in more than an individual member could derive on its own, that is, a system is greater than the sum of its parts. Moreover, it is reasoned that a system (or nonsystem) can be integrated classically with another system (or nonsystem) through their respective, relevant interacting objects. Integration occurs at both the conceptual and corporeal levels through these objects. For engineers and management, integration is planned to be two objects combined and made operational through various interfaces representing connectivity and flows of energy, matter, material wealth, or information. The result of integration is more than mere extensions of physical boundaries as is with synthesis. Integration brings full extensions of functional and behavioral boundaries, in addition to physical boundaries. Synthesis can be thought of as weak integration, which is to say, consolidation of objects, but shy of integration. Synthesis is a required step, a step that is often a final stage for many objects. The example of a simple, swinging pendulum illustrates synthesis. A string tied to a fixed pivot point swings a mass tied at the other end. Emergence results as a tight string due to the connectivity, high coupling, and high cohesion. Synthesis has revealed the value of physical boundaries and the boundary conditions. And integration occurs with an observer who needs time-keeping and whose problem is one of not knowing how far a ship has traveled on sea, that is, navigation. Integration is quite different from synthesis to demonstrate the simple, but elegant, demonstration of cyclic, gravity-enabled motion. Following the footfalls of Derek Hitchins, bottom-up integration (that "mechanistic, building-block approach, as opposed to a holistic, organismic method, and, as such, is unable to accommodate the internal subsystem trades necessary to satisfy overall system constraints, . . ." (Hitchins 2003)) fails to achieve a system in an efficient fashion. Integration requires synthesis to advance to a ProtaSystem and holism to achieve the rights and features of a system. ProtaSystems do not exist without synthesis; systems do not exist without synthesis and integration—the difference is in the degree of interactions across the boundaries of the objects. Systems engineering requires both synthesis and integration to deliver product and service systems.

Work of the Systems Engineer

The customary work of the systems engineer is to grind through the issues of integration diligently, identifying the number and types of interfaces, the quantities and frequencies of exchanges to assure the mechanical and electrical connections are established, the data types and flows are identified, and the expected behaviors are planned into the work to meet stated requirements. Trying to accomplish integration in this manner is a difficult and problematic task. Since projects rarely fail during start-up or system design, integration is usually relegated to early planning with the bulk of the work scheduled for later in the project. The first opportunity to observe substantial and measurable progress is during the development phase. It is here that the first objects are built and tested. If those objects do not perform in an acceptable manner, then the program schedule and budget should be considered at risk. The second opportunity to note significant progress is during integration. Integration refers to the phase of bringing objects together specifically to enable functions (or their decomposed subfunctions). Nearly half of the development budget can be spent on integration, with 80% of development and integration costs associated with software (Maier 2006). These percentages are typical of systems engineering projects, although the actual amounts vary widely. But as a rule of thumb passed along by systems engineers, there is a consistency across many projects. When planning for integration, the allocation of time and budget for integration often varies between 25% and 40%, indicating that the lessons learned is useful for rough planning. Depending on the amount of software (for an average-sized project), the percentage of its allocation can be greater than 80%, depending on the approach taken to develop, integrate, and test software. It seems straightforward to envision integration as the means for consuming a large portion of expenditures to deliver a product or service. Integration of objects that are insufficiently mature to have small variances in their performances (assuming that performances of any sort are achievable) are saddled with unreliable demonstrations of functionalities. For software systems that are considered very large, the lifecycle cost of the system is nearly equal to the combined integration, rework, and maintenance expenditures. These are estimated to be greater than 90% of the total cost of ownership (Jones 1994; Donaldson and Siegel 1997). Regardless of the size of the project, a significant portion of the systems engineering costs and the lifecycle costs are wrapped up in integration work.

Systems engineers promote structure for processes, functions, and physical domains, and as such, integration is formally inculcated into systems engineering processes. The nature of integration changes over the lifecycle of the product, beginning with the bringing together of ideas to form a concept, moving through development object by object: by aggregating parts into units, units into components, into subassemblies, assemblies, and subsystems, then into operations with other systems, and finally eventual disposal. Each stage of the lifecycle brings about integration.

Of importance to the systems engineer is that domain-specific knowledge needs to be integrated with systems thinking (Troncale 1977). Both domain knowledge (e.g., engineering) and systems reasoning are essential to establishing a means from which to appreciate systems engineering processes, as these processes are reflective of both. Yet, thinking in systems is not found naturally in domain knowledge. The specifics of the domain knowledge can be all consuming without having to consider issues that seem to be beyond the immediacy of dealing with a particular issue. However, it is sometimes exactly those issues that seem to be beyond the defined boundary of the problem that become problematic over the lifecycle of the new product or service. It is not enough to just design and build a product or service. The consequences of the product or service must be considered and incorporated into the systems engineer's work.

No human-built product or service is without interactions with another object. Many products and services do not have a great number of interactions with other objects and thereby lend themselves well to some degree of insularity. The boundaries of these insular objects are not complicated with interactions from other objects, and if their interactions are not integrative with other objects, their independence is maintained. Insular products and services do not face the encumbrances and difficulties of large-scale or otherwise complex systems whose interactions may have grievous and deleterious effects on systems that impact on our survival or goodwill. Systems engineering and its equivalent thinking* in other disciplines and fields are well equipped to deal with boundaries and interactions. It is the abstraction of thinking across domains that distinguishes the systems engineer from the engineer or domain specialist; the systems biologist who manages to unravel seemingly disparate processes in living systems by enigmatic patterns, the

---

* That of bringing about the conceptualization and actualization of structures of thought, equipment, or notions.

systems sociologist who recognizes behavior patterns as symptomatic of structure and processes, the systems psychiatrist who sees integrative medicine as more effective (Lake 2007), and the systems mathematician who appreciates the equal sign as signifying relations rather than equality.

Systems and Engineering

To engineer and build a system means to know what it takes to define a system and to make a system. But just defining and building a system are insufficient in the main to deliver the system that is needed and will be used. Defining a system is inadequate if the stakeholder needs do not address the real problem (as opposed to merely addressing symptoms of the real problem). And, building a system that merely addresses symptoms of the real problem only frustrates users, as much time and expense is wasted dealing with symptoms. A common example is purchasing medicines that placate the common cold. They do virtually nothing to cure the ailment, but only ameliorate the symptoms. We feel both frustration in not being able to do what we want to do and misery which accompanies illness. Defining a system that is needed is the first stage of systems engineering work. Some people know they have problems and also have the wherewithal to ask for help. Their needs may be sufficient to warrant paying someone to push technology and incorporate it into a product or service. Even if the needs are sufficient, such a product or service may be unobtainable. Systems engineers are skilled in thinking in systems and engineering to work alongside those people who have the needs to solve complex problems. Key stakeholders are those who represent the totality of the people who have various needs associated with the product or service that is to be built by systems engineers. The stages of work are stated generally and simply as (1) describing the problem in sufficient detail to manage the development effort; (2) designing a set of alternative solutions that reflect the functionalities that are wholly responsive to solving the problem within the limitations of budget and schedule; (3) architecting the artifacts to be consistent with the various relations and performances; (4) developing and integrating the one solution that is consistent with the requirements, specifications, constraints, and conditions imposed by the system boundaries, boundary conditions, skills, technologies, and policies; (5) verifying that what was conceived and built as the system solution indeed matched the requirements and specifications of the stakeholders; and (6) validating that the delivered system solution satisfied the stakeholder needs in solving the defined problem. The systems engineering development process is highly interactive, sometimes rigorously recursive, and necessarily iterative. It is highly interactive because multidisciplinary teams of people and subject matter experts are called on to participate; it is rigorously recursive for the planned revisiting of work previously attempted to reflect the newly acquired information and knowledge for planning and staging new work; and it is necessarily iterative to satisfy the persistent need to double check the work, clarify issues, voice concerns, and fix what needs to be fixed so that the cumulative work reflects the benefits of all the subsequent results.

Systems engineering can be extremely frustrating for a domain engineer, expressly focused on these three habits that are inculcated in the systems engineers thinking: (1) interactivity implies frequent, but not necessarily substantive, communications (often with nonengineers) to tease out subtleness buried in the proposed concept of operations, or implied by requirements, or ensconced as basic tenets of the problem statement; (2) rigorous recursion exposes the misunderstandings and omissions typical of developing and integrating something that has not be built previously; and (3) iteration determines the eagerness to correct all problems and the tenacity to do the right things so that the product can be built and delivered.

Charter of Systems Engineering

The charter of systems engineering is to create and express ideas and integrate components into systems that are referred to as products or services. These products and services are most often presumed enigmatic or incomprehensible by some methods, means, or fields of study. Each domain of study is built on the premise of boundedness, framework for measurements and interpretation, and theory. The essence of systems engineering is to unbound the seemingly bounded, broaden the concepts to beyond recognition, open the solution domain to include the ridiculous, and consider the issues and problems in an abstract space rather than as they are posed or presumed to be real. No other discipline or field carries with it that worldview. The rationale and purpose are clear to the systems engineer, but enigmatic to others. Specialty engineers (e.g., mechanical, electrical, and materials) often find dealing with systems engineers frustrating and annoying. Why discuss issues and factors that are not germane to the problem at hand? Why spend any time on any issue that does not go to solving the problem? Why "waste" valuable effort on talking about solutions, when simple trade-offs analyses will surface the salient relations from which to make decisions? Why indeed? The roots of systems engineering supposed the problem was either incorrectly defined, in which case the solution might be perfect, but inappropriate to solving the needs of the stakeholders, or correctly defined, but all that was needed to solve the problem was unavailable to those charged with solving the problem. The distinguishable difference between systems engineering and other engineering disciplines or fields is the context of the thinking—that of thinking in systems (always thinking in systems). Every problem needs to be thought of as a system; every stakeholder is considered as a system or system of systems; every solution reflects not only the problem with which it is matched, but also its implementation as a system or system of systems. Systems engineering exists only if its implementations can result in the integration of its artifacts into the requisite product or service.

No single discipline has developed the tools to engineer multidisciplinary products or services. Systems engineering is more than engineering. Systems engineering is the nexus of bringing together the variety and breadth of disciplines and fields required to accommodate the needs and priorities of objects (e.g., people, organizations, and the environment) put at risk during the product's or service's lifecycle. Each object put at risk has a stake in the lifecycle of the solution and are referred to as stakeholders. Stakeholders may be key stakeholders who impose requirements or are affected directly from the building, delivery, or use by the primary users. By definition, all stakeholders have needs that can be expressed as requirements. But nearly all stakeholders are undeterminable at the onset of the work, that is, the conceptualization that eventually will result in a set of requirements that will drive systems engineering and integration will themselves help expose additional stakeholders and new requirements. It is the role of the systems engineer to elicit requirements, and by doing so identify the hundreds of people, organizations, and situations that will be affected by the proposed system over the system's lifecycle.

Lifecycle Considerations

The allure of using systems engineering for solving vexing problems is determined to a great extent by three issues: (1) how comfortably the solution reflects lifecycle needs; (2) the broader context in which the design is considered to have utility; and (3) the flexibility to incorporate cross-disciplinary views. These three issues are captured in a lifecycle thinking in systems.

Lifecycle needs are often mentioned in the same vein as low-cost solutions that deliver high performance. Yet the realities of development within the constraints of budget and schedule often imply and impose "hidden" requirements on the system design and architecture. These hidden requirements are likely visible during the product or service lifetime, and are indeed traceable to the original specification documents. Specifically, the requirements indicate what the stakeholders need to solve their problem. Specifications are written by the project team to guide the project work. And while the specifications embody the letter and spirit of the requirements, the interpretation of these specifications and the decisions made by the individual engineers may result in more or less than what was stated in the requirements. Sometimes these differences between what is delivered and what was required are subtle, but oftentimes not. For example, specifying the number of lines of code for a module may improve the overall system performance, but make the development time longer. So a typical rule is not to limit the developers in too many ways, thereby permitting greater flexibility to achieve early testing of the work. Once the module demonstrates proficiency in satisfying the specification "passing" the tests, and being verified as both responsive to the needs and suitable as per the specifications, little concern is shown for reworking the module to some additional specification (in this instance, the number of lines of code). The developers and planners assume that if there is a performance issue that surfaces during integration, then work can be directed at that time to deal with the issues. The particular module that is referred to in this example may not be the one which is targeted for rework later on. In other words, the decisions by the engineers are considered to be "good enough" if the tests, verification, responsiveness, and suitability are demonstrated. It is very difficult to specify all that is important for the system when the system is not demonstrable, when all the functions and their performances are known, and when all the losses that incur due to achieving those performances are measured. What lurks in the "completed" and "acceptable" modules only shows their combined action *after* the system is integrated and most likely *after* it is in the hands of the user(s). It is important for systems engineering and systems integration to estimate not only the performances of the system's functionalities, but also the losses that incur to achieve those performances (Chapter 1, Principle 7). The lifecycle issues that impact on the users are most often of the type that come from implementing the specifications.

Lifecycle can be seen as a structured progression from an initial beginning state to an end state, often thought of as from inception (beginning of life) to disposal (end of life). Lifecycle is not comprised of sequential or successive processes. Yet, lifecycle discussions are appropriate to all processes and activities. It is instructive to consider the lifecycle of the problem, the stakeholder needs, the development effort, the product, and the product uses. If either the lifecycle of the problem exceeds that of the need or the need exceeds that of the problem, the problem has been solved differently than expected. Either there is no longer a need to solve the problem (i.e., the problem has changed from that originally defined) or there is no longer a problem that needs solving (i.e., the problem has vanished due to circumstances). In both cases, the problem should be redefined to determine the stakeholders who need to solve the newly defined problem (or if there is a newly defined need that addresses a new problem). If the lifecycle of the product is greater than that of the need, the product is overdesigned or the market changed. If the lifecycle of the need exceeds that of the product (solution), and the problem remains the same, there is market opportunity for an enterprise—the hallmark of a successful lifecycle product.

A product's lifecycle refers to the overlapping, concurrent stages through which the product passes from its earliest beginnings through design, development, production, use, its discontinuance, and settlement of the final legal actions. The concept of lifecycle divided into stages with processes embedded within each stage is typical of systems engineering. At the end of each stage, a decision is made whether to go on to the next stage or continue the processes within the current stage.

Inception includes the earliest conceptualization of the problem that needs to be solved or awareness of the opportunity that ignites compelling interest in stakeholders. Disposal is the scrapping of a product as a consequence of it having unacceptable or no utility for its intended or emergent purposes. The key drivers of a product's progression through its lifecycle result from the trade-offs between economics and fear. In the case of commercial products, the economic success factors are revenue and profitability; the fear is loss of market share, declining image, and deterioration of brand.

The arrows of lifecycle success converge to deliver the lowest lifecycle cost alternative for a required system effectiveness by avoiding the congeries of losses throughout the lifecycle. But lifecycle success does not favor any one stage under the guise of politics, opportunity, innovation, selfishness, or profitability. Such preferences for expediency may encourage the development of a product that is a widely accepted; perhaps even considered respectable by seemingly rational benefits (e.g., the internal combustion engine promotes modern industrialism). But losses accumulate from both manageable and unmanageable factors. And unintended consequences may result from either category of factors. Success, therefore, should not be predicated on ignoring or not exploring any particular stage(s) or activity, but rather on the totality of losses accrued to the set of all stages—the lifecycle.

Lifecycle Success

Our past and our prospective plight are in the hands of dreamers and pragmatists. Dreamers set the boundaries of our future, concerned with a world that could be. Pragmatists build our next reality, driven by the success or failure of their work. Neither is a credit to humankind *if* they solve the wrong problem, or *if* they try to solve the right problem in the wrong way. Their solutions may be clever, apt, and ingenious—boffo exemplars of accomplishment—but they may have failed a crucial test: lifecycle success. Success over a lifecycle is indicated by (1) an aggregation of activities, processes, or results that accomplish their purpose, (2) the synergism of events that produce an effect greater than the sum of their individual efforts, (3) the state of prosperity or good fortune that is consequentially related to one's intentions over an extended time to attain eudaimonia, or (4) the semipermanent winning record that is envied. By the first definition, success is found through the processes of achievement, those acts that, when mixed or summed, combine and compound in a way that is strictly in definite proportion to the total of the constituent results. In other words, while the activities and results may have intertwined, there is no synergy—no greater effect than the sum of their individual effects. We term this process of interchange and interplay as the interaction between elements. An element is an artifact or atomic unit that is one of the individual parts that comprise an entity, a cognitive process, an operation, or an object within a system. It is the fundamental unit which is perceived, known, or inferred to have existence (living or nonliving). Elements that are interactional may be capable of acting on or influencing each other under various conditions. However, these elements do not have such undue influence so as to impart information, energy, material, or wealth as part of their interaction. In other words, elements may interact, but if they leave no imprint or lasting effect, their interaction is not causal. Causality requires that another condition be met. The first enactments that lead to success emphasize the aggregative effects, whereas the property of emergence is embodied in the second portrayal of success. The interaction between objects takes care to distinguish between interaction and integration (which relies on emergence).

Lifecycle success is challenged by development and integration: (1) figuring out what product to build; (2) determining which parts are required; (3) deciding on which parts to make or buy; (4) putting the parts together; (5) keeping the parts together so they perform as expected; and (6) then disposing of the parts when they no longer are useful to society.

Lifecycle Stages

Within the systems engineering parlance, lifecycle can be described as stages that characterize the product as it transitions from a concept to its final disposal (and the last lawsuit). The stages are defined for an individual project as agreed upon by the stakeholders. The stages represent work that moves the product or service through milestones, each milestone showing progress toward the delivery of the requisite product or service. Systems engineering lifecycle models are specified, tailored, and used to pace the events leading through development and operations. In a general sense, these stages can be characterized as follows:

- *Conceptualization*: The first stage concentrates on two activities— defining the problem faced by stakeholders and determining their need; conceptualizing the concept of solutions (CONSOLS) and the concept of operations (CONOPS). Conceptualization is done to attract financing and set a preliminary course to explore the stakeholder's problem, need, and solution space. Conceptualization increases the likelihood of appealing to the profit or opportunities that may interest investors. As more time is invested in conceptualization and searching for financing, the conceptualizer's risk increases for not being able to realize a return on investment. However, failing to attract needed capital may or may not increase the risk of success. Conceptualization is often presented as a case that justifies the product (e.g., business case, business plan, or proposal).

- *Financing*: The second stage focuses on financing—determining which financing source has sufficient interest, capital, and expertise to consider and possibly fund the effort to accept or amend the proposed solution to satisfy market demands. Financing is secured to reduce the risk of not having sufficient resources to complete and sell the product. Outside money is exchanged for a say in decisions and a part of the results of all activities (make equivalent to other areas, including process decomposition to first level).

If financed, the project begins work to prepare a design that captures the preliminary set of top-level (most general) requirements. These requirements derive from the various stakeholders and represent their cardinal interests and is responsive to the various delineations that bound the solution(s) as well as the conditions under which the boundaries impact on the design. Additionally, a CONSOLS (describe how it reduces risk) and a CONOPS (describe how it mitigates risk) are prepared that relate the design, the operational environment, and the key performance parameters that distinguish the solution from other existing and proposed solutions. Preliminary design reduces the risk of failure by (1) providing confidence to the developers and financiers; (2) showing transparency in the work and the design, so that "independent" analysis and evaluation can be accomplished before additional money is spent; (3) uncovering areas of work that necessitate further study; and (4) addressing specific issues that were identified earlier as items of risk.

- *Preliminary design*: The third stage transforms the revised CONSOLS and CONOPS into requirements at appropriate levels of detail to produce a design that can be evaluated for risk, utility, and market acceptance (or mission dominance). Planning is performed.
- *Detailed design and architecting*: The fourth stage completes the design by inductive reasoning, innovation, creativity, reductionist logic, modeling and simulation, and prototyping. Alternative architectures are formulated. The detailed design and architecting follows from the preliminary designs to provide additional detail, derive requirements from the stated requirements, and further specify tasks and their outcomes. Sometimes protoproducts or protoservices are developed to verify the stakeholder requirements and needs are met (but as important, to validate that the team can in fact build a model of something that represents a key function that is

expected to be delivered after integration and test). In the case of entrepreneurs, the funding sources will nearly always insist on seeing an early prototype as a means of validation of the business model, technical capability, and proof of market viability (as determined by testimonials from potential customers and users). At this stage, the risk of not meeting the requirements is reduced, but not eliminated.

- *Development*: The fifth stage describes the bulk of development. Development personnel and support staff are hired or contracted, resources are allocated, parts are procured, and units and subassemblies are built and tested.

- *System integration*: The sixth stage blends with the fifth stage as development transforms parts into subassemblies that are integrated into subsystems and then into the protosystem. The protosystem can be accepted by the acquisition stakeholder as the deliverable product or service.* The protosystem can also be a production prototype that the customer and user should then contract to improve reliability, engineer for production or manufacturing, or use as a system that has limited use in a reasonably benign, not critical environment.

- *Manufacturing or production*: The seventh stage deals with the engineering of the protosystem to specify and deliver a manufacturable end item product that satisfies stated requirements. Manufacturing is the remaining test that the stakeholder requirements can be met with more than a one-of-a-kind build. Batch or series construction deals with using noncustom methods applied to problems that would have been solved previously by hand-crafting custom solutions. Supplying a more standardized product or service further reduces the risk that more than one product or service can be built to satisfy stakeholder's requirements.

- *Purchasing*: The sixth stage involves the offer to sale and the acceptance to buy. To purchase means to recognize the problem; to identify a need; to determine if the problem and the need are applicable to current and future intentions of the acquiring organization; to gather information to increase organizational knowledge; to negotiate; to agree on price, terms, acceptance, warranties, product support, maintenance; and delivery options and constraints. Purchasing is the first test of the user's acceptance of the product. That the product is believable and should be accepted by the buyer reduces the risk that no one will need it.

---

* There is sometimes a concerted effort that builds only a few protosystems that have robust properties and traits according to a set of requirements that imply use in an operational setting and environment. In those cases, the development effort must also include "design for operations" which adds additional requirements and testing.

- *Operations*: The seventh stage spans integration into existing operations for use, maintenance, support, training, upgrading, modifying, documenting, and reporting. Acceptance by more than one buyer and the user is another benchmark of risk reduction. Having multiple and different customers buying shows that the product or service is satisfactory (by its first impressions) given some variability of conditions and circumstances. Use is the putting of the product into service for a particular purpose. Most products or services serve two purposes: to render its inherent or natural purpose and to assist the user in making or carrying out decisions (if different from its purpose). Use involves integration of the new product or service into the existing organizational procedures, culture, social, and behavioral environments.
- *Disposal*: The eighth stage addresses the final disposal of the product or winding down of service and all related issues.

Each stage of the lifecycle is highly organized with checks and balances to reduce the risk of finding problems in subsequent stages. One of the checks is to verify that the activities and results of the work are in fact fully responsive to the requirements and needs of the stakeholders. Through the early stages of work, verification is most often comparison of ideas and prototypes with written documents that either explain or describe the customer and user's intentions and requirements. Other checks include the processes of systems engineering that stipulate the starting points for the stages along with the criteria for leaving that stage and progressing to the next stage. Often, rework is required on previous work, but the impetus is to move on to the next stage as each stage reveals more data and information about the product and service. For example, at disposal, one should expect to have learned much of the product's or service's properties and traits, some of which are converted into knowledge.

All previous stages contribute to reducing the risks of disposal, with particular emphasis on amount of money and planning that is incorporated into the design and use.

Conceptualization reduces the risk of financing; financing reduces the risk of preliminary design; preliminary design reduces the risk of detailed design and development; detailed design and development reduces the risk of manufacturing or construction; integration reduces the risk of not delivering what is expected, which in turn reduces the risk of purchasing; and which reduces the risk of use and the risks associated with disposal. The risks of disposal may include lawsuits due to harm or safety concerns (e.g., radiation effects and inhalation of asbestos), and unintended consequences that arise due to social behaviors (violence associated with certain types of entertainment).

The lifecycle stages can be thought of designed to reduce risk of later stages. The greatest uncertainties that threaten lifecycle success (i.e., bring about risk) are discovered and dealt with by the activities that are carried out during the period of each stage. The transitions between stages are planned points of stability. Major decision points typically follow milestones at the end of each stage. Milestones are designed to show stakeholders that the product is progressing through a stage and at an agreed juncture, and the product is deemed ready to enter the next stage. Therefore, the majority of product uncertainties are determined by observation and analysis during the course of a stage, with only a small fraction identified at the entry or exit point of a stage. In systems engineering parlance, the lifecycle stages are captured by the systems engineering process models. Each stage of a process model is bounded and delineated by processes that reduce the uncertainties in a systematic, highly regimented environment. The principal risk of lifecycle success is that of not achieving the desired level, kind, and degree of integration.

Lifecycle success results from managing a product's lifecycle to achieve the lowest lifecycle cost for the desired level of product utility at a risk that is deemed acceptable to the stakeholders. Those put at risk or who are positioned to lose or gain over the lifecycle of a solution are called the stakeholders. The key stakeholders are those who are put at significant risk during any one lifecycle stage. Stakeholder interests relate to one or more stages of the product's lifecycle. In the main, stakeholders fall into one or more of four groupings—those representing the product, the customer or user, afflicted parties or organizations, and the Earth's biosphere and its purlieu. Building successful products depends on satisfying this quadruplet of stakeholders. Success is the attainment of a goal, if and only if, no stakeholder is burdened with losses that are destabilizing or from which there is no recovery.

Lifecycle investigation—inquiry into the staging of activities, their recurrence, concurrency, and progression—involves lifecycle assessment and lifecycle analysis. Lifecycle assessment develops knowledge of the consequences arising from each stage, whereas lifecycle analysis examines the activities and their relations in comprising the stage. Lifecycle investigation is a framework for two critical economic analysis tools—management of product worth and of product risk.

The activities in each stage of the product's lifecycle and the manner in which they are performed can add to the product's worth by lessening or eliminating risks. While there is an abundance of risks, lifecycle investigation is focused principally on one risk that is omnipresent and persistent. That is the fundamental risk of integration. In the broadest sense, integration is the combining of objects, whether two thoughts form an idea or take action, two people form a team, two units perform a test, two subsystems build a system, or two systems complete a system of systems.

Lifecycle Measures

A measure is a basis for comparison derived from a single, repeatable process of assigning numbers to phenomena according to a rule. Lifecycle measures have traits of being relevant and appropriate. Allowing for the possibility that not all measures capture key factors that are directly causal, we define key measures as also contributing in a central way to the essential character of the phenomena—that which is directly causal. Measures must be quantifiable with some precision. Good measures have relatively high accuracy, that is, low variance and high precision. Measures are distinct qualitatively and quantifiable as an attribute of a phenomenon or matter (IEEE 1991).

Measures need not be applicable to all parts or the whole of the system. Optimizations do not occur through analyses of measures.

Lifecycle measures that focus on low-level determinants may be useful for estimating other development projects for budgeting, scheduling, and planning. Comparisons with other projects: An example of such a lifecycle measure is the number of source lines of code of software. Over the course of the product or service lifecycle, the number of source lines of codes grows. A comparable measure of the number of source lines of code for a specific function for a project is the absolute number, growth rate in the number, the ratio of the growth rate for one stage of the lifecycle versus another, the rework number, and the same measures from one project to like-kind projects. Other lifecycle measures focus on the mean time between failure, while still others are concerned with a measure of effectiveness (which is only determinable at the system level). For smaller projects, these lifecycle measures are quite useful. However, for larger, complex projects lifecycle measures are often inadequate due to the quite dissimilar nature of the prior projects. Comparisons of new projects with non-like-kind projects are problematic. There is another way, but it is also fraught with uncertainty.

Measurements, measures, frameworks, theory, variables (and their dependencies), metrics, and causality are all essential ingredients for comparing projects for both estimating purposes (in the case of new development efforts) and planning purposes (in the case of operational issues).

An introduction to concepts of measurement suggests that there is an inherent error in all that we measure. Measurement is the interpretation of observations, where the interpretation requires a context and a conceptualization of meaning. The interpretation is expressed through a framework— the relationships, dimensions, interfaces, form, and fitness that act according to the accepted standards. The framework is the logics of a scale by which to compare various constructs. In total, the essence of a theory is made simple and comprehensible through a framework. A theory, broadly defined and widely recognized, is expressed in cognitive substance that explains phenomena and guides our actions and experiences. Reasoned and rational measurement premises a serious-minded, deeply rooted theory. At the heart of any scientific explanation is a mechanism, the cardinal enactment of a function; or an activity, the central workings of a process. Measurement is fundamental to comparing measures.

When investigating potentially causal factors, it is posited that the mechanisms and key activities that characterize various acts (i.e., events or experiences) are essential to developing an understanding about the joint occurrence of events or experiences.* Mechanisms are enacted (i.e., whether by action or activities) through variables. Measures are the independent variables that are reference points from which other items can be evaluated. Measures are fundamental to comparing projects.

Variables take on different numerical quantities or states of existence. Variables differ from constants, and constants are stable within a defined construct. Variables indicate the linkages between events and experiences and help indicate the possible causal relations. Discovering indicators and isolating variables are an integral part of the scientific method.† The investigation into the relationships between events or between experiences is organized and progresses through method—the intervening substance and way of going between principle and practice (Pawson 1989). Mechanisms convert or arrange inputs into outputs basically mapping a set of characteristics into outcomes. A mechanism converts inputs (independent variables) into outputs (dependent variables) (Reskin 2003). Constructs that relate multiple independent variables are referred to as metrics. Mechanisms and metrics are important for comparing projects.

The relations between the variables may suggest possible deterministic qualities that influence the behaviors that are observed, perhaps considerably. Under certain conditions, a variable may be restricted in a pair-wise fashion with other variables to ascertain if their interactions are prima facie suggestive of an inevitable consequence or deterministically causal by their portrayal. By careful design of experiments or well-conceived *gedanken* experiments‡ the independence of a variable can be investigated in this fashion. The nature of a system or a system component can be construed as being dependent on the variables that affect its make-up and design. However, to quantify the system aspects and features, variables that are independent of other influences (i.e., other variables) form the groundwork to describe the system operations, be combined for accurate representations of the system (i.e., models), and be predictive of situations in which the system is planned to operate. Discovering and isolating independent variables or a set of variables whose combinatorial actions and influences act in concert as a single independent variable is a primary task of the systems engineer, systems architect, and systems integrator. The dependencies of variables are important for comparing projects.

---

* The precedence for this position on mechanisms is based on the procedures of a physicist who would first observe and then conjure a law of the basic or underlying phenomenon. The law would be predicated on a mechanistic perspective that linked the observed behaviors with a principle of action. The implication is that a simple causal relationship exists and is descriptive of what is observed in nature. The causal relationships derive from variables that capture these relationships.

† Method organizes and represents the structures, artifacts, concepts, means, and context into the endeavor and exercise of the work.

‡ Thought experiments.

For the purpose of comparing non-like-kind projects, typical of many new, complex endeavors, integration offers a unique insight into many measures of development. Specifically, the events of integration represent all that has transpired during system design, architecting, and development. The sequence of events can be represented as a Markov chain where the amount of rework that is necessary for two objects to demonstrate requisite functionality is a strong measure of the progress toward final systems integration. For these purposes, the amount of rework includes all the iterative actions found predominantly before integration as well as the recursive thinking that often occurs once integration begins. If there is a substantial percentage of rework during physical structures, and hardware and software development, then of the primary suspect causes (i.e., ill-defined requirements and poor functional decomposition, or poor mapping of functions to physical entities), partitioning problems are often the most likely to persist until integration reveals the consequences of poor partitioning. This is not to say that the other root causes of rework have been eliminated. It is only that requirements and functional analysis can be reasonably checked and reviewed to discover and correct issues. However, poor partitioning with mapping to physical objects is significantly more difficult to detect and will present problems such as different coupling or cohesion than is expected across the physical interfaces between the two objects undergoing integration. Coupling and cohesion due to poor partitioning of lower-level functions are not normally completely thought through either at the system design or shown in architectural views. Rework to correct partitioning or its symptoms, coupling, and cohesion can occur during integration rather than earlier after development testing. Since functions are demonstrable when two objects are integrated, the function either is not shown or is demonstrated with poor performance. In either case, rework is required for one or both objects. There is no reasonable, clear dividing line between development and integration and arguably development and systems integration are often one task after the next as they intertwine during the building of objects. But there does come a point when the elementary objects are presented and tested, then brought to the next object for integration. The framework of integration points out the key measures of coupling and cohesion as symptoms of improper partitioning. The measure of reworks provides insight into the intricacies of the development stage—an often enigmatic and vexing morass of uncertainty and risk for developers.

The practicalities of using rework as the only indicator of progress or for estimating purposes is much dependent on the ability of the workers to show consistency in their actions so that their rework as a percentage of work completed can be represented as a distribution function based on the type of work, the stage of development, their years of experience, or some other measurable variable. We refer to these factors as the unencumbered measures. In addition, the influence of management or some other identifiable issue to coerce the workers to deviate from their "natural" tendencies and thwart the unencumbered measures mentioned has a masking effect on the meaning of the measure. The degree to which that masking effect can be quantified in a period and averaged somewhat mitigates the masking effect if that period is meaningful in the context of the overall usefulness of using the measure as the primary indicator for determining progress or for estimation purposes. Because although tests can be developed to "measure" the proficiencies of the engineers for their statistical weightings, coercive effects complicate the use of rework as a measure. Such tests of proficiency, for example, may be applicable to software, but not to building physical entities. Whereas source lines of code or function points (to mention a few) have been used to estimate software work efforts, using rework metrics may provide supplemental, supporting evidence that proves useful in some circumstances. Previous works on physical structures and computer hardware also have various methods and measures to estimate work. Combined with other such measures of rework, the totality of measures may be sufficient to perform the necessary comparisons for planning.

Lifecycle Measure: Time

A lifecycle stage is characterized as a distinct period within a sequence of stages. Each stage is a conglomerate of periods of activities. Time can be considered a limitation that bounds the period between the notional onset of life and the somatic moment of death of an entity. Time can also be regarded as a constraint that restricts what is available within the defined limits. Each stage can be constrained in different ways, for example, by budget, schedule, available skills, and allocations of office space (for example).

A condition for lifecycle success requires the individual summation of each type of constraint not exceed the limitation imposed on the type. In other words, lifecycle success means not overrunning budgets or not exceeding timetables to deliver the expected product. With regard to a limitation, time refers to that of the lifecycle, and in the context of a constraint, time refers to that apportioned to each stage (or that which is allocated to an activity within a stage). Time is a consummate, independent measure of a cycle.

Lifecycle Measure: Cost

Cost is a measure of what is expended to attain or accomplish something. Cost is the total of expenses that it takes to provide the totality of a product or a portion of the product, within the context of the product's lifecycle (e.g., a stage or total lifecycle). Cost can also be construed as a constraint within defined limits of an overall budget that is imposed on a stage or an assemblage of stages. In this context, cost takes on the notion of "budget"— the allocation of cost for a particular time or purpose. Budgets can be allocated to each stage and expenditures monitored against those budgets.

Cost, as a constraint, is apportionable by stage or over the product lifecycle. When viewed over the product lifecycle, cost has been referred to as the total cost of ownership. Lifecycle costs consist of the costs associated with the processes (and it can be said, with the results of the processes) within a cycle or the product lifecycle. Lifecycle success means not overrunning cost budgets ascribed to delivering the expected product. Since cost is deemed to a substantial measure of lifecycle success, its management is an unassailable measure of a cycle or aggregation of cycles. Cost is a consummate, independent measure of a cycle.

Lifecycle Metrics

Combinations of measures whose relations have meaning within a framework are termed as metrics. Metrics, themselves dependent, relate multiple independent variables to facilitate the quantification of a particular characteristic. In addition to the properties indicated for measures (germane and quantifiable with small variance), good metrics should (1) be complete as an indication of a particular system-level attribute (to be meaningful), (2) have verifiable physical meaning (to avoid misuse), and (3) indicate the degree of satisfying an ideal performance. A metric is a quantitative assessment of the degree to which a system, component, or function possesses a given measure. For example, metrics can be used to quantify productivity and efficiency.

Unlike measures that need not be applicable to all parts or the whole of the system, metrics do. Optimizations occur through analyses of metrics (not measures). Metrics are the criteria that contribute to decision making.

Lifecycle Metric: Money

Money functions as a sanctioned, legitimate means of exchange (Newlyn 1978). Investments are represented by money or other exchange of intellectual capital. For a monetary type of investment, labor and requisite skills can be acquired to carry out the appropriate processes within the lifecycle stages. Depending on the activities chosen and the manner in which those activities are performed, a product may be engineered, managed, and maintained to achieve desirability and usefulness. The total lifecycle investment can be considered a limitation that bounds the amount for the project and a constraint regarding the rate of expenditure for each stage.

But while the notion of investment appeals to the human nature of wanting to get the most for the money, it is instructive to think of the product as a means to store the value of that investment. Through a medium of exchange, money is converted into a product and participation in the outcome of the enterprise that produced the product. The participation may be as the buyer and user of the product (as in the case of governments contracting for goods) or as a partner in the enterprise that is intended to benefit from the sale and profitability of the product. In either case, the lifecycle success of money is through an investment that returns benefits (usually preferred in monetary terms) through planned uses of the product's inherent functionalities. So, unlike the independent measures of *time* and *cost*, investment through the construct of money is a dependent metric of stored value for a given product performance, based on several measures, including cost and performance.

Lifecycle Metric: Performance

Performance is the consequence of accomplishing work, the outcome of an event. Performance is a metric of functionality, a dependent variable comprised of measures that give rise to a product's utility. Performance is multidimensional, having meaning only within a domain in which its measures are continuous and quantifiable (Euske and Euske 2002). As such, the mechanisms that deliver functionality do so based on an input (independent variable(s)) while the performance (output metric) represents a measurable set of dependent variables. The context of input and output (Reilly and Reilly 2000) and continuity of the measurements of all the variables must be considered when developing an understanding of performance. An example of performance is the average speed (km/h) at which a vehicle of mass (kg) travels between two locations, a stage, or the lifecycle of all travel between an initial and final location. The ratio of three measures—distance (e.g., km) times mass (kg) divided by the time (e.g., h)—is a performance metric. In the case of an internal combustion engine, the lifecycle success of the performance metric could be referenced to the quantity of gasoline consumed during the travel. Therefore, performance is a metric that is relative to the measures of time, mass, and distance. These measures are impacted upon by temporal changes and events (Phelan 1993) in various reference frames: vehicle, operator, and environmental. The mechanism(s) that converts the energy stored in the vehicle's gasoline into speed may change over time (due to wear or nonoptimum tuning), by environmental events (headwind and rolling friction), or due to operator effects (strong acceleration that averages over time and distance to a constant velocity, but at a lower rate of efficient conversion of gasoline).

Lifecycle Metric: Complexity

Complexity results from emergent properties of integrated objects, number and types of processes, and the number, types, and frequency of interactions between and within processes.

Lifecycle Sense

We observe the beginning and end of things to appear to be both a natural occurrence and one that we contrive by our own intentions. But it seems that lifecycle is more than a structured progression; however, we apply our intentions. Rather, lifecycle presents as a piece-wise continuous succession marked by recursive, iterative processes and not that which is posed as discrete, independent stages where processes are kempt and respectable. No stage of a lifecycle or its totality satisfies this Utopian model. Yet, a lifecycle perspective makes sense in this regard if we allow discrete activities to recapitulate and update previous findings. Posing lifecycle as the primary orchestration and organization of systems engineering in itself is not new. The systems engineering process models do in fact regiment the overall view of a product in terms of lifecycle stages. But what is new is the discussion on lifecycle through the portal of the fundamentals of integration. Lifecycle is active within the framework of integration, rather than being the framework in which integration is performed. While it seems intuitive that integration occurs throughout the lifecycle, it is more accurate to consider lifecycle as a continuum of integration. In fact, lifecycle is the result of integration. Without integration, no process would show emergence, and product functions would be self-governing and separate. This autonomy would perhaps reinforce interactions, but would by definition not result in integration. Every process, every function, and even the physical space that encompasses the domain of interest exist only because of interaction and integration, both necessary for a system. Integration requires interaction, but interaction does not imply integration. Lifecycle is merely the temporal interpretation of integration.

Introduction to Defining the Problem

Systems engineers attach great significance to defining terms. Every project has a litany of words and phrases that have specific meaning to people working on the project. The explicit and expressed meaning of a word needs to be clear and definitive about its uses, its opposites, and its causes, as well as be descriptive of the thing (Swartz 1997), its model, or its representation. The results of well-defined (i.e., nonoverlapping, nonunderlapping) terms will be fewer errors made in judgment, fewer mistakes made due to miscommunication, less cause for variation from expectations, and agreed limitations that will be shown respect. Typical of the interaction between buyer and seller, defined terms are an agreement, a normative means by which to work within a zone of comfort. Outside the definition is uncertainty, and inside the definition is assurance that others will appreciate and understand the meaning. You might say that without proper limitations and constraints, the words you use might create a problem for others. But this characterization of a problem incorrectly attributes the problem to the notion that by not using an agreed-to (or prerationalized) definition, a problem is created. This situation by itself does not pose a problem. In fact, the listener might simply have thought the word was clear, inferred meaning from context, guessed, or ignored the word's meaning and follow along with the discussion to take away what could be gleaned. The problem may manifest itself later when the word is used in a different or mistaken manner and a part is built that does not pass a test or meet a requirement (whether or not the part(s) passed their tests). Analyzing the part's failure might focus on various aspects of the design of the part, the building of the part, the testing or test set-up, milestone reviews and oversight, or some aspect(s) of communications. The consequence of a failed test means rework and analysis, which adds to schedule and costs. If these difficulties are assumed to be part of development, and taken into account when determining budgets and schedules for deliveries, then there is no problem. A problem only exists when there is a difference between what can be done and what you need to do, and you do not know how to achieve what needs to be done. Defining the problem is a process "... which transforms an indeterminate situation into a pattern of factual data ..." (Hall 1962).

Defining the Problem

A problem is devised in relation to a need or want of a solution that accomplishes something that cannot be done due to some objective reason(s), that is, availability, technology, science, opportunity, resources, or desire. A problem is relative to circumstances, meaning a problem for one person may not be a problem for someone else. Problems are relative. Therefore, systems engineers (and others) who aspire to find a set of solutions to a problem must be mindful that there are problems great and small. Great problems demand an apt consideration of the problem space, that is, the problems that exist in association with other problems. Such association might be thought of as a nesting of problems (mix of heterarchical and hierarchical relations), a strict hierarchy of problems, or an affinity of like-kind problems.

Nested Problems

Nested problems are problems within problems with a relation to the other problems through an abstraction. The relation between the problems is amorphous, having more to do with the abstraction that categorizes the problem than the relation between individual problems. Nested problems are typically related to some abstract object, whether physical or intellectual. An example of a nested problem within the domain of physical objects is that of a house. The abstraction of the house could be considered as a shelter or living environment. As a shelter, the house (with its myriad of functions that support "to shelter") and the boundary of *house* are well beyond the physical limitation of the bricks and mortar. Supplying water to the house includes the systems of source, storage, and distribution. The problem domain for the abstraction shelter includes all problems associated with each of the systems that interact with the house. Problems with the source (e.g., insufficient rain and runoff) have an impact on house and living in the environment of the house. So the abstract problem of shelter subsumes many other problems due to location, circumstance, specific design details, and habits of the inhabitants. Nested problems intertwine like a cauldron of cooked and drained spaghetti. Parsing the topology of nestedness can be taken in many dimensions. Nested problems have different relations depending on the relations between the dimensions in which the problem is expressed. Figure 5.1 relates problem A (the abstraction of house) to the reservoir that stores water (Problem A_4) to the distribution of water from municipal storage to the house (Problem A_1) to piping the water to the rooms in which water is desired (Problem A_2) to the restricted use by the inhabitants (Problem A_3). In this example, the connection of the municipal water pipe to the house is indicated as (Problem A_5), while the source of water (rain) is disconnect (Problem A_6).

Hierarchical Problems

Problems of a hierarchical nature are related by the need to associate one problem with another in a structure of subsets. Each problem is considered to be a subset of another problem, with the highest level in the hierarchy representing the totality of the subset problems. All the problems in the hierarchy are related by a common factor that weaves a common theme throughout their relations. Such a hierarchical structure of problems provides a convenient

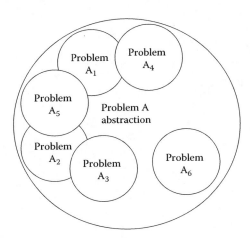

FIGURE 5.1
Nested problems related through abstraction.

place for all problems related by the common factor. The problems represented in a hierarchy are deduced from the highest level, with each successive level a deduction from the level above. Unlike nested problems that can have a mixed relation between problems (relation by hierarchy or heterarchy, hierarchical problems are only related through decomposition). Take for example, a problem with the foundation discovered after the building of a house. The floors of the multistoreyed house slope slightly toward the front of the house. The doors do not close, a gap showing at the floor and apparent, because the doors either swing open or close in the unlatched position. With the walls out of plumb, doors and windows do not function properly. The foundation is a problem from which many structural subproblems arise. Figure 5.2 illustrates the hierarchy of problems with the house. At the top level, Problem A (the foundation) is decomposed into problems caused by the foundation: Problem A_2 (floors), Problem A_1 (walls), Problem A_3 (doors), and Problem A_4 (windows).

Like-Kind Problems

Problems of a like kind are related by similarities but not by circumstance. The appearance of a problem, its resemblance to another problem, and the problem's general attributes, traits, or properties are sufficient to group one problem with another completely unrelated problem. For example, a problem with the performance of a car from one manufacturer has many anecdotal similarities with the performance of a car from another manufacturer. The manufacturer neither uses the same parts nor has the same processes and procedures, yet both cars seem to exhibit the same type of performance problem. Uncontrollable acceleration of both vehicles, separated by thirty

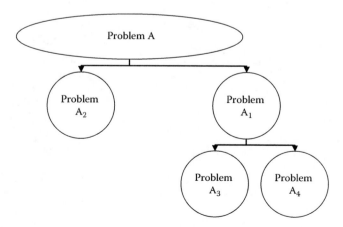

FIGURE 5.2
Problems within a hierarchy.

years, from manufacturers on opposite sides of the globe may not be tightly coupled situations; however, there may be a commonly practiced design flaw, a set of parts with similar heritage, or a particular method and approach to installation or testing that masks a potential problem. The categorization of such a problem would be "uncontrollable acceleration." Actual cause(s) under investigation, perhaps, will never be known.

Problem Domain Analysis

Over the lifecycle of a problem, some things or people are affected and they reside within the lifecycle boundaries of the problem, while others remain outside those boundaries. Those within the lifecycle boundaries are termed as stakeholders (lifecycle stakeholders). Whether the problem domain is characterized as nested, hierarchical, or like kind, the domain will involve stakeholders. As with systems engineering, an integration perspective relies on defining the problem in a most important way: the problem stipulates the solution. It is the role of the systems engineer and the systems integrator to provide that solution. If the problem is ill defined, erroneously defined, or undefined, the solution has no meaning. In other words, the work, resources, and skills were misused. However, if the solution is defined in terms of the problem domain, much insight is gained into the type of problem that needs to be explored to define the problem with which systems engineering and systems integration must deal.

The usual early work in systems engineering revolves around the triad: stakeholder, problem, and need. A typical sequence is (1) ask the stakeholder(s) what their needs are, surmise what the problem might be, discuss the problem to gain concurrence from the stakeholder, and then declare the problem is (fill in the blank); ask the stakeholder(s) what their problem is, surmise their needs, discuss their needs in terms of the problem to gain concurrence (or reach consensus), and then declare what the problem is (fill in the blank); and have the stakeholder(s) tell their problem, what they need, reach consensus, and then declare the problem is (fill in the blank). All too often, very little time is spent on determining either what the problem is or what the problem means. Defining the problem means more than just defining the problem. Rather, defining the problem means exploring the domain of the problem to determine what type of problem (nested, hierarchical, or like-kind) needs solving.

If the discussion with the stakeholder(s) indicates a nested problem, then the systems engineer needs to identify both the top-level abstraction and the particular aspects of the nested set of problems that apply. Then, narrowing down the applicable nested set will expose the causal problem that needs to be addressed. As part of that analysis, conceptual solutions must be posed, a general systems design developed, a concept of operations presented, a high-level architecture presented, and perhaps an initial feasibility study must be completed. In this manner, the extent of the problem domain can be compared within the nested problem space to assure that the correct problem domain is identified. Nested problems require the most analysis and care to discern the actual problem and the nature of the problem.

If the discussion with the stakeholder(s) indicates a hierarchical problem, then the systems engineer needs to identify the root (or top-level) problem in the hierarchy. Asking the question *why* at each level of the hierarchy exposes the next higher level of the problem set. For example, *why* do I have a problem arriving at work at 8:00 a.m.? (I do not get up early enough to avoid the traffic.) From this point, the questioning can (and should) proceed in any number of directions. Why do I not get up earlier? (I stay up too late and I need 8 h of sleep.) This question and answer suggest that the problem might be related to trying to accomplish too much or taking too long to complete certain tasks. The problem could be phrased as *taking too much time to complete tasks delays the start of required sleep, thereby causing me to be late to work.* The solution sets are plentiful:

Objective (1): Eliminate some of the previous day's activities, and go to bed earlier.

Objective (2): Learn to complete the work faster.

Objective (3): Train to get by on less sleep.

Objective (4): Get help with the tasks.

Objective (5): Negotiate with the boss to have some tasks completed by others.

Objective (6): Some of the tasks are irrelevant; do not even start them.

Objective (7): Change jobs.

And further, why is there so much traffic when I want to commute by bicycle? (Everyone needs to be at work at 8:00 a.m. as this is the current schedule for all the business in the area.) This question and answer suggest that the problem might be to encourage public transportation, lobby for bicycle lanes, work from home, or go to the beach (or fishing) and work outdoors. What problem domain is important enough for the stakeholders to reach consensus? Defining the problem is itself problematic. Identifying a symptom and offering a palliative solution are quickly revealed as eye candy. Considered analysis upfront avoids false starts to the effect that the real problem is discovered after exposing the sham during reviews.

Characteristics of a Problem

Identifying the problem is as much about understanding the context of the problem as it is about knowing what the stakeholders believe they need. It is as much about knowing the constraints and limitations of the boundary conditions and boundaries as it is appreciating the uses of the product or service. It is also as much about the functions as it is about knowing what decisions the users expect to make after they put the product or service into operation. In order of importance, the defining characteristics of the problem can be expressed by (1) the decisions that the users will make either because the product or service exists or because in anticipation of the working with the product or service; (2) the limitations imposed by the physical, functional, and behavioral boundaries; (3) the constraints of cost, facilities, and other nonlabor issues and resources; (4) the implications of thinking in systems about the future of the proposed product or service; and (5) the revised needs of the key stakeholders, as determined after the stakeholders have examined and discussed the trade-offs imposed by the previous three items listed.

Scope of a Problem

Within the boundedness of the problem, the scope of the expected development work can help define how much of the long-term solution will be delivered in the first tranche of value. The determination of the scope of the work can be expressed by (1) the limitations imposed by the total development costs, the deliverable functions, design and architected performance(s), and the levels of quality expressed as the losses to achieve the deliverable system performances; (2) the constraints imposed by the development schedule, systems engineering skills, and management skills; (3) and the degree of integration. The scope issues are a constant reminder that development uncovers many issues that were not envisioned at the beginning of the project. Since both boundary discussions and scope issues are contractual by their nature, the discussions with key stakeholders are a convenient, albeit proper means to deal with such issues.

Nature of a Problem

An example that points out the specific nature of a problem is that faced when the project is on a tight schedule and a wrong decision in the beginning days of work becomes expensive to correct later and co-opts the schedule with undue additional work. Your mathematicians describe a means of evaluating two conceptual designs developed for a project in the first week of work. Both designs seem to address the needs of the customer and appear workable to the engineers. However, there is no consensus or rationale for a consensus that favors one design over the other as the designs are only representative of a set of vague requirements from the customer. The designer has inferred much and liberally made decisions to eliminate uncertainties. Early designs are conceptual in nature and lack the refinements that are reflected in later-stage design. The schedule of this project is compressed more than the systems engineers found comfortable, but acceded to their project manager's requests to streamline the development process. A part of streamlining was to commit early to a design that met the requirements. Typically, a design matures concurrently with the results from engineering models and detailed analyses that expose additional requirements. As there are relatively few tools that help shorten this transition time from concept to a robustly considered scheme (Abeln 1990 as referenced in Giachetti et al. 1997), a decision must be made with some uncertainty to avoid time spent in rework. The technique offered by the mathematicians is to apply fuzzy logic (a form of set theory introduced in the 1960s by Dr. Lotfi Zadeh (Zadeh 1965)). The mathematicians plan to model the design variables and specifications as a means of characterizing the differences between the two designs. More specifically, the designs are based on different assumptions that are shown by differences in the objects and interactions. Without a way to determine the implications of these assumptions, there is little on which to base a choice. Unlike the work presented by Zimmerman and Sebastian (Zimmerman and Sebastian 1994 as referenced in Giachetti et al. 1997) to represent imprecision, the project's mathematicians believe that fuzzy logic can assist in the analyses of assumptions. By representing assumptions as parts of one set (i.e., one design) versus another set (i.e., the other design), and allowing for assumptions to have partial membership in both sets, the designs can be instantiated in terms of their dependencies on assumptions. Making a wrong choice (if in fact there is a wrong choice) dooms the project to not deliver. Making a right choice merely moves the possibility of making additional wrong (or right) decisions in time. Presumably, if wrong decisions are made later in the project, the impact will be less than making a wrong decision earlier. The systems engineer rightly recognizes there is no room for material error—a catastrophically wrong decision at the beginning of the project is a problem.

Defining the problem can be depicted as summarizing the situation, elaborating on a particular aspect of the situation, and explaining the situation with relevant details. The forms of stating the problem (summarized, elaborated, and explicated) are detailed as follows:

- *Summarized*: The difference between making a decision without analysis and knowing that the design best suited to meet the schedule needs to be chosen, and you have no viable plan to succeed indicates you have a problem.

- *Elaborated*: The difference between making a decision without analysis (and without carefully soliciting and considering all the opinions expressed by the project team) and knowing that the design best suited to meet the schedule needs to be chosen (because difficult times lie ahead for the entire project team should a catastrophically wrong decision be made too early in development), and you have no viable plan to succeed (should there be inordinate delays in the work) indicates that you have a problem.

- *Explicated*: The difference between making a decision without analysis (in this case mathematical) [the part: *what can be done*] and without carefully soliciting and considering all the opinions expressed by the project team (people who will in great part determine the success of the project and your decision) [combined as the part: *what you need to do*] and you know the design best suited to meet the schedule needs to be chosen because difficult times lie ahead for the entire project team should a catastrophically wrong decision be made too early in development and you have no viable plan to succeed (should there be inordinate delays in the work) [the part: *you do not know how to achieve what needs to be done*] you have a problem. The problem involves not having enough schedule to make catastrophic mistakes (in this case early on).

Domain of a Problem

The domain of the problem is succinctly characterized as a trait of the team's work and the project context rather than an intrinsic property of the project. The team takes responsibility for determining the problem within the structure and culture of the project. The problem does not present itself in isolation. The problem is inextricably tied to the stakeholders, one of which is the project team. But the problem is neither an inherent part of the project nor the context of the project. In short, the project cannot be blamed for a faulty problem statement, but the team can. Therefore, it is vital for the team to diligently work to better the definition of the problem, to focus effort on satisfying the needs of the stakeholders so the problem can be mitigated, mollified, or solved, and to create a collaborative environment for all stakeholders to share openly. If the problem was always thought of as success or failure (representing an intrinsic part of the project), then every decision would be in the vein of fatalism (an unavoidable necessity). Were fatalism the operative doctrine, one would have not proposed to tackle the work (and by any measure should not be a part of the team).

Systems Engineer's Perspective of a Problem

It is instructive to point out a key difference between systems engineers and project managers. Indeed, the difference is so pronounced that it is the rare individual that should perform both roles. In the example of the project where the systems engineer needed to decide between two conceptual designs, the project manager advocated compressing the development and delivery schedule, resulting in acquiesce by the development team (including the systems engineer). The sensitivity to a shortened schedule heightened the importance for the systems engineer to take particular care, as there was no time to mature a design as might be planned. The project manager's intuition instilled sufficient confidence to bring up the idea of compressing the schedule. But the systems engineer should not and cannot rely on intuition for such matters. A systems engineer is inclined to analyze the details and view the project as a system of systems. A manager may rely on the analysis provided, but often has a good intuition for what the situation warrants to initiate the project. Without the decision to compress the schedule, the work may have been awarded to another group of developers. The first step is to secure the work under conditions that are conducive to its successful completion. Since the project team agreed with the project manager, no problem occurred. Isenberg (Isenberg 1984 as referenced by Busman 2008) suggests that there are circumstances in which intuition is used. These include sensing when a problem exists and circumventing detailed analysis and move expeditiously to posit a solution. The approaches of the project manager and the systems engineer are quite dissimilar, but their goal of a successful project is paramount. The thinking of the manager and the systems engineer are not only different, but they must also be different to provide the checks and balances necessary to navigate project pitfalls. Interestingly, systems engineers with different backgrounds interpret problems differently—applying different sets of criteria using different approaches (Bernstein 2001). After some discussion, most systems engineers will agree on the problem (and the related subsets of needs that helped determine the problems).

Stakeholder's Perspective of a Problem

A problem is devised in relation to a need that accomplishes something that a key stakeholder cannot do because of some objective reason(s), that is, availability, technology, science, opportunity, or resources, or condition (limitation or constraints). A problem is also considered with regard to circumstances. In relative terms, a problem for one person may not be a problem for another. Since problems are relative, the solutions are also relative (often based on the predecessor systems that existed before the stakeholders deemed that they had a problem).

For systems engineering, stating the problem was the founding notion. Beginning with a key stakeholder who had a need that was causing a problem (that could be defined) suggested a tangible benefit from finding a solution. The engineering mind (and most other disciplines are challenged intellectually to solve problems, whether self-imposed or otherwise). The academic literature and the public media bait the researcher with problems faced by society, military, and government. Problems seem to be omnipresent, with everyone gaining fame and fortune by solving "big" problems. Industry rewards people who solve problems, so that revenue and profits can increase. The betting public rewards companies who solve problems and gain market-share. Waring nations are rewarded by winning battles and war(s) by solving the problem of defeating the adversary. Problem solving is tantamount to one of the best roles which inspires movies and best-selling books. Heroines and heroes might ask, "What is wrong with solving problem?" The answer would seem to be: some problems need to be solved now, some later, and some not at all. Not all problems have an urgency to find a solution. For example, military systems that take 10 years or more to build do not represent an urgent need. Instead, such a development schedule suggests that new technologies are being inculcated through the design into the envisioned product or service. If the lifecycle of a new technology is estimated to be less than the time it takes to design and build the system with today's technology, and the system is needed for at least another lifecycle of the proposed new technology, then the system should be built. The system design should begin with the lowest-risk technologies and over time be upgraded so that additional capability can be included as new technology is introduced and proven. The result would be a planned upgrade using new technologies, maturation of which results in the system have a potentially longer use in operations. Longer use, however, does not automatically signify that the newest technology reaches the operational environment first. Rather, the technology makes it to the field when it is ready. Once the new technology is proven and fielded, the existing systems (with the "older technology") can be upgraded to improve performance(s) across the family of systems that use the new technology. Contrast this upgrade scheme with waiting longer to field the first unit and losing out on the increased performance for a number of years. But it is much more than merely losing out on some years of enhanced performance during the formative years for the product. It is also the missed opportunity to develop the supply chain, logistics, maintenance, support, and training that is often more expensive to set up and sustain. Early fielding provides many benefits (both in improved capability and costs of operations). Moreover, the total cost of development for a limited series production is significantly less than an extended development period. If greater production is the intent of the acquisition, then once the production prototypes are built (i.e., the limited series production units), manufacturing engineering can present a more mature manufacturing item for scaled-up production.

Defining the problem in systems engineering is akin to forming a question that needs to be answered for integration. Integration relies on bringing together objects that satisfy the metaphoric question: What does it take to satisfy the needs of a specified group of stakeholders? From an integration perspective, the needs of stakeholders must be considered both from the limited perspectives of the desired parts and from the whole. This implies that the stakeholder requirements for physical, functional, and behavioral objects must also be reflective of the lifecycle issues that will confront and pace the whole. The question domain that describes the distinctive nature of product or service features desired by the customer(s) can then be posed and answered through integration. Does the integration of two physical objects result in the required functionality and the resultant behaviors? Does the test that is planned for a physical object represent the questions that eventually must be answered if the unit fails. Of course, a functional requirement being met is the important determinant for systems engineering. But for integration, requirements have already been established. As such, integration concerns how the requisite parts form a whole. When an object fails a test, it is the failure of the performance of a function. The first and driving thought is not that a functional requirement is shown to have a degraded or out-of-specification performance. Testing is but a small part of assessing the efficacy of integration. Testing needs to be done as an audit to determine what functions are demonstrated to some level of proficiency; what rework is necessary to achieve the requisite level of performance; what system behaviors reveal about the intended uses of the deliverable system; and what areas of risk may impede completing the product or service. Testing highlights the test object's performance at a particular instance under certain conditions. The process of integrating objects provides opportunities to test the performances of additional functions, each object showing its worth in transforming EMMI. Assessing a system's performance is more than an aggregation of discrete instances of tests. The end-to-end assessment is indicative of combined demonstration of transformed EMMI into an output that represents the totality of system behaviors and enactments. Assessment of a system involves integration, testing, verification, and validation. Integration provides the sequencing of objects to be tested. Testing indicates whether the functions are operative.

Verification and a Problem

Verification shows the mapping of the functions to the specifications, and ultimately to solving the problem. That is the role of verification. Verification is the means of establishing that the requisite functions of the product or service have been provided. Verification is referenced to the specifications that guided the engineering activities as well as the requirements which guided the design and architecture. Integration concerns itself with answering the questions that plague engineers and systems engineers—how will the performance requirements be met. Integration is more than simply providing object-to-object integration (i.e., providing functions). Integration must also satisfy the performances that are required for each of the functions. Validation illustrates the limits of the performances of the design, system architecture, and implementation.

Integration and a Problem

Formulating a question requires that systems engineers who take responsibility for performing appropriate analyses act independently to make necessary decisions, acknowledge perspective and context, and are accountable for the consequences of the question. That accountability is focused on due diligence in preparing the integration plan, which results in the questions that need to be asked and later in the answers. In addition, the development of the integrated solution (or the answers to the questions posed) must satisfy good engineering practice.

Characterizing the Need

In developing a theory of integration, three needs arise: first, the salient factors (e.g., assumptions, independent and dependent variables, measures, and measurement) need to be identified, classified, and categorized; second, a model must be posed that relates the facts, based on certain specified assumptions; and third, the hallmarks of explanation and prediction must be laid out. Each of the three needs address several problems. For every problem there may be multiple needs. A need is a condition requiring relief. A need needs to be resolved and managed. Needs have characteristics that distinguish them from wants. Needs map to the intentions of stakeholders through the stated or unstated wants. Stakeholders have need(s) and want(s). The systems engineer differentiates needs from wants by reflecting the needs in the design and architecture baselines, weighing whether needs or wants are disguised and miscategorized.

A need has measurable requirements. Systems engineers define needs for developing the product or services, while the program manager satisfies the needs to support the project. The user of the new product or service defines their needs. Users have needs. A need can be for an object's properties, traits, or attributes. The systems engineer's role is to assure that all objects are appropriately specified, that is, the need(s) are met.

Stakeholders

The word "stakeholder," one who has a stake in the outcome, is most typically an entity (a person either acting alone or representing an organization) who can influence the conceptualization or funding of the development project, or the product's or service's acceptance, operations, or disposal. A stakeholder is anyone who significantly affects or is affected by decision-making activity the influences the product or service. In a broader sense, it is someone with an interest or concern, and specifically someone at risk due to the product or service.

Stakeholders have needs, as such it is important to capture their needs so that the systems engineer can incorporate, acknowledge, or choose to ignore specific requirements. Discovering stakeholders early in the development work is less disruptive than later. Identifying and analyzing the needs of stakeholders is referred to stakeholder analysis.

Stakeholder analysis is a methodology for identifying stakeholders and analyzing their underlying value and interests in the system. The methodology involves several processes and tools that cater to discovering types, significance, and value of stakeholders. At risk are the consequences of not uncovering the current and potential future interests and objectives of affected parties. Conjugate benefits include (1) a better appreciation of the complexity of the system and the undertaking; (2) understanding of the stakeholder influence(s) and how to manage those influences; (3) a more thorough examination of multiple-use objectives; (4) identification and resolution of potentially conflicting requirements; and (5) exploration of architecture alternatives. Additionally, stakeholder analysis encourages a forum to improve mutual understanding about issues, ideas, and solutions that represent potential stakeholders not included or perhaps not even yet considered stakeholders. Stakeholder analysis aids discovery of new stakeholders and their requirements. This more extensive involvement increases the long-term stability of the system's appropriateness and applicability to changing situations. The methodology outlined here becomes increasingly important for systems of greater worth and complexity.

Stakeholder Analysis

Stakeholder analysis is the systematic gathering and analyzing of qualitative information to determine whose interests should be taken into account when developing and/or implementing a policy or program. The sheer number of potential stakeholders that can influence system development can be quite large. Therefore, instead of posing the question "Who should be considered a stakeholder for a system?" an alternative question is "Who should *not* be considered a stakeholder for a particular system?" Answers to these two questions help further identify stakeholders.

Consequently, stakeholder analysis is an examination not only of the individual stakeholders but also of how their motives, interests, and values affect system development. In conducting a stakeholder analysis, a clear purpose must be defined in the beginning or the analysis could lose focus and direction resulting from the large quantity of stakeholder inputs.

There are five major steps in stakeholder analysis: (1) identification of potential stakeholders; (2) classification of potential stakeholders; (3) determination of potential stakeholder and system relationships; (4) determination of key system stakeholders; and (5) definition of stakeholder requirements.

The list of potential stakeholders begin quite naturally with the customer(s) and user(s) who have supplied requirements for the development project. Expanding the list by referrals is the straightforward means to complete this first step.

The next stage in the identification process is the creation of scenarios that require potential stakeholder interactions. These scenarios may help identify additional stakeholders overlooked during the initial brainstorming session. The scenarios should involve aspects of the system under development. Each scenario is then adapted using events that give rise to the reason behind the scenario. These adaptations take the form of parameter changes related to timing, location, participants, or other pertinent factors that alter the assumptions or initial conditions. Additionally, the analyst explores alternatives in the scenarios based on "what if" situations that represent different courses of action (i.e., the result of different choices). Each adaptation will drive a different system response. By examining the different responses from these variations, one will observe (or in some cases, discover) the stakeholders that interact with the system.

Finally, a master list of potential stakeholders is compiled from the results of the brainstorming session, augmented with the lists generated from examination of the scenarios.

Any stakeholder analysis will result in a degree of uncertainty with regards to the problem that needs to be solved and the requirements for a corresponding set of solutions. A possible defect is missing a stakeholder of consequence. Furthermore, participation by a group of people and representatives of organizations or entities brings with it a set of unique and dynamic characteristics. As a result, personalities and agendas are complicating factors that must be endured, although they sometimes impose difficult limitations and constraints.

After the stakeholder analysis is completed, it is useful to evaluate the specific methods and their consequences through which the system can be thought of and analyzed. The comprehensiveness and usefulness of the stakeholder analysis is revealed in discussions with stakeholders to explore the earlier-stated requirements to ascertain if those requirements are certain in the minds of the stakeholders, reflect the needs of the stakeholders, and may be indicative of other requirements that have as yet been unsaid (whether missed or simply omitted). For example, changes in a policy may have very visible impacts on the design of the system. If the stakeholder expects changes in policies from time to time, then it may be important to consider desensitizing the system design to small changes in policy using the stakeholder sentience analysis as a feedback measure. Further, changes in policies do occur whether they can be planned for in advance or not. Knowing which policies the system design (or architecture) is most sensitive to allows the systems engineer to consider defining back-up options should key policies change. These back-up options would not become requirements, only discussion points from which scenarios might be developed to investigate the sensitivities of planning on the futurity of events. Rather than a prediction or a projection,* the futurity

---

* A prediction is a probabilistic statement that something will happen in the future based on what is known today. A prediction generally assumes that future changes in related conditions will not have a significant influence. In this sense, a prediction is most influenced by the "initial conditions"—the current situation from which we predict a change. In contrast to a prediction, a projection specifically allows for significant changes in the set of "boundary conditions" that might influence the prediction, creating "if this, then that" types of statements. Thus, a projection is a probabilistic statement that it is possible that something will happen in the future if certain conditions develop. The set of boundary conditions that is used in conjunction with making a projection is often called a scenario, and each scenario is based on assumptions about how the future will develop.

of events helps determine the interests of the stakeholders and their needs at some point in the future.

Classification of Potential Stakeholders

Classification of potential stakeholders proceeds using the following steps: (1) determination of the system boundaries, (2) classification of potential internal stakeholders, (3) classification of potential first-order stakeholders, and (4) classification of potential second-order stakeholders. First, to define the system boundary, one must understand that it can be somewhat ephemeral in nature. That is, the incidental interactions between stakeholders, the elements and domains that characterize the system, and external interactions with other systems and stakeholders will change over time and therefore change the system boundary. Scenario building and analysis are convenient means to explore the role of stakeholders at various stages in the lifecycle of a product or service.

Those stakeholders that interact only with internal system elements or with other stakeholders are classified as internal stakeholders. Those stakeholders that are in direct *contact* with the system but do not have direct *interaction* with the internal stakeholders are considered first-order stakeholders. Second-order stakeholders are defined as those stakeholders who are connected indirectly to the system via interaction with first-order stakeholders. Both first- and second-order stakeholders are classified as boundary stakeholders because they interact with external entities across the system boundary. Therefore, the group of internal and boundary stakeholders comprise the set of valid system stakeholders (Ku 2007). After classifying the stakeholders, it may be necessary to prioritize them based on when they influence the system. Determining the relationships between the potential stakeholders and the system is an initial (and critical) step in prioritizing the stakeholders. The purpose for prioritizing the stakeholders ensures that vital inputs (stakeholder problems, needs, and requirements) are utilized to develop the functional analysis, and thereafter, the system architecture for the human capital management (HCM) strategy. Drawing from the pool of potential stakeholders established during the previous steps, stakeholders are grouped into different system roles, which assist their prioritization and facilitates the selection of appropriate stakeholder inputs.

Stakeholder analysis helps identify the key system stakeholders—those stakeholders who help form the acquisition and development, and then take delivery of the product or service. Determining which stakeholders have significant roles during development and which are focused on those aspects of defining the project are most likely to either be involved early on or be represented for the early discussions regarding requirements. All stakeholders must be represented during requirements analysis.

As with any set of requirements, not all stakeholder needs can be met. There will always be some requirements that are not included or changed significantly from that which is desired by the totality of stakeholders. Not all stakeholders will be happy with the compromises or the outcomes of the requirements analysis. And fundamentally it is impossible to build a system that does no harm. The reason for such a strong statement is rooted in the recognition that when systems are built for commercial purposes they are meant to compete in a lively and profitable manner. Other companies lose marketshare and often substantial money. The stakeholders include not only the developers and the customers and users of the new product or service, but also the competitors who may be affected by loss of sales. All have a stake in the requirements for the new product or service—with the competition wanting less functionality and lower performance which is completely at odds with those stakeholders who want greater functionality, greater performance, or a lower cost. Every system has stakeholders who at odds with the requirements. Acquirers set the domain of the key stakeholders. Military planners must consider the "competitive" environment for the products and services they have developed. Adversaries are indeed key stakeholders, but not in the same way as the developers. Consequently, systems are affected by the key stakeholders during acquisition and development in a way that grants them a higher importance and greater influence than those stakeholders—the competitors and adversaries who hold completely opposite views on the requirements.

Stakeholder importance is a qualitative measure based on the product of the number of interactions a stakeholder has with other stakeholders, and the worth of these interactions as determined by the worth activation function—the measure of performance multiplied by the loss incurred if the performance deviates from a target value of desired EMMI, then divided by the expenditure of EMMI to achieve that performance. From the work of Ku (2007), the importance of a stakeholder is based on the number of interactions each stakeholder has with all other stakeholders (internal, external, first-order). The more direct an interaction a stakeholder has with others within the system, the more likely it is that the stakeholder's actions will affect the whole system rather than individual subcomponents of the system.

Unlike stakeholder importance which is quantifiable in terms of EMMI, stakeholder influence is a qualitative measure based on the types of relationships the stakeholders have with the system domain (internal, first-order, or second-order) and the duration of these relationships throughout the product's lifecycle. The higher the risk of gain or loss a stakeholder has with regards to the system domain, the greater the influence that stakeholder may have over the system. Therefore, internal stakeholders may have greater influence than first-order stakeholders may have. In turn, first-order stakeholders may have greater influence than second-order stakeholders may. In addition, the duration of the relationships has a bearing on the stakeholder's influence. If an internal stakeholder only interacts with the system during the concept development phase, but a first-order stakeholder interacts with the system well into the deployment phase, the first-order stakeholder may have a greater influence on the system than the internal stakeholder may. Both the type and duration of stakeholder and system domain relationships contribute to stakeholder influence (Ku 2007).

The selection of key stakeholders is based on the product of the stakeholder's importance and influence. From these factors, the stakeholders are ranked as primary, secondary, and tertiary entities based upon thresholds determined by the analyst(s). Primary stakeholder needs have direct input into the development of the system's functional analysis and the overall measure of effectiveness model. Secondary stakeholder inputs have limited weighting in the development of the functional analysis and the overall measure of effectiveness model. However, these stakeholders will be incorporated to the maximum extent possible within system boundaries, as described in subsequent sections of this chapter. Tertiary stakeholder inputs are considered beyond the scope of this analysis and will not be incorporated into the functional analysis and the overall measure of effectiveness model.

The final step of the stakeholder analysis is the definition of stakeholder requirements. This step is closely related to the stakeholder requirements definition process described in Revision 3 of the INCOSE Handbook, which states: "The purpose of the Stakeholder Requirements Definition Process is to elicit, negotiate, document, and maintain stakeholders' requirements for the system-of-interest within a defined environment" (INCOSE, 2006, p. 4.2).

After identification of the primary, secondary, and tertiary stakeholders, problem statements can be developed. Langford et al. (2007) defines a problem in the following terms: "Whenever there is a difference between what can be done and what you want to do, and you do not know how to achieve the desire, there is a problem." For every stakeholder problem, several stakeholder needs can be identified. A need arises from a condition faced by the stakeholder that requires a solution to alleviate it.

Once stakeholder needs have been documented, they are used to derive stakeholder requirements, which are essential for guiding system development and serve to frame the project scope (INCOSE 2006). These requirements drive the development of the functional analysis, the overall measure of effectiveness model, and system architecture. In addition, the stakeholder requirements are used in gap analysis to determine the desired state sought by the stakeholder ("where we want to be") and, in conjunction with the perceived existing state, establish the gaps to be addressed by the system solution.

After the stakeholder analysis is completed, it is useful to evaluate the specific methods and their consequences through which the system can be thought of and analyzed. For example, changes in certain policies may have very visible impacts on the design of the system. It may be important to consider desensitizing the system design to small changes in policy using the stakeholder sentience analysis as a feedback measure.

From a systems point of view, a stakeholder is an object of the system. A system is a set of objects that are either dependent or independent but yet interacting pair-wise—temporally or physically—to achieve a purpose. Likewise the objects that interact with other objects outside the system form the boundary of the system and are "boundary" objects. Objects that only interact with other system objects (and have no interactions outside the system) are "internal" objects. Both internal and boundary objects are system objects. Boundary objects are also objects in the system with which they interact. These definitions include both the permanent and episodic interactions among objects of a system, systems of systems, or a system of systems. Thus, the lasting and occasional interactions, as well as emergent properties and behaviors of a system, are driven by the object within the system that are driven through their EMMI from both other internal objects as well as external objects. These interactions result in the transfer of EMMI. The transfer of EMMI can result in various behaviors, for example, cooperative, competitive, enhancing, enabling, destructive, and degrading.

Complexity

Complexity, scope, and extents underlie the challenge of detecting, engaging, and integrating stakeholder issues into phases of the lifecycle of the system. Complexity stresses the constructs of systems engineering practices and modern management techniques. Complex systems have a great variety of interactions, which transcends physical, information, and social interfaces. The system complexity thus augments the management challenge because of the large number and various types of system elements and stakeholders. In this book, complexity is reflected by the number and types of interactions that lead to integration of the elements of a system or among the systems of a system of systems. Since an element of a system may also be or represent a stakeholder of the system, increasing the number of stakeholders increases the complexity. Managing complexity or managing stakeholders thus amounts to managing the value and the worth activation function, and therefore the number of interactions.

For issues that have numerous, multiplicatively tortuous boundaries (i.e., physical, functional, and behavioral) and complicated boundary conditions, systems engineering seems to be the only discipline that has the tools and methods to deal with the issues that we herein define as *complex*. Complexity has been said to be an important concept, perhaps as important as the concept of a system (Klir 2001). Since a great number of books, scholarly works, and musings have been offered by ponders of complexity and its implications, there is not much more to say about the subject of complexity. It would seem that complexity should be simple, rather than complex. The reason for this is there are a great number of things and issues that appear complex scaled from the atomic level to the universe. There seems no limit as to what can be complex. Complexity is omnipresent, commonplace, almost to the point of being pointless because it is ordinary by its nature. But in the ordinary, the humdrum of existence, the exceptional transpires. As phrased by Klir, "Complexity (in the epistemological and methodological sense) is thus associated with systems, that is, some abstractions distinguished on objects that reflect the way in which the objects are interacted with" (Klir 2001). Complexity is the result of interactions between objects. Yet, more interactions do not necessarily increase complexity. Complexity can increase and decrease. Complexity can be seen as relative to the level of abstraction in which one views two objects. For an atom, the constituent parts of protons and neutrons may not be complex at that level of abstraction. But examine the constituent parts of the proton and reveal details that are as yet unexplained. Scale up from an atom to molecules and then subsequently to a human-built system that interacts with other systems. Regardless of scale, we find complexity. Interaction is the defining process of complexity. Interaction is the catalyst that results in integration. Integration is the process of systems and of systems interacting with other systems. The fundamental mechanism that drives complexity is interaction across the three types of boundaries that lead to integration of the elements.

Process Models

In a simple and inaccurate form, systems engineering process models relate work that needs to be done to a set of stages that map to the systems engineering progression of defining the problem; designing and architecting solutions; developing and integrating objects; and testing, verifying, and validating to satisfy requirements. Again in a simple and inaccurate form, process models indicate what should be done next, and how long it should take. In addition to process models, maturity models have been developed to describe various aspects of systems through the development process. These maturity models include parameters such as capability and integration. Each model is specialized to deal with perceived risk, verification of specifications or requirements, domain interactions, or a catch-all of multiple factors. Process models help to orchestrate the work as well as communicate what type of work should be performed to reach the next milestone. Selection of the process model for a systems engineering effort depends in part on the preference of the acquirer, the skills of the developer, and the particular key limitation (e.g., cost, budget, or performance) that is to drive the work.

Process models differ from acquirer to acquirer. Within the U.S. government, multiple process models are used. Within industry, home-grown process models dominate the work, many adapted to the particulars of the project or as an accommodation of the business enterprise. All have a high-level graphical presentation that defines the organization of processes. All relate that organization to milestones, progress reviews, and decision points. All imply coordination, process architecture, requirements management, and goals and objectives. Some process models are designed to work with particular standards, while some standards help to orchestrate process models. There are literally dozens of process models and many hundreds of references (IEEE 1220-1998 1998; Martin 1998; ANSI/IEEE 2000; Sage and Rouse 1999; Sheard 2001; ISO/IEC-15288 2002; INCOSE 2010), including a systems integration process model (Jain et al. 2010). All these models can be adapted to virtually any project given enough understanding of the relations and time to communicate both the intentions and the methods that will be used for all stakeholders.*

Scalable Process Models

The effectiveness of a process model is at worst dependent on a preference for one model over another and at best based on the circumstances surrounding the project. While the process model must be tailored to the needs of the acquirer (e.g., in their reporting to others, in their review of the work, in their perception of risk, and in their decision to continue support) and must also be matched to the skills of the project team, the process model should always serve to improve and facilitate the problem solving and decision making, rather than in itself become the problem that must be worked around. Assuming that the needs of the acquirer and the project team are met, success with the chosen systems engineering process model depends on satisfying two conditions. First and foremost, the resources needed to succeed with the project must be available at the start of the project (GAO 2001). Second and nearly equal, the process model must be scalable. The meaning of scalable for process models is the relation between the time it takes to complete the project using the systems engineering process model and the size of the problem that is to be solved (given that the two conditions are satisfied). If the requirements are known in detail and the only derived requirements are those that expound on the detail provided by the acquirer, then a process model that begins with certainty (e.g., the iterative waterfall model and like-kind kin†) will provide varying degrees of success). All waterfall derivatives are scalable. The more requirements are detailed up front, the more definite the outcome of development. The characteristic of scalability that is important for systems engineering is the capability to cope and perform under changing requirements. Requirements that

---

* Stakeholders include the project staff.
† Example derivatives of the waterfall process model include the spiral and the vee process models.

up front are ill-formed, incomplete, or deemed unnecessary may have importance as the systems engineering activities begin to discover and detail
trade-offs and derive new requirements. The iterative nature of systems
engineering inculcates the maxim, "If at first you don't succeed, try and try
again." The trade-off for systems engineering is to live with an error and deal
with it later. Intentionally letting an error pass is unacceptable because something that is known to not work properly will bedevil integration work.
Functionality and performance will suffer.

Scalability is about doing what needs to be done with more people or
being able to do more with one person. Scalability in the first instance
(more people doing what needs to be done) is enabled by providing services to support the people. Scalability in the second instance (enabling
people to do more is through efficiencies). The systems engineering process
model needs to work effectively in both instances. The desired scalability
is achieved when an economy of scale is reached. This economy is related
by the output level of the team per unit cost increasing at a nonlinear rate.
If the resources and the process model are as described above, an effective
scalability is achievable. Economies of scale result from both members of
the project team working in concert as well as the development objects
being synergistically integrated. One of the key factors in achieving such
an economy of scale is to instill a recognition of scalability by accessing the
processes through synthesis at the unit level first in development, and
again early in the integration work while at the same time supplanting
iterative actions with recursive thinking. One of the key factors in stimulat-
ing scalability is through the use of tools that are not limited by their access
to a single instance in a database. Multiple users of the same data through
single-threaded access thwart people in achieving economies of scale and
are therefore scalable. This way of thinking about scalability is the same as
enabling greater capability with respect to multiple simultaneous uses of
common tools and data. When magnified across the development or integration work, an improved use of resources results. In essence, the net-
working of process through common access provides scalability from very
small projects to diverse, multiple domain programs (NIST Special
Publication 1108 2010).

Checklist for Scalability

Satisfying the two key conditions (resources available at start of project and
scalability) is essential to realizing an economy of scale. The checklist for
scalability is as follows:

- Mandates, preferences, or desires of the key stakeholders (e.g.,
  acquirers and users)
- Experience of the project manager and the business enterprise

- Skills and experience of the systems engineers and the project team
- Time limitation to complete the project
- Budget limitation and schedule of payments for the project
- Economies of scale due to common access to network tools and processes
- Applicability of approvals and milestones to the pace of development and integration
- The level of functionality, performance, and quality desired
- The scope of the project (determines the set of heuristics necessary to make decisions)

Without satisfying the two conditions for a successful project (i.e., resource satisfaction at start and process model scalability), the project should be considered to have an undeterminable risk. Moreover, the view is that once a systems engineering process model has been adopted, it should be sacrosanct. However, if the conditions warrant and the initial development work determines that a change in the systems engineering process model would help focus the work on a particular methodology or emphasize an aspect of the work that is discovered to be most important, then consider tailoring the model. Again, that change in the systems engineering process model needs to be vetted carefully against the needs of the key stakeholders, and strict unanimity is paramount.

Testing

Testing objects and integration are often indicted as being either related or similar in nature. "Test and integration," "integration and test," or "integration test" appear frequently in reports, systems engineering documentation, and in presentations by stakeholders involved with an acquisition activity. Integration and developmental testing or integration and unit testing are key to building a system. Integration has been thought of as "... progressively linking and testing ..."* The common usage of the word "integration," while suggestive of bringing objects together to build a system, should not be thought of as building a system. Rather, the process of "integration" is more accurately described as putting parts together in a particular order and fashion to demonstrate the requisite system functionalities. Whether these parts form a system (or not) has nothing to do with the process of putting them

---

* Institute for Telecommunications, United States Department of Commerce.

together, but rather is a separate evaluation after the completion of the work. Integrating objects merely brings their boundaries into some sort(s) of conjunctive relation(s) through the exchange of EMMI. However, the mere conjunctive relation(s) of individual parts, units, components, subassemblies, assemblies, or subsystems does not satisfy the requirements for a system unless all the various kinds of objects are operative. The goal of a project may be to build a system, but a system is not a system until it satisfies the requirements for a system. Integration is the means for building a system, as it is the means to put anything together. The process of integration is necessary to build a system, but it is not sufficient.

It is a learned response, whether by a previous stressful experience (Kinnaird 2003), by warnings, or by procedures that systems engineers have turned into the habit of testing what is determined that needs to be tested. Sometimes items are included in that determination that has no more than a reflexive stimulus (Gabora 2001) to evoke testing. The impetus for testing an item should be based on something other than fear or instinct. As the practice of systems engineering has evolved and taken its lessons from engineering, the basis for a test is problematic. Certainly, we are unwilling to test every part (e.g., every resistor, every wire, every washer, and every transistor under every condition that is likely to occur in conjunction with a finished product or service). Instead, systems engineering and projects have built-in formalisms to provide a level of assurance that certain standards (ISO 9000 series) for the improvement of organizational performance are followed for production of individual parts.*

Testing is a process to determine the difference(s) between an object's properties, traits, and attributes under certain conditions in a given set of circumstances with that of a representation (or test model) of what is desired. The representation includes the test setup; the test procedures; the test plan; the test personnel; the test objectives; the data analysis (and tools); the theory in which the measurements are planned, executed, and interpreted; and the biases (of all parties). It is tacitly assumed that all factors not included in the representation (or test model) are factually extraneous (and therefore not significant either to a specific test or to a concatenation or totality of tests). Great care must be exercised in planning and executing the testing of an object. Testing is a means of comparison: comparing an object to the test model. Tests do not prove anything; they only show a correspondence to an expected result (Aerts 1983).

What to test is extremely important for any project. Testing impacts on the schedule, use of resources, budget, and final performance(s) of the product or service. The high costs and impacts of remediating defects discovered after the product or service is in operation cause great consternation when deciding what to test (Boumen et al. 2006). From an integration perspective, the

---

* International Organization for Standardization, About ISO introduction http://www.iso.ch/iso/en/aboutiso/introduction/index.html, website modified February 16, 2004, accessed April 18, 2004.

purpose of testing is to show that the functions are required by the final product or service are substantiated in a manner that invokes the desired behaviors of the users. Therefore, testing requires more than the test model that is typical of most projects today. To assure that the required functionality is present, testing must not only confirm the mechanical, electrical, chemical, biological, computer hardware, and computer software issues, but also the fitness for use and the sociological aspects of the user behaviors. There are likely no quick fixes to the product or service after it has been placed into operation, so the operative moments to deal with these issues is during the system design phase when the subsystems are being identified and the requirements are being allocated to components. System design involves evaluating off-the-shelf components, evaluating alternatives, determining the selection criteria, analyzing and allocating requirements from the system requirements, and identifying the interfaces. Components are defined in the system design in terms of their functionality and performance.

System Design

System design is the opportunity to conceptualize the user's needs by answering the following example questions:

- What decisions does the user need to make?
- How much information does the user need to make those decisions?
- From where will the information come?
- How much detail does the information need to assist the user?
- How does the user determine what information is needed?
- What level of trust does the user ascribe to the information?
- How should the information be represented?
- How should the information be presented to the user?
- What is the time line for the user's decisions?
- Is there a natural sequence in which the information should be provided to the user?
- What are the differences between the users and their decisions?
- What procedures should be built into the product or service that are most natural to assist the user in making decisions?
- What are the functional characteristics of the product or service that the user requires to complete the user's tasks?
- What are the performance(s) of the functions that are expected? How do those expectations compare with the requirements?

System design focuses on the functional nature of the interactions between the product or service and the user. While system design does not impose an architecture (which establishes the relations between the structural components, e.g., physical entities, computer hardware, and computer software), system design poses alternatives that could be considered as solving the stated problem to some varying degrees. The match of the system design alternatives to the most effective solution to the posed problem is a matter for analyses and evaluations which then become some of the guidance for the architectural alternatives. System design will tend to provide a general-level perspective of the product or service through varying levels of detail down to the component level. The iterations of design will result in the allocation of requirements first to subsystems, then to assemblies, then to subassemblies, and then to components.*

An item that is a unit (the lowest level of an object that results from work) can be tested. The decision to test at the unit level is fundamentally based on the model of testing used on the project. Based on a study of development teams across several industries, Wheelwright and Clark (Wheelwright and Clark 1992) suggest that a design–build–test strategy seems to be more effective in creating an air of confidence in people's work. This confidence is clearly ill-founded as substantial amounts of rework are necessary throughout a development cycle. Consequently, projects must develop a rework strategy with procedures, inspections, and accommodations for revising work plans, schedule, milestones, and budgets. Selecting an aggregation of units and components for testing at a higher level eliminates low-level tests. The counterargument from the engineer is that testing provides visibility into how well the work matches with expectations. The point is that expectations are often incorrect and while correcting work to match with expectations is satisfying, it is ruinous with regard to schedule and budget. From an object's point of view, its design and implementation are only testable in conjunction with another object, whose combination results in a function. If both objects were built and tested to expectations, but the desired function was not demonstrated, then one or both objects have a problem with mechanism(s), outputs, or inputs. With perfect execution of the work (as assumed in this discussion), the design is faulty. Detecting faulty designs quickly is essential for remaining within schedule and budget constraints and argues for building to functionality. The argument that without perfect execution the functions were not demonstrable is specious. Poor execution of the work to build an object that is deficient in some way may only reduce the performance value of the functions, but may still demonstrate the function. The key issue of what to test is not determined by testing all that can be tested, but rather testing what needs to be tested. Sometimes it is simply better to plug the

---

* Projects vary as per their terminology for different levels of work. Here, components are meant to be an element of a larger whole, that is above the individual part and the aggregation of parts into a unit. Components are aggregations of units.

lamp cord into the circuit to test the lamp, rather than testing all the parts individually. If the designer fails to allocate a requirement to a component and that component is tested, the requisite functionality will not be demonstrable (i.e., an essential requirement is missing). If, however, that requirement is allocated, but poorly implemented, the functionality will be shown, but at a reduced performance. Arguing for a strategy of design–build–test may encourage or inspire engineers, but it is not a logically defensible position. An example to illustrate the point begins with a group of engineers who build and install electric vehicle recharging stations in anticipation of widespread acceptance and mass production of electric vehicles. Testing of the components was extensive, as was the finished station. The test plan was derived from the Systems Design documentation which included a comprehensive set of requirements. The requirements document was carefully considered by the urban transportation specialists, policy makers, city planners, the recharging station engineers, and representatives of the electric vehicle industry engineering groups. Due to a last-minute problem with a new battery design based on lithium-ion energy technology that provides the prime energy to move the vehicle, the development team working on the electric vehicle's battery system revised their plans and reverted to a more reliable design (but one known to have a tendency to catch fire when overheated due to overcharging or high ambient heat). While the vehicle met the form, fit, and function of the requirements documentation, the hazards issue was dealt with as a safety issue and not one as a functional requirement for the recharging station. Instead, the system of systems design had this safety issue allocated to the electric vehicle. The designers of the electric vehicle dealt with this issue in the user's instruction manual both as a warning and as a set of procedures. Since the allocation of the safety issue regarding the electric battery was to the electric vehicle and not to the recharging station, the recharging station was not robustly designed for a vehicle fire of the sort produced by an overheated lithium-ion battery. Extensive safety testing of the lithium-ion battery was carried out by the electric vehicle team for mishandling, electrical malfunctions, overheating (due to various conditions) explosions, inundation by water, and mishandling. According to the Electric Power Research Institute, "whenever there is a concentrated quantity of stored energy, the possibility always exists of creating high temperatures that can lead to combustion" (Eckroad 2002). Regardless of the intentions of the various groups of engineers, the belief in the strategy of design–build–test resulted in two components that were tested extensively and expensively. After the second fire destroyed the second charging station, three lawsuits were filed against all parties of the electric vehicle-charging station system.* There are several ways in which the problems that resulted from the systems engineering process could have been addressed before delivering an operational system. But the point was that extensive testing of

---

* This is a fictitious example based on two real events, both of a similar nature and outcome.

a flawed design is wasteful. To circumvent a design problem, the integration plan needs to more than simply define the sequence of activities that will be accomplished to integrate components into subsystems and subsystems into a whole. The plan is important by two measures: what function(s) are to be demonstrated first, then second, and so forth, and how the users respond (user behaviors) to their use of the function(s) (Federal Highway Administration 2009). If integration planning relied only on architecture, behaviors are process related but not user related. Process behaviors capture the system performances of the product or service, not the behaviors of the users. The system may perform in such a manner as to inhibit the user from accomplishing their task in the requisite period of time. Architecture merely reinforces the product or service performance, not the combination of user and product or user and service behaviors. Naturally, the systems engineer endeavors to design the system to accommodate and provide for user behaviors through architecture, functionalities, performances, user interfaces, and physical adaptations. But since there is no one item that captures the behaviors, the only means of validating a product's or service's fitness are through modeling, simulation, or actual use of the product or service. Integration planning should provide for at least one of these three means of validation. Testing by itself is not validation.

Architecting

Architecture is different from design, as different as marketing is from sales. In some ways, an architect is like the salesperson who readies a pitch to reach a deal. The pitch is derived from a plan which is in line with policies set down by the position of product (the buyer's perception of the product) in the marketplace, by the product's design, and by the manner in which the product is thought to be useful. The architecture carries with it the organization of the product or service (embodied in its objects and their interactions) to provide the user with various functions. The premise of the architecture is the flow of EMMI that satisfies the needs of the user, reinforces the desires of the seller, is in agreement with what the customer expects, and is sustainable during its lifecycle.

Whereas design has more to do with setting up the problem, architecture must solve the problem. Architecture has more to do with the ways in which the purpose of the design is to be achieved than with the selection of the optimal means for realizing that design. Architecting brings order to misleading, ill-fitting, and confounding data; at-odds opinions; differing values; and problematic convergences. Architecture must sort these, implement the decisions, and show that the resulting compromises satisfy the key stakeholders. Designs that are not well defined, problems that are misstated, and needs that are unspoken merely frustrate the architect. But architect is not deterred in the search for more than one set of solutions that satisfy the problem. The system design reflects innovation and creativity. The architecture must be clever and robust.

When building an architecture, there is no route from beginning to end, no hint from the design process as to the best perspective from which to reckon with an architecture, no sequence of steps that can be followed, and no list of rules to guide the work. Architecting is engineering and systems engineering in that the architect puts things together. Design identifies those things and therefore the end of design is when architecting is finished. Change the design, and most probably change the architecture. Change the architecture and potentially change the design. The end of designing is the end of architecting. But it is not until the end of architecting that the design can sustain its first verification with requirements. "Engineering aims for technical optimization, architecting for client satisfaction" (Maier 2002). Design leads and is interactive with architecting through the product or services physical objects, functional characteristics, and the behavioral responses of the user. Architecting is one of the many feedback loops to improve design. In this iterative manner, architecting is much like systems engineering, moving forward and stepping back to fix and refine, ever keeping the presence of a solution that is more highly matched to solving the problem. With rework, the evolving product or service improves at the unit level. Architecting is that space somewhere embodying the art and science of solving a puzzle.

It is the aim of architecting to recognize that the problem is only *the* problem which needs to be solved when viewed from a particular perspective. What one person views as a problem may be different from that of another person. The properties, attributes, traits, context, boundaries, and boundary conditions of the problem suggest the types of solutions that can apply. The types of solutions also suggest the type of problems that can be solved. The constraints of time, money, skills, policy, or rules further constrain the nexus of problem and solution. Architecting needs to be responsive to all of these.

The architectural views are the greater variability cast by the perceptions of attributes, context, boundaries, and boundary conditions of the problem. The wide range of possible solutions is often noticed in industrial and commercial enterprises. So, it is the job of the architect to pose the brilliant solution that captures the essence of the design objectives given the resources, limitations, stakeholder sensitivities, and constraints. Architecture brilliance is portrayed by the simplicity, coherency, and robustness as seen through the manageability of the objects and their mechanisms to solve the problem easily, within budget, on schedule, and with all functionality, requisite performances, and expected quality.*

The architect's strategy may notice the breadth of the principal features of the system design and use that as a key from which to tie other factors

---

* At a minimum.

together. Then, the architect may develop one area in particular to sufficient detail to surface as many details as possible. Others may develop all areas concomitantly in parallel, developing similar levels of detail, expanding and detailing iteratively at each level of abstraction. Then, when the underlying structures are worked through, the definitions and contexts are defined and clarified further. Regardless of the proclivities of the work, a structure of abstractions from the top level leading to lower-level detail is exacted. Much like the process of decomposition, the nature of architecting is often to move from top to bottom, from bottom to top, and from abstraction to detail and back. It is the iterative process of decomposing the design space and recomposing the elements that forms a preliminary architecture. To use the iterative process successfully, the architect must conceptualize a schema in the domain at each of the levels of abstraction that captures the exact behaviors needed by the product or service. This schema is an internal representation of each level of abstraction that includes what (and does not) belong, the degree of misalignment of the layer of each abstraction that is acceptable, the organization of the concepts (partitioning of the objects, functions, and behaviors), and the actions that can potentially revise the schema (and under what conditions those revisions may take place). To be successful, the architect must have content knowledge, structural knowledge, domain-specific strategy, general searching strategy, general representational strategy, and general abstraction strategy.

Architecture describes what the system does and generally how it does it. It reflects the optimizations and trade-offs that support the key operations. It identifies the processes to be performed by the subsystems and components, defines the flows of information and interfaces between the elements, and signifies the priorities. Architecture is explicitly concerned with the views of what and how things are done in the context of the domains. The domain of a relation is the set that contains all parameters that identify the members of a relation. The domain is defined as the sphere of activity that includes the physical entities, functions, processes through their relations, and context. Domain analysis is defined as determining the (1) operations, (2) unit modularity of data and associated processing (data), (3) properties and abstractions, and (4) appropriate partitioning. Domain analysis provides a representation of the requirements of the domain. The domain model identifies and describes the structure of data, flow of information, functions, constraints, and controls within the domain.

For a domain, the system architecture view shows how multiple systems link and interoperate. Additionally, the system architecture describes the internal construction and operations. These descriptions should include the physical connections, location and identification of the key nodes (or points of interaction), and the component performance parameters.

Architecting focuses on the major elements—the elements that are structurally important to achieve the end-to-end functionality desired by the system solution. Architecting is choosing the small set of mechanisms, patterns of behavior and operations, and styles that are consistent with the needs (and desires) of stakeholders. Architecture is a tool that allows us to tame complexity. Architecting is not the whole design, but architecting is design. (Design is not architecting.) Architecting is infrastructure, the foundation of being able to move data. Moving data require paths (connectivity) and interfaces that link the path to various nodes of processes or mechanisms. The architecture consists of the network of connections, and the movement of data (energy, material, money) through the network. Good architectures are usually formed by a team of people who are committed to a solving an agreed-upon problem, who embrace the same performance goals, and who hold the same approach to resolving the vexing issues confronting the characterization of the solution.

System architecture precedes hardware and software architecture. At the system's level, architecture reveals its relations outside the boundaries of the product or service. Inside, architecture is an end-to-end (boundary-to-boundary) solution. Integration is the key issue that distinguishes architecture from design. Integration requires the forethought to prepare the structures so that data can flow to and from nodes where it is needed at the appropriate time. Architecture can and must be validated.

Much like designing, architecting brings order to misleading, ill-fitting, and confounding data; at-odds opinions; differing values; and problematic convergences. Designs that are not well defined, problems that are misstated, and needs that are unspoken merely frustrate the architect, but do not deter the search for more than one set of solutions that satisfy the problem. The system architecture views show how multiple systems link and interoperate. Additionally, the system architecture describes the internal construction and operations. These descriptions include the physical connections, location and identification of the key nodes (or points of interaction), and the object performance parameters. Architecting focuses on the major objects—the objects that are important to achieve the structurally supported end-to-end functionality envisioned by the system solution. Architecting is choosing the key set of mechanisms, the desired patterns of behavior and operations, and the styles that are consistent with the needs (and desires) of stakeholders. Architecture is a tool that allows us to tame complexity. Architecting is infrastructure, the foundation of being able to move data. Moving EMMI requires pathways (connectivity) and interfaces that link the path to various nodes of processes or mechanisms. The architecture consists of the network of connections, the movement of EMMI through the connections of objects.

Integration is the key issue that distinguishes architecture from design. Integration requires the forethought to prepare the structures so that EMMI can flow between objects where it is needed at the appropriate time.

Validation

Validation is an assessment of the operational system that exposes and quantifies the systems' limitations. The intent of validation is to determine if the user's needs are satisfied for different uses (often referred to as scenarios). When the functions are provided, the physical entities are adequate, and the user's behaviors are as needed, the product or service is deemed fit for the uses intended by the set of requirements. The concept of validation suggests that requirements can be mapped into physical, functional, and behavioral needs of key stakeholders. This mapping is indeed essential to confirm that the product or service satisfies the key stakeholder's requirements and is found to be acceptable. Acceptance of a product or service is both formal and definitive. The formality can be set out by contract stipulating terms and conditions for acceptance, criteria that must be met, and other formal representations as to fitness for use, warranties, and exclusions. Acceptance is definitive from both the buyer's and the seller's perspective. On acceptance, the buyer assumes a responsible role for the use of the product or service. Of course, the seller may have both contractual and ethical obligations, in addition to continue interest in assisting one of their customers.

From the perspective of integration, validation is the confirmation that integration had satisfactory results. Should there be problems discovered during validation, they are more often of two cases: either a functionality is performed inadequately for the expected use or there are inordinate losses to achieve the desired level of performance. In both cases, the functionality is present and the issue is one of performance. By the time of validation, the issue of functionality has been adjudicated and settled: either the functionality is there or it is not there. The customer decides what is acceptable or not and works with the developer to either wait for the full cadre of functionality or move to validation without some aspect included (perhaps with a waiver to permit the addition of a missing function after validation). Missing functionality implies missing object(s), missing (or inadequate) mechanism(s), or inappropriate EMMI. With functionality present, system performance may be degraded due to effects that were not adequately accounted for in the scenario or the product or service failed to achieve what was necessary and sufficient. The remedy may be to "tune" the performance(s), recognizing that there were trade-offs made during design and development. Validation is also the process of demonstrating the effectiveness of the new product or service. There may be several measures of effectiveness (common ones include cost, temporal responsiveness, and resilience). Validation is direct evidence that the new product or service meets the requirements through its design, architecture, and implementation. Validation is ongoing throughout the development phase beginning with engineering models, then prototypes, and ending with production models and early manufactured items that are limited to a few in a short series.

References

Abeln, O. 1990. *CAD-Systeme der 90er Jahre—Vision and Realitat.* VDI-Berichte Nr. 861.1.

Aerts, D. 1983. *The Description of One and Many Physical Systems*, Brussels: Vrijo Universiteit.

Aerts, D. and Gabora, L. 2005. A theory of concepts and their combinations I. *Kybernetes* **34**(1/2): 167–191.

ANSI/IEEE 2000. Architecture Standard 1471-2000. *Recommended Practice for Architectural Description of Software-Intensive Systems.* New York: Institute of Electrical and Electronics Engineers, Inc.

Bernstein, J. I. 2001. *Multidisciplinary Design Problem Solving on Product Development Teams.* PhD thesis, Technology, Management, and Policy Program. Boston: Massachusetts Institute of Technology, 259pp.

Busman, R. 2008. Intuition and the systems engineer: Learning from management. *INCOSE 2008 International Symposium: Systems Engineering for the Planet.* Amsterdam, the Netherlands.

Boumen, R., de Jong, I. S. M., van de Mortel-Fronczak, J. M., and Rooda, J. E. 2006. Test time reduction by optimal test sequencing. *INCOSE 2006—16th Annual International Symposium Proceedings: Systems Engineering: Shining Light on the Tough Issues,* Toulouse: International Council on Systems Engineering (INCOSE).

Christie, T. 2006. What has 35 years of acquisition reform accomplished? *Proceedings of the United States Naval Institute* February: 30–35.

Donaldson, S. E. and Siegel, S. G. 1997. *Cultivating Successful Software Development: A Practioners View,* Upper Saddle River, NJ: Prentice Hall.

Draucker, C. B., Martsolf, D. S., Ross, R., and Rusk, T. B. 2007. Theoretical sampling and category development in grounded theory. *Qualitative Health Research* **17**(8): 1137–1148.

Eckroad, S. 2002. *Handbook of Energy Storage for Transmission or Distribution Applications.* Palo Alto: Electric Power Research Institute, 300pp.

Ender, T. R., McDermott, T., and Marvis, D. 2009. Development and application of systems engineering methods for identification of critical technology elements during system acquisition. *7th Annual Conference on Systems Engineering Research 2009 (CSER 2009).* Loughborough: University of Southern California.

Euske, L. and Euske, K. 2002. *Theoretical and Conceptual Issues.* Cambridge, UK: Cambridge University Press.

Federal Highway Administration. 2009. *Systems Engineering Guidebook for Intelligent Transportation Systems.* United States Department of Transportation. Washington, DC: U.S. Government, 323pp.

Finkelstein, F., Land, F., Carson, E. R., and Westcott, J. H. 1988. Systems theory and systems engineering, *Measurement & Technology,* IEE Proceedings, **135**(6): 401–406.

Gabora, L. 2001. *Cognitive Mechanisms Underlying the Origin and Evolution of Culture.* PhD thesis, Center Leo Apostel for Interdisciplinary Studies. Brussels: Free University of Brussels, 275pp.

GAO 2001. *Best Practices: Better Matching of Needs and Resources Will Lead to Better Weapon System Outcomes.* Washington, DC: U.S. Government Accountability Office, 80pp.

GAO 2005. *Defense Acquistions: Assessments of Selected Major Weapon Programs.* Washington, DC: United States Government Accountability Office, 150pp.

GAO 2008. *Defense Acquisitions: Assessments of Selected Weapon Systems.* Reports: GAO-08-467SP March, U.S. Government Accountability Office.

GAO 2009a. *Defense Acquisitions: Assessment of Selected Weapon Programs.* U.S. Government Accountability Office.

GAO 2009b. *Defense Acquisitions: DoD Must Prioritize Its Weapon System Acquisitions and Balance Them with Available Resources.* Washington, DC: United States Government Accountability Office, 19pp.

Giachetti, R. E., Young, R. E., Roggatz, A., Eversheim, W., and Perrone, G. 1997. A methodology for the reduction of imprecision in the engineering process. *European Journal of Operational Research* **100**: 277–292.

Groah, J., Joel, S., and Blake, T. 2007. *Shock Wave Interactions in General Relativity: A Locally Inertial Glimm Scheme for Spherically Symmetric Spacetimes.* New York: Springer.

Hall, A. D. 1962. *A Methodology for Systems Engineering.* Princeton: D. Van Nostrand Company, Inc.

Hitchins, D. K. 2003. *Advanced Systems: Thinking, Engineering, and Management.* Norwood: Artech House.

Hitchins, D. K. 2007. *Systems Engineering: A 21st Century Systems Methodology.* West Sussex, England: John Wiley & Sons, Inc.

IEEE 1220-1998 1998. *IEEE Standard for Application and Management of the Systems Engineering Process.* 1 May, Institution of Electrical and Electronics Engineers.

IEEE 1991. *IEEE Standard Glossary of Software Engineering Terminology.* New York: Institute for Electrical and Electronic Engineers.

INCOSE 2006. *Systems Engineering Handbook: A Guide for System Life Cycle Processes and Activities.* INCOSE Systems Engineering Handbook v.3. San Diego, International Council on Systems Engineering (INCOSE), 4.2 of 24.

INCOSE 2010. In C. Haskins (Ed), *Systems Engineering Handbook.* San Diego: International Council on Systems Engineering (INCOSE), 382pp.

Isenberg, D. 1984. How senior managers think. *Harvard Business Review* November–December: 81–90.

ISO/IEC-15288 2002. *System Lifecycle Processes.* International Standards Organization.

Jain, R., Chandrasekaran, A., and Ozgur, E. 2010. A systems integration framework for process analysis and improvement. *Systems Engineering* **13**(3): 274–289.

Jones, C. 1994. *Assessment and Control of Software Risks,* Englewood Cliffs, NJ: Prentice Hall.

Kinnaird, B. 2003. *Use of Force: Expert Guidance for Decisive Force Response.* Flushing, NY: Looseleaf Law.

Klir, G. 2001. *Facets of Systems Science.* New York: Kluwer Academic-Plenum Publishers.

Koestler, A. and Symthies, J. R. (Eds) 1968. Beyond reductionism. *The Alpbach Symposium.* Hutchinson of London, 438pp.

Ku, P. 2007. *Commanding the Global Fleet Station and the Joint Sea Base.* M.S. thesis in Systems Engineering Analysis. Department of Systems Engineering and Department of Operations Research. Monterey: United States Navy Postgraduate School, 95pp.

Lake, J. 2007. *Textbook of Integrative Mental Health Care*. New York: Thieme Medical Publishers, Inc.

Langford, G., Raymond, F., Thomas, H., and Ira, L. 2007. *Gap Analysis, Rethinking its Conceptual Foundations* (Report Number: NPS-AM-07-051). Monterey, Graduate School of Business and Public Policy, Naval Postgraduate School.

Maier, M. W. 2006. System and software architecture reconciliation. *Systems Engineering* 7(2): 146–158.

Maier, M. W. and Rechtin, E. 2002. *The Art of Systems Architecting*. Second edition. Boca Raton, FL: CRC Press.

Martin, J. N. 1998. Evolution of EIA-632 from an interim standard to a full standard. *8th Annual International Symposium of the International Council on Systems Engineering*, Vancouver, Canada, INCOSE.

Mattice, J. J. 2005. *Hubble Space Telescope: Systems Engineering Case Study*. Center for Systems Engineering. Wright-Patterson Air Force Base, Air Force Institute of Technology, 90pp.

Newlyn, W. T. 1978. *Theory of Money*. Oxford: Oxford University Press.

NIST Special Publication 1108 2010. *NIST Framework and Roadmap for Smart Grid Interoperability Standards*. Release 1.0. United States Department of Commerce. Washington, DC: National Institute of Standards and Technology, 145pp.

Pawson, R. 1989. *A Measure for Measures: A Manifesto for Empirical Sociology*. London: Toutledge.

Phelan, P. 1993. *Unmarked: The Politics of Performance*. New York: Routledge.

Poh, K. L. 1993. *Utility-Based Categorization*. PhD thesis, Department of Engineering-Economic Systems. Stanford: Stanford University, 238pp.

Przemieniecki, J. S. 1993. *Acquisition of Defense Systems*. Washington, DC: American Institute of Aeronautics and Astronautics, Inc.

Reilly, G. P. and Reilly, R. R. 2000. Using a measure network to understand and deliver value. *Journal of Cost Management* November/December: 5–14.

Reskin, B. F. 2003. Including mechanisms in our models of ascriptive inequality. *American Sociological Review* 68: 1–21.

Sage, A. P. and Rouse, W. B. 1999. *Handbook of Systems Engineering and Management*. Hoboken: John Wiley & Sons, Inc.

Schlager, K. J. 1956. Systems engineering—Key to modern development. *Institute of Radio Engineering* EM-3: 64–66.

Semler, R. 2004. *The Seven Day Weekend*. New York: Penguin Group.

Sheard, S. A. 2001. Help! How do I make my organization comply with yet another new model? *Proceedings of the 11th Annual International Symposium*, Melbourne, Australia, INCOSE.

Spetzler, C. 2003. *Garbage In, Garbage Out: Reducing Biases in Decision-Making*. Executive Briefing Series. Menlo Park, CA: Strategic Decisions Group.

Stem, D. E., Boito, M., and Younossi, O. 2006. *Systems Engineering and Program Management: Trends and Costs for Aircraft and Guided Weapons Programs*. Santa Monica: RAND Corporation, 199pp.

Swartz, N. (1997, November 8, 2010). *Definitions, Dictionaries, and Meanings*. Retrieved July 19, 2011 from http://www.sfu.ca/~swartz/definition.htm.

Taguchi, G. 1986. *Introduction to Quality Engineering: Designing Quality into Products and Processes*. Tokyo, Japan: Asian Productivity Organization.

Troncale, L. 1977. Linkage propositions between fify principal systems concepts. *North Atlantic Treaty Organization Conference Series: International Conference on Applied General Systems Research*. New York: Plenum.

von Bertalanffy, L. 1968. *General System Theory: Foundations, Development, Applications*. New York: George Braziller.

Wheelwright, S. and Clark, K. 1992. *Revolutionizing Product Development*. New York: Free Press.

Zadeh, L. A. 1965. Fuzzy sets. *Information and Control* **8**: 338–353.

Zimmerman, H. J. and Sebastian, H. J. 1994. Fuzzy design-integration of fuzzy theory with knowledge-based system-design. *IEEE International Conference on Fuzzy Systems*, Orlando, Florida.

6

Systems Integration Management

Engineering practices reflect advances in technology to satisfy the demands for products. This situation is quite similar to other fields that have a goal to satisfy, a problem to solve, or a need to fulfill. There are stakeholders whose requirements are expected to be met, procedures that are set up and followed, and results that are measured in some fashion. According to the *Oxford English Dictionary*, something devised or contrived for bringing about some end or result is defined as a project (1290 AD). The view that the project is a social structure that results in questions, investigating, questioning, studying, examining, and building things is the focal point for systems engineering management. "An engineering project starts because of a social need for particular service" (Gosling 1962).

The process of performing management has been described as achieving objectives by influencing others (MacKenzie 1969, 1988). The processes of management have been variously described (Steiner 1969; Kerzner 2009) and are often thought of as the means of systems engineering to carry out its tasks. Similarly, systems integration requires a form of management that guides the workers according to a set of processes. While the "to manage" process can be broken down into many subprocesses, this presentation uses "to plan," "to organize," "to direct" (or "to command"), "to control," "to communicate," and "to team-build." The convention used in this text signifies processes by double quotation marks, in contrast to the single quotation mark used to indicate a function. These six processes (and their respective decomposed subprocesses) adequately represent all the major processes carried out as part of management. Within these six processes* of "to manage," the many other subprocesses that make up the first-tier decomposition of the six major processes (listed above) are meant to be independent of each other and concomitantly their respective decompositions into lower-level processes are meant to be each hierarchically structured to reveal more detail, while maintaining that same independence. The top-level decomposition of "to manage" is shown in Figure 6.1 and the detailed structure of "to manage" is outlined in Appendix 1.

It is particularly important in performing systems integration work to distinguish carefully between individual process and individual functions. Combining, mixing, aggregating, or otherwise erasing the atomic nature of

---

* An alternative view of management is to describe the roles taken on by the people on a project. "A role is an organizational identity that defines a set of allowable actions for an authorized user" (Jansen 1998).

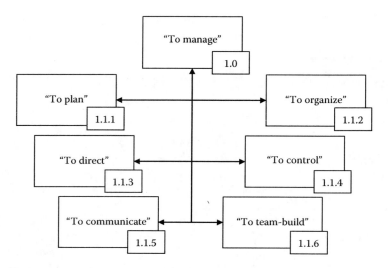

FIGURE 6.1
"To manage" process decomposition.

processes or functions obscures interfaces, masks boundaries, and prevents access by users. For integration, hiding an interface between objects prevents integration at the exact level where the interaction between objects must occur for efficiency, thereby forcing an interaction at either the wrong level of abstraction or forcing an interaction with more objects than required. The impact of not exposing the proper interfaces, the interfaces that minimize the number of interactions that are required to transfer the requisite EMMI, between objects at the correct level of granularity is the number one encumbrance on effective integration (Park et al. 1997). Granularity can be thought of as the partitioning of what is included in one object as opposed to including it in another object. Granularity is the amount of action that is ascribed to object(s) that have been grouped by the system design, by the engineer following the specifications, by the convenience as determined within the purview of the engineer, and by the practicalities of the physical entities.

Granularity

The degree to which an object is composed of discrete objects each separably enabled by mechanisms relates the minimal logic that can be partitioned. The sense that it makes to associate one object with another is justifiable by the resultant dependencies of the aggregate of discrete groupings; the summative impacts on the other objects; and the logicalness of the completed partitioning. The reason for one grouping of objects versus another is more than a matter of convenience. Various logics apply, including compliance issues (e.g., standards, policies, and requirements); preferences based on logistics or support; socioeconomic rationale that makes sense on a normalized or weighted basis; or other justification that is rational and defensible based on the rationale. Granularity is the parsing of objects (e.g., requirements) on the same level of abstraction to differentiate one (in this case) requirement from another. For the top-level process that describes "to manage," each of the portioned subprocesses is determined to have an approximately equal amount of influence on an organization, that is, having equivalence with regard to the totality of what is covered. There are also similarities in the relations, each relation being associated with a common mechanism at the top level of abstraction to a common set of procedures. In other words, the reasoning and rationale applied to partitioning objects (that are enacted to form each of the subprocesses) should accommodate the needs of the stakeholders, the limitations set by the boundaries, the constraints established and imposed by the architecture, and the anticipated changes deemed most likely as the product or service is used in an operational environment. Granularity is said to be flexible (Kaindl and Dvetinovic 2008) and if that flexibility is removed too early in the development process, integration is made more difficult. By not removing essential flexibility through iterative design and development, integration is made easier. The ease of integration is managed by sensitivity and attention to this issue from the systems engineers, the engineers, as well as the project management. A portrayal of granularity is shown in Figure 6.2.

Figure 6.2 illustrates a hierarchical graphical representation of granularity and abstraction. For granularity, the partitioning of the system of systems (in this case) into two systems (1.1 and 1.2) indicates that system 1.1 is an integral whole, as distinct from system 1.2 (also an integral whole). While there may be overlap in subfunctions, the design and architecture of system 1.1 is unique and different from the design and architecture of system 1.2. If systems 1.1 and 1.2 were designed and built prior to their integration into a system of systems, their respective lower-level subfunctions may only be accessible through interfaces that have already been designed and built. If one or both of systems 1.1 and 1.2 are being designed and built, then the separability of the two systems as well as their interoperability will become the major design and architecture drivers. These design drivers need to be managed (i.e., planned, organized, directed, controlled, communicated, and carried by consensus (team-building)).

The notion of overlapping processes or functions is a significant, germane issue for granularity. Were a function to overlap with a similar, if not identical, function, in an adjacent set of objects, then there may be confusion as to which of these identical functions should be available to the user. In the least, there is a redundancy that may or may not be intended. Should there be an underlap between functions, that is, a part of a function is missing, or a whole function is missing, then that missing function will inhibit a

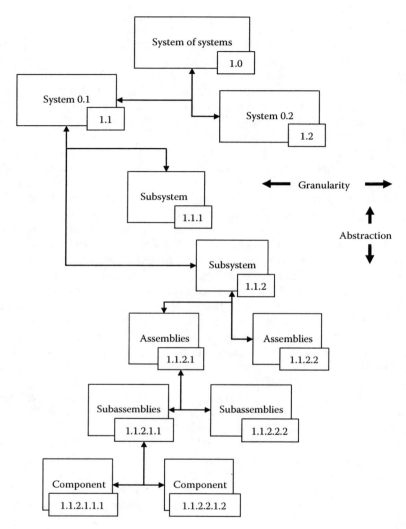

FIGURE 6.2
Granularity and abstraction.

local function to execute (or be available) as well as a system-level function (or thread). Overlapping and underlapping functions (or processes) signify problems in the design or construction of objects. These problems may or may not be detected during design reviews or walkthrough once development has begun. If missed before testing, they can appear as a performance measurement that is out of range set by the specifications. All such overlaps or underlaps are detected during integration when local and end-to-end system-level functions are to be demonstrated.

Granularity and Integration

Granularity deals with the organization of EMMI as mediated by the aggregation of objects (and their mechanisms). Granularizing (or partitioning physical entities, functions, and behaviors) into various domains is the most important task for the designer and architect after all the requirements are captured and characterized. Since it is exactly the iterative nature of systems engineering that helps to surface and describe the system requirements, partitioning of objects is consequently iterative. If the granularity is too broad (too coarse, or too wide in a hierarchical graphical representation), access to the specific mechanism may be encumbered by other objects that must be enacted. If the granularity is too narrow (too fine, or too narrow in a hierarchical graphical representation), access to the desired set of mechanisms may be encumbered with multiple enactments or the desired mechanism (and object) may be missing. While it is quite difficult to know in advance what the optimum granularity should be, building in process managers (for processes) and a common physical object structure (for functions) provides flexibility without burdening either the design intent or the testing. A process manager is a module of hardware with software that is an interface between objects. A process manager (Pikula and Siemion 2007) creates a new process instance based on the activity model embedded in an object or a group of cooperating objects. That activity model maps the input or output EMMI to the object's internal processes. Each new process instance maintains the current state of the processes for external and internal actions so that individual processes can be executed simultaneously. After one object completes its operation, the process manager determines which mechanism will execute next based on the state of the process instance that is completed and that remains to be completed. Therefore, the suite of applications used by the process manager can be enacted independently without regard to the sequence of steps defined in the activity model or in the overall system of systems operations. Process integration ensures application operational independence and allows the process manager to be domain (or object) independent because the process manager only needs to interpret the basic constructs that form an activity model. Testing of a process manager is the same for all process managers, since they are designed to be identical in both hardware and software. It is in this manner that the granularity of processes need not be determined as precisely as needed for efficient integration. Building flexibility through process managers overcomes the designed interfaces between objects. Again, progressive refinements to design and building objects are fundamental to the systems engineering process. In contrast, a goal of integration is to work with a design that has included sufficient flexibility to allow recursive integration, without iteration. Here, recursive refers to enabling a mechanism without redesigning either the interface for input or output, or the mechanism.

For ease of integration, the goal of granularity is to partition objects so that each is composed of simple, independent groupings (Stevens et al. 1974). However, the initial emphasis needs to be placed on the design of the objects (i.e., a system that is being built from scratch), while second best is to orchestrate the design through interoperability of objects that have strong similarities across systems (i.e., a system of systems). Managing the interoperability for integration requires clear partitioning in objects that can be considered both as a whole (resultant output is reflective of a single unit of functional operation) and as a part (required input is reflective of a single unit of functional operation). The object's roles of a whole and a part are focused on the aggregate performance of a set of mechanisms that enable a single function. These roles can be represented in a hierarchical view with the subfunctions aggregating into a function (e.g., Figure 6.1). To simplify integration, to provide the requisite interoperability, and to achieve lowest possible expenditures of labor necessary to realize the deliverable product or service, engineering efforts need to focus on (1) simple, independent objects, (2) a minimum number of interfaces, (3) a minimum number of connections across the interfaces, and (4) achieving the set of object behaviors that are required. Managing the integration efforts needs to focus on this engineering thinking in addition to the systems engineering thinking that is associated with recursion. The two challenges of managing integration are to first, recognize and manage to achieve the easiest path to integration and second, to provide the necessary resources and leadership to stay on that path.

Abstraction

A close second, and similar, abstraction to a higher level than is necessary may mask the detail needed for a mechanism to be effective in transforming input EMMI into an output. At a higher level of abstraction, the existence of lower-level details may be acknowledged, and if they are, then an additional exchange of EMMI will be necessary to extract what is necessary for the mechanism to complete its transformation. If the lower-level details are not acknowledged, not known, or obscured, then several additional exchanges of EMMI may be needed. In both these cases, granularity and abstraction interfere with the efficient transformation needed to present the requisite functions. Here, abstractions are referred to constructions of varying degrees of details shown "... by taking an exemplary case or instance and removing detail" (Machamer et al. 2000). Abstraction is the result of redefining a previously constructed schema into a new set of schemas—by extracting common features from specific instances, merging, and replacing with another that has less detail, but yet embodies the general notion of what is missing along with what remains.

For example, the process of "to manage" and the function of 'to manage' extend across the entire work domain that encompasses the integration activity. The process of "to manage" overlaps exactly with the function of 'to manage'. Both are measurable, but in different ways. The process of "to manage" is measurable by comparison with another process of "to manage." By that, the differences between two processes are objectifiable by comparing like-kind or similar processes, whereas the difference between two functions is demonstrable through comparison of their performance(s). Both processes and functions have mechanisms, inputs, and outputs, and losses that result from achieving their outputs.

"to manage": "command" and "control"

In the case of one of the decomposed processes "to command" (or "to direct") shown in Figure 6.3 and "to control" shown in Figure 6.4, the hierarchy of subprocesses highlights the differences between "command" and "control" (or "direct" and "control").

Take for example "provide resources" (1.1.3.2) in Figure 6.3 and "to report" (1.1.4.5) in Figure 6.4. If the concepts of "command" and "control" were considered to be the same (i.e., the process "command" and the process "control" are always used together without a means of differentiating them (Steinbit 2002)), then all of the "command" hierarchical subprocesses will be mixed with the "control" subprocesses in levels of abstraction that may be different (i.e., different levels in the hierarchy, the vertical graphical representation) and with partitioning of when a process begins and ends (i.e., the horizontal

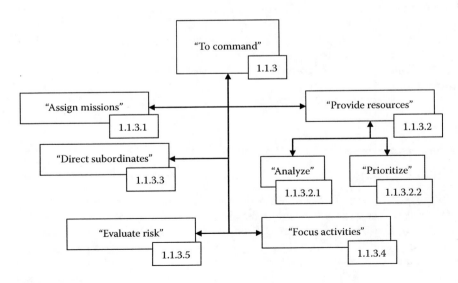

FIGURE 6.3
"To command" process decomposition.

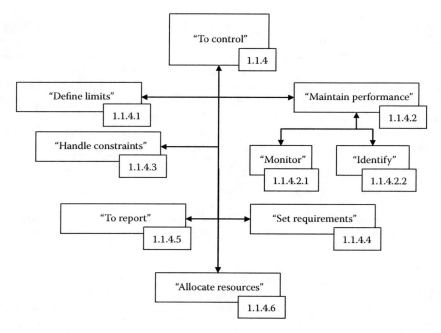

FIGURE 6.4
"To control" process decomposition.

graphical representation) may be bounded and mixed different than if the two processes of "command" and "control" were kept separate and distinct (Bornman 1993). When integrating a system that is predicated on commanding and directing (or controlling), it is essential to separate the functions so that the proper allocation can be made to physical entities and so the users of the system can be exposed to the most effective functions to carry out their work.

Project Management

The project may be charged with organizing simple tasks and tasks of significant complexity; it may be established within an organization or may arise as the confluence of like-minded people; it may be temporarily started and stopped or proceed to its logical conclusion; or it may be part of a program that involves many projects. Regardless of its context, constraints, or structure, a project is temporary. It may be convenient to classify projects in other ways to investigate a particular aspect or nuance that has particular meaning in a specific context, but the fundamentals of a project by definition are pervasive across these various circumstances.

After World War II, scientists were motivated to push technology as they explored the limits of the known physical laws. Progress seemed only limited by imagination, yet the appetite for innovative products that could be produced inexpensively seemed insatiable. Management techniques, applicable since the industrial revolution, were being revised. The design of new products was no longer governed by a single physical law or simple set of rules. The confounding factors were the number of items in the product, the myriad types of and relationships between these items, and the number of these items that required choreographing to achieve the desired functionality. The management systems that were used to develop such products were not designed to deal explicitly with these confounding factors. Therefore, visibility into the progress of work was limited, delays (due to changes in work, insufficient skills of workers) were common, and predictability of status and progress suffered. Systems engineering grew out of the need to build products whose elements were inextricably intertwined.

The systems engineering view required its own brand of management, not unlike what most others use, but postured to reinforce the focus on defining requirements, system design, architecture, and integrative processes. Some books on project management sometimes reference systems engineering (Kerzner 2009), many systems engineering books included section(s) on managing for systems engineering projects (Sage and Armstrong, 2000; Hitchins 2007), and a few systems engineering books are devoted entirely to the subject (Forsberg and Mooz 1996; Blanchard 1998). One can easily fail to count or acknowledge all that has been written about project management. However, the author knows of no writings that focus on the management of integration. Aside from the litany of books, handbooks, organizations, professional journals, practicing managers, and consultants, new concepts arise with the intention to replace, yet still add, only to add to the funambulation* that ropes terms together into the web of ideas we term as project management.

For developing a product or service using systems engineering, the focus is on managing the systems engineering process. The systems engineering management plan (SEMP) lays out the plan, procedures, and the representations (or models) that are necessary to describe, provide, and be the documentation, milestones, reviews, and steps for the project team to carry out. However, invariably the section on systems integration begins with the word integration, follows through with words discussing integration, and then ends with the word integration. Nowhere is there a discussion of any detail about what management of integration means or even what is entailed. Typically, integration is viewed as a phase of work that must occur as development ends and prior to completing the product or service work.

---

* Tightrope walk.

Integration as a Recursive Process

Fundamentally, building a product or service is an integrative process; however, testing is not an integrative process. Planning for integration should be based on developing system-level functions that are in-kind built up by their subfunctions. The first step is to focus on those subfunctions that when linked together with other subfunctions form an end-to-end thread that stretches across the entire system. That end-to-end threat is referred to as a system-level thread (or "thread"). Each time a subfunction is linked to another subfunction (recall each function is the result of the integration of two objects), the performance and quality of the linked subfunctions are measured. Measures of performance and measures of quality are stipulated as part of the test plan. The integration plan identifies the objects that are to be linked together to provide the subfunctionality necessary to constitute a thread.

Measures of Integration

The measures of quality are premised on two factors: quality is the achievement of a level of acceptable variability of each measure of performance, and the variability in performance is representative of the user's perception of quality. For example, a vehicle is said to be moving at 60 km/h (a measure of the vehicle's performance). The accuracy of the measurement (its bias due to systematic errors) is 0.003% of the measured performance; the precision of the measurement (its uncertainty due to random errors) is 0.0002% of the measured performance. The manufacturer of the vehicle offers an automatic speed control device as an accessory. The variance (the spread of the measurements for both accuracy and precision) matches a Gaussian distribution. Based on the customer's (user's) requirements, the manufacturer will customize the speed control device by setting its upper and lower limits. When the vehicle speed is less than the lower limit or greater than the upper limit, the speed controller will adjust the speed of the vehicle whether moving on an incline (positive or negative slope) or on level ground. Assuming that the precision and accuracy are measurably the same, the adjustment of the upper and lower limit of performance is now only determined by the customer's willingness to pay the price asked for by the manufacturer. The smaller the variance (i.e., the closer the lower and upper limits are to each other) around the target value for the vehicle's speed, the higher the price. However, the price is not a linear function of the distance between the upper and lower limits. As the distance between these limits approaches the exact target value for the vehicle's speed, the cost is significantly higher than near the limits (which are further away from the target value). See Appendix 2 for a derivation of the loss function that describes the quality based on a variance from a target value of performance. Choosing a loss function depends on whether the use is for making predictions, making estimations, estimating risk, defining optimal testing procedures, or making sense of optimizations (Hennig and Kutlukaya 2007).

Quality

Quality can be thought of in any number of ways, all leading to the notion of great importance to the users of products and services, but defined inconsistently (Reeves and Bednar 1994). Here we distinguish quality in the narrow sense of only that which is measurable through an association with a function.* The interpretation that follows concerning quality is that quality is a property of a function. As a property, quality is then deemed as conformance to performance(s) for that function as objectified through a set of specifications. Functions have performances (a minimum of one per function), and each performance has a quality (a minimum of one quality measure per each performance). In this manner, the functions of a product or service are completely objectified by performances and qualities. When goods are within specification they are considered to be of high quality, contrasted with low-quality goods that are determined to be outside the bounds of specification. To be within specification means the product or service should function as expected, while outside specifications signifies potential problems such as increased wear, unreliable operations, or inoperative functions. Such a notion of quality is often typified by the statement: the quality is remembered long after the price is forgotten. However, Taguchi (1986) proposed a view of quality that relates to cost and therefore a loss measurable in monetary terms. This loss accrues not only to the designer, developer, and manufacturer but also to the customer, user, and broadly to society as a whole. In aggregation, these entities represent the 'seller' in a buy–sell relationship. For the seller, their loss is accumulated, throughout the product lifecycle, from conception to include adjudication of the last lawsuit, or the last simulacrum of support. Whereas Taguchi offers that seller's losses are incurred up to the time that the product is shipped, we take a broader perspective and extend the seller's losses beyond the shipping event to include support,

---

* It is often said that quality of a product or service is due to having a certain set of features. When a feature is "missing" (referenced to the desired set) the product can be considered to be of lesser quality. Applying this logic for quality within the structure of functions, performance, and quality is interpretable as a missing feature has a quality of zero (reflecting no performance and no function). Therefore, the quality of a product or service with a "missing" feature is of lower quality than a product with a "full" set of features.

maintenance, and service. According to Taguchi, after the product is put into use, it is society and the customer who bear the cost for low quality. However, we distinguish between the customer's and society's responsibilities in contrast to those attributable to the seller. Even while the seller's purview can be narrowed to the Taguchi limit by eliminating all interactions with the customer after the shipping event, our general perspective considers real costs as well as the costs of negative customer reaction. Sellers sometimes have difficulty in capturing and accounting for money and money-equivalencies spent by customers to deal with problems, dissatisfaction, and inconvenience associated with low quality. The approach here of developing and applying a quality loss function allows quantitative evaluation of losses caused by variations in behaviors and limits of performance specifications as they consociate with various functions of the product (Choi and Langford 2008).

Since Taguchi's quality loss function is primarily focused on manufacturing, a more general formulation of loss functions is needed beyond his quadratic formulation to assist in analysis and decision making that must be made during other lifecycle stages of the product, including conceptualization, research, development, integration, operations, maintenance, and disposal.

The requirement is for a quality loss function that is applicable for the lifecycle of a product or service. Further, we must tackle head-on the definition of product function and associated performance(s). Taguchi uses a quality loss function to evaluate product performance relative to the performance specification. When applying Taguchi methods, performance evaluation becomes a most demanding task that challenges managers to distinguish carefully between processes and functions, without providing the benefit of firm definitions and theoretical standing. The diversity of definitions of performance, the complexity of measuring performance, and the scarcity of generally accepted performance measures compound these problems.

In general, there are *at least* seven distinct, although not necessarily mutually exclusive, performance measures used in practice. These are effectiveness, efficiency, quality, productivity, quality of work life, profitability, and innovation. Thus, performance can be defined as a concept that takes on different meanings in different situations for different organizational systems (Kaplan and Norton 1992, 1993, 1996; Kumpe and Bolwijn 1994). For example, efficiency, quality, time, innovation, and contribution to profit are often used as performance or effectiveness measures. Taguchi used quality (i.e., loss) as a performance measure for evaluating product performance. When referring to EMMI, performance can be defined as the net work accomplished during a period t. For example, the net work accomplished during a period t is equal to the amount of work "completed" minus the amount of rework required to finish the amount of work that was thought to be completed. The units of loss can be in units of energy (joules or electron volts), mass (commonly, kilograms or pounds), material wealth (converted into local currency, e.g., dollars), or information (*bits* as adopted by Shannon for information (Shannon 1948)).

The quality loss function developed by Taguchi (1990) is used to describe quality in terms of smaller-the-better (STB), larger-the-better (LTB), and nominal-the-best (NTB) characteristics. An STB output response results when it is desirable to minimize the performance, with the ideal target for performance being zero. Examples of STB output responses are the wear on a component, the amount of engine audible noise, the amount of air pollution, and the amount of heat loss. The LTB output response reflects cases when it is desirable to maximize the result, the ideal target being infinity. Examples of LTB output responses are strength of material or fuel efficiency. The NTB characteristic results when there is a finite target point (or domain of cooperative agreement) to achieve, often associated through a negotiated outcome. In this case there are typically upper and lower specification limits on both sides of the performance target, representing the maximum or minimum acceptable bounds for the parties of the negotiation. Examples of NTB characteristics are the plating thickness of a component, the length of a part, and the output current of a resistor at a given input voltage.

A great many papers relate to the quality loss function largely from only one side of the quality characteristics (Kapur and Wang 1996; Chung and Chao 2005; Yahya and Chanwut 2007). Here, we derive a quality loss function for broader applicability in managing quality characteristics, regardless of domain and characterization, regardless of input and output, and irrespective of preference or specifics for any discipline or field. We are particularly interested in loss functions as a means to determine the effectiveness and efficiency of integration.

There is some common ground that reconciles traditional and Taguchi views of quality. Quality is viewed as a step function such as a product or service is good or bad. In reality of course, there may be a preponderance of characteristics that in aggregation transition from acceptable to unacceptable (or vice versa), but the general sentiment assumes that the product or service quality is uniformly good between the lower specification and the upper specification, and bad outside these limits. Even traditional decision makers and those using Taguchi's loss function will make the same judgments, as they both will set upper and lower limits for acceptability. However, the limits may not be equally distributed from a target value that is deemed the best trade-off between good and bad. If decision makers consider both the position of the average and the variance, and if the averages are equal and/or the variances are equal, then the traditional decision maker and one using Taguchi's loss function will make the same decision. Typically, the traditional decision maker calculates the percentage of defective units over time, when both the average and the variance are different. Both the average performance and variation from a target value are measures of quality (Taguchi et al. 1989).

Further, Taguchi formulates and it is widely held that the customer becomes increasingly dissatisfied as performance departs farther from the target value for performance of a function. His extensive work with manufacturers over the last 30 years suggests that a quadratic curve best represents a customer's dissatisfaction with a product's or service's performance. The customer's view of the product is restricted to the operational and disposal stage of the product or service lifecycle. A straightforward and accurate means of representing a quality characteristic is through a function that uniquely defines the relation between a loss in EMMI and the deviation of the quality characteristic from its target value (Taguchi et al. 2005). For products or services that are in operation, the quadratic form of loss functions matches well to customer satisfaction (Taguchi et al. 2005).

A Taylor series expansion is often used to approximate a function as a polynomial of terms whose first terms turn out to be reasonably close approximations to that which would otherwise seems mathematically complicated (Mason et al. 2003). The first derivative of a Taylor series expansion taken about the target value is a quadratic curve when the target value is set to zero. The curve's minimum (or nominal position) is centered on the target value, which (Taguchi et al. 1989) has shown to provide the best performance in the eyes of the customer. However, identifying the appropriate performance measures as well as selecting the best target value can be challenging. Designers sometimes offer their best guess. The quadratic form was chosen by Taguchi because it was both simple, and as it turned out, useful. Further, after the Taylor expansion, higher powers in the series change the loss at the target value by a very small margin, and for practical purposes can be ignored within experimental error. Symmetric formulations of loss functions are assumed to be approximate and accurate to a first order. This assumption is shown to be accurate since the result of development is indeed what is placed into service by users, that is, that which is equivalent to a Taguchi validated quadratic form of loss function. The general loss function discussed in this book, provides the quantitative means to evaluate integration from conceptualization through disposal by adapting the order of the loss function to the desired phase in the product's lifecycle. Asymmetric loss functions are most useful when integrating systems into a system of systems. The reason for this situation (as distinct and different from that of integrating a system) is the requirement for reversibility of actions to allow a system to remain a system when it is no longer a part of a system of systems.

The loss function offers a way to quantify the benefits achieved by reducing variability around a target performance value. It can help justify a decision to invest further to improve a function that is already capable of meeting specifications, but there exists a requirement to achieve the same (or better) performance at a lower loss (e.g., in energy, matter, material wealth, or information). According to Taguchi, the objective of minimizing the loss to a customer was to improve product quality by minimizing the effects of variations in its performance while striving to achieve the performance target value. The narrower the performance limits, the higher the quality (and within the same design space, the higher the cost). However, achieving higher quality does not need to come at the expense of eliminating the causes of that variation. Eliminating the causes within an existing design must be invoked through solutions that are already included in the mechanisms, processes, and physical entities. The designer's and the architect's intentions are to design robustness into the product to mitigate excessive variation in achieving performance(s) so that value can be imparted to the customer without an associated loss (Yao et al. 1999).

The generalized loss function covers all quality characteristics such as NTB, STB, and LTB (Taguchi 1986, 1990). One party to a negotiation, conflict, or point of view determines that more performance is better (considered as LTB strategy) while the other party (in opposition in some way) considers that STB demands on performance is required. The LTB (e.g., a buyer's strategy) benefits from larger values of performance, *m*, coupled with lower loss. Alternatively, the STB (e.g., a seller's strategy) faces higher losses from delivering a higher performance, *m*. For example, a seller might want to deliver greater product performance but is unwilling to accept increased costs which when passed on through higher pricing to the consumer may result in stiff competition that may lead to a reduction in market share. The buyer might desire and come to expect greater product performance for lower pricing as competitive factors, innovation, and new technologies offer sellers a means to satisfy that need. The primal relation between a buyer and seller is depicted in Figure 6.5.

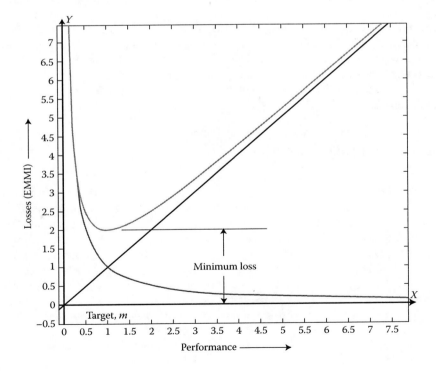

FIGURE 6.5
Generalized loss function relating performance to quality.

A simple, yet instructive, portrayal of a loss function is to view the seller as having a decreasing value function as performance increases and the buyer in the opposite position of having a decreasing value function as the performance decreases (Langford 2009). This can be represented as $L(y) = k_0 + k_1 y + k_1 m^2$, where $L(y)$ is the combined loss of the buyer and seller, and k_0 is a constant equal to $-2k_1 m$.

Between any two contra-posed positions (whether weakly or strongly held), the loss function indicates the minimum loss that can result from the positions (assuming that both sides are willing to not gain any advantage over the other). In essence, a loss function is useful in determining the amount of EMMI that is used (i.e., lost) to achieve various levels of performance for individual or aggregated functions. For integration, the functionality of a system can be measured by both the performances of the functions and the losses that are attributable to achieving those performance. There is a direct correlation between a loss in EMMI and "utility," where utility is a measure of relative satisfaction (e.g., in economics or systems engineering). By applying the loss function to the management processes, the workforce can be monitored in real time, the projection of predicted work can be evaluated against current status, and metrics can be established to analyze the impacts of applying resources (e.g., engineering labor) to particular problem areas. By applying the loss function to the integration activities, the various functions can be evaluated by both their demonstrations of performances as well as their losses in achieving those performances. This determination of functional effectiveness helps to define and design tests, revise integration sequencing, and outline validation schemes to facilitate better determination of the usefulness of a product or service.

Types of Quality Loss Functions

For each quality characteristic, NTB, STB, and LTB, there exists some function that uniquely defines the relationship between economic loss and the deviation of the quality characteristic from its target value. Taguchi has demonstrated through practice the quadratic representation of the quality loss function to be an efficient and effective way to assess the loss due to deviation of a quality characteristic from its target value. For a product with a target value m, from a customers' perspective, $m \pm \Delta_0$ represents the deviation at which functional failure of the product's or service's component occurs. When a product is manufactured or a service is provided with its quality characteristic at the extremes, $m + \Delta_0$ or $m - \Delta_0$, some measure to counter the loss must be undertaken by the customer.

Following the simplified loss function $L(y)$ (average loss) with the characteristic of NTB is the combined buyer–seller dynamics that can be described as follows:

Nominal-the-best

$$L = k(y - m)^2 k = \frac{A_0}{\Delta_0^2} \tag{6.1}$$

where k is a proportionality constant and can be described as the cost of each unit (returned, modified, reworked) divided by the range limits of process variability divided by 2, y is the measure of performance (e.g., output) for a given function, m is the target value of y, and A_0 is the loss per unit that encompasses the lifecycle of the unit that must be expended to mitigate loss (e.g., countermeasure). The loss function can also be determined for cases when the output response is an STB response. Following the same procedure as for the case of NTB, where the target value for performance is zero, the loss function is described as follows:

Smaller-the-better

$$L = ky^2 k = \frac{A_0}{y_0^2} \tag{6.2}$$

where A_0 is the consumer loss and y_0 is the consumer tolerance.

For an LTB output response where the target is infinity, the loss function can be written as follows:

Larger-the-better

$$L = k \frac{1}{y^2} k = A_0 y_0^2 \tag{6.3}$$

Outline of the General Quality Loss Function

To achieve the desired level of quality and to determine the target value for a product within each stage of a product's lifecycle, stakeholders pose the question—how much loss can or will be incurred? To address this question, a general quality loss function must be developed—one that accounts for the changes in the allowable variance from a performance's target value. We introduce a shape parameter that governs the amount of losses as a function of the product's or service's lifecycle and present a function which covers all three quality types from the perspective of a product's or service's stakeholders. Through this effort we propose a general quality loss function covering all lifecycle phases. It can be shown (Appendix 1) that given the following notation and assumptions, a general quality loss function reduces to STB, LTB, and NTB forms shown in Equations 6.1 through 6.3.

The general loss function is shown below:

General quality loss function

$$L_n(x) = -2C_s m^n + C_s x^n(1 + m^{2n}x^{(-2n)})$$ (6.4)

Further, the relationship between proportionality constant C_s (under STB) and for C_1 (under LTB) is as follows:

Proportionality constants and target value

$$C_1/C_s = m^{2n}$$ (6.5)

where C_s is a proportionality constant of stakeholder's loss per response of quality, if the type of quality characteristic is STB, and means a proportionality constant of developer's or manufacturer's loss per response of quality, if the type of quality characteristic is LTB, and x is the response of quality; $L_n(x)$ is the total quality loss per piece in case of shape parameter n and quality response x; and L_n is the expected quality loss per piece in the case of shape parameter n and quality response x.

With the total quality loss ($L_n(x)$) consisting of the stakeholders' loss plus unknown losses, and if the level of quality equals the target value of the quality (i.e., m), the total quality loss is to be zero (or the minimum loss that is inherent in the system); the parameter n, the shape parameter, represents the adaptation of the loss function to the specific uses intended by the circumstances in which the loss is to be determined; and the minimum value of a shape parameter is close to zero and the value of the shape parameter in the concept refinement phase of the acquisition phases varies from 0 to 1.

The loss function can be used to assist in decision making by considering the impact(s) of one strategy versus another. Decision making is foundational for determining the requirements for functions, the performances of those functions, and the losses incurred to achieve those functions. The loss function is a mapping of a decision into an integration framework that reflects the consequences of that decision.

Integration Strategy

Integration activities need to be sensitive to the differences in style and techniques that pervade the social structures and strategies of the development work. If the integration effort is primarily equipment engineering, then a plan-driven strategy (Smartt and Ferreira 2011) is most typically found in the workplace. A plan-driven strategy emphasizes well-defined roles and responsibilities, partitioning of the effort into planned work packages, and implementation through repeatable processes. Integration for simple, engineered products or services relies on the plan-driven strategy. It is effective most of the time. Repair people work on networks, products are manufactured, services are provided, upgrades are installed, and routine operations are brought together with commonplace and habitual practices. However, for products and services that require systems engineering techniques and management, systems integration is fraught with difficulties and confounding predicaments.

It might seem troubling to learn that on some of the smaller systems engineering projects, integration plans are considered optional; for the larger more complex projects, integration plans are prepared as a matter of contractual obligation. One should expect these integration plans to be well thought out and rationally prepared ("because it is not the plan that is important; it is the process of planning"). However, in terms of overall planning effort, the integration plans only lightly reflect the expectations of the planners. The widely held belief with large integration efforts presumes significant delays due to uncompleted developments on components and subsystems. The integration plan is based on design documents, architecture, and best-case scenario sequencing of components and subsystems. The integration plan is notionally a plan, but most often laid down without the capability of management, systems engineering, and engineering to be able to achieve what is laid down. Instead, the integration plan needs to represent the extent of human activity that is organized to perform certain tasks at certain times, following specified procedures that are derived from policies and rules, resulting in a model or representation of the formative acts of cognition (used to think about the plan and planning) and the physical and intellectual portrayal of both the acts of cognition and the human activities that carry out the procedures envisioned by the cognitive acts. These physical and intellectual representations are the best efforts to match the cognitive and procedural acts. Representations (or models) are the paper drawings, the planning document, the written procedures, the reports on the activities, the schematics and specifications that will guide the work from which the physical entities will be built, the presentations, and all that is corporeal concerning cognitive and procedural activities. In the integration framework, this discussion is captured as the subjective frame. The integration plan is thought of, written, and presented in a tangible form (e.g., paper, digital account, or in various formats for presentation). The process of planning is fundamental to project management and absolutely fundamental to systems integration.

Recursive Nature of Systems Integration

Planning is a periodic, recursive process. As such, the results of planning for integration must necessarily be recursive. If something is objectively true, there is an implied impartiality that fairly weighs all facets. However, systems engineering is the process of transforming objective needs through subjective processes to achieve an objective result. The subjectivity comes from planning and the results of planning. Recursive implies that the answer is not known, but rather is discoverable in the process of objectifying the requirements and procedures into objects that satisfy the needs of key stakeholders. Recursiveness is the property of a process that is identifiable by an event (i.e., interaction between objects), deemed causal for that event, and that has predictive qualities for the next event or chain of events. Recursive processes are not iterative. Iterative processes are typified as refinements on existing objects. Recursive processes are typified as refinements on future objects.

Here, the issue of causality is not statistical, but rather based on experimentation with the objects' mechanisms to ascertain the relation between the adjustable parameters of the objects. Statistical models are incomplete because they are recursive (Heckman and Vytlacil 2006).* There are two such types of recursive processes: (1) self-recursive: those that are determinable by internal processes (e.g., modeling, simulation, integration, and testing) that reveal or associate with patterns that help the decision makers move toward an integrable set of objects) and (2) open-recursive: external processes that reveal or associate with patterns that are not considered to be part of the work domain of the project. Examples of self-recursive processes are patterns of improved performance due to a particular physical configuration or the need to change an existing (or in-progress) object to better reflect the needs of future integrable objects. Self-recursive processes benefit from the interactions between objects as they often produce patterns that are suggestive of what needs to be accomplished next. For both types of recursiveness, the operative model is that of a social network (Fowler et al. 2009), where the work of the project development team is controlled to a great extent by the subjective procedures that were used as well as the extrospective analysis provided by key stakeholders. The nature of recursive processes lies in their dynamics which tend to point out the next best step by exposing the consequences of the previous steps. Unlike iterative processes that focus on redoing what was done previously, recursive processes point to the future, the next event in integration that is now less problematic than before. Systems engineering process models are both iterative and recursive (Sopha et al. 2010), where the great reliance on iteration is meant to improve on what has been done previously. Systems integration is recursive, in part due to the nature of systems engineering that has been applied to the system design and specifications for objects. So, by their natures, system integration is quite different from systems engineering. The management of systems integration is quite different from systems engineering also.

---

* The scientific model embodies nonrecursive causal models (Heckman and Vytlacil 2006).

For integration efforts that involve significant software work, a more agile approach is sometimes adopted (Smartt and Ferreira 2011). The hallmark of agility is to adapt (self-organize) the engineering activities to compensate for the delays that have arisen during development. Both the plan-driven and the agile-driven strategies to integration illustrate the inadequate predictability of increasingly complex system developments. The lack of the systems engineering tools and their abilities to adequately plan and diagnose the development effort for integration problems is readily apparent in the strategies and methods currently used. For simple engineering integrations, the plan-driven strategy suggests that complexity (multiple interactions across the three kinds of boundaries: physical, functional, and behavioral) is a key driver. When software is added to the project, neither the engineering tools nor the systems engineering tools appear adequate. Integration is considered to consume the major portion of expenditures and is held as the most likely cause of project problems.

To implement a strategy of planned integration while maintaining the agility to adapt the sequencing of components and subsystems, the strategy should be considered a policy that delineates all the actions required for integration in the integration plan. Integration planning should be mandatory, not optional based on a perception of simplicity or fitting within a "normal" routine. The reason: interactions can lead to integration—some interactions are desired, perhaps some are not.

Integration Planning Concepts

Planning integration means more than just allocating time to various activities or sequencing those activities. Planning is predicated on the development team working in an incremental fashion to (1) build the objects that are specified at the onset of the specification stage in the lifecycle process model, (2) evaluate the functionality and performance(s) of those objects through testing and modeling, (3) predict the adjustments that need to be made to the objects associated with the next event (e.g., testing, simulation, and modeling), (4) broaden the scope of what is learned to all threads of objects that are planned for development, and (5) determine the limit for applying what is learned. Planning for integration is the distributing of tasks that reflect what is known to be necessary but includes the ability to adapt to what is sufficient. As tasks are assigned (according to a model of integration), the task manager is responsible for ensuring that subtasks are allocated and completed successfully. The task manager must collaborate with other task mangers so that the lessons learned from the patterns of integration can be shared. Each task and subtask should have a deliverable, and each task should have an entry in a log that is maintained current on a daily basis. The accuracy of planning will be improved by having a reference log with adequate details to chronicle the work activities and events. The significance of this log is understood by the fundamental notion of planning: planning is only about the future. The job of the task manager (and the supervision of the management, in general) is to focus on establishing the procedures for readying all objects for integration based on the strategy of reducing a multiobject problem to a two- or three-object problem. This strategy is referred to as the recursive strategy (Zhang and Norman 1994).

Planning deals with the future of present decision. According to Steiner (1969), planning is a process which begins with objectives; defines strategies, policies, and detailed plans to achieve them; establishes an organization to implement decisions; and includes a review of performance and feedback to introduce a new planning cycle. According to Drucker (1959), planning is a continuous process of making present risk-taking decisions systematically. It involves the best information about the future, it organizes the efforts needed to carry out decisions, and it provides for measuring the results of the decisions against the expectations. Planning includes who, what, where, when, why, and how to be done in the future. For integration planning (as for all planning), planning is not forecasting.* Planning is the charting of the course to find the most probable future events. Planning is not what you would like to do, want to do, or in fact, need to do. Planning is doing what you will do to make happen what you must do. Integration planning is the process of establishing realistic objectives and the strategy to bring objects together in a systematic, logical fashion to show functionality that demonstrates (1) subfunctions within a thread of a system function, (2) multiple subfunctions that interact with each other, and (3) end-to-end (system) threads that illustrate the integrative effects on system behaviors. In the short term (before verifying functionalities), planning focuses on anticipating the unforeseen. The more plan Bs (i.e., backup plans), the better in the early days before object-to-object integration. This notion that planning can help deal with problems is philosophically based, borne out in anecdotages, and now steeped in the honored position of common practice. For integration planning, it is more than meets our perceptions. In a systems engineering environment replete with changes in requirements, missing requirements, and requirements that are left unsaid (and perhaps left undone), integration planning must deal with more than just the complexities of the project (which includes the people and the results of their work), and integration planning is the single-most visible element of how well the project is progressing. All the indications of well-being that occur offer false hopes and deceptive elusions of milestone completions. It is not until integration that the true nature of the work is exposed and put to the first real tests or viability. The planning of integration must endeavor to find the most probable course of future events that will result in achieving the project objectives.

---

* Forecasting attempts to project the probable future events, often based on modeling or simulation (e.g., weather forecasting).

The problem that must be solved with integration planning is to innovate and find the unique events that will change the probabilities of success. Planning helps expose those events so that the opportunities can be capitalized into resources and activities to endure the hardships and anguishes that arise from the inherent nature of building something that is new and as yet unexplored. The focus of planning is on what should be done now to make desirable things happen given the uncertainties that will show up in the future. "What futurity do we have to factor into our present thinking and doing, what time spans do we have to consider, and how do we converge them to a simultaneous decision in the present?" (Drucker 1964).

Integration planning is not a schematic for the future; not a set of functional plans; not a rigid series of tasks that must be attempted because they were stated in a planning document; and not a complete set of all that is known about what is to be integrated. But integration planning is an effective means of scoping what is to be accomplished given uncertainties (uncertainties that will be resolved over a series of events or alternatively a period of time). The important point about integration planning is that it can be accomplished either in event-space or in the time domain (i.e., temporal space). Event-space captures the events that need to occur to accomplish a set of tasks. For integration planning, event-space is most useful as the combination of objects that provide various functionalities is for the most part known after the preliminary design is completed. More often than not, the integration plans are time-based and laid out as if the objects will be completed at a particular time. If one of two objects is completed and the other is not, then the event-based planning has plan B (contingency) tasks to combine the work forces of the completed object task with that of the uncompleted object task to work together to demonstrate their object–object functionalities. This may not be the case for time-domain planning, which may (1) try to integrate the completed task with another completed task (based on the two assumptions: any progress is making progress and the team completing their object can then go on to work another object (and logically, sometimes to help with the object that was supposed to have been completed), and (2) may embark on tasks that are in fact counterproductive to the overall system integration effort. It is particularly distressing to see integration activities that focus on getting any objects to "work" with any other objects through the use of simulated interfaces. This practice requires the construction of simulations based on the "completed" object and presumes that the completed object has achieved its (near) final composition and outputs. This practice helps to reinforce the iterative aspects of existing integration thinking. Build it close to what is expected, then modify it to work with another component, then modify it again to work with yet another component, and so forth, each time moving the components toward an anticipated "integrated" whole. However, this chase to integrate is often met with a seemingly unending set of changes that doom a project to cancellation, as the key stakeholders lose confidence in the ability of the integration effort to result in a product.

Integration planning sometimes assumes a model that covers the tasks (and their subtasks or activities) that are planned to carry out integration. Terms such as big-bang integration (where all components are brought together at once), bottom-up integration that focuses on a functional approach, modularity integration (Burkatzky 2007), top-down integration, agile integration, pipeline integration (Lewis 2006), are prevalent approaches to integration, and the list goes on (Burkatzky 2007). The sheer number of integration models is suggestive that no one model has been widely accepted. An integration model (similar to systems engineering process models that describe the steps and the milestones that must be met to move from one step to the next) defines the nature of integration and the mechanisms for integration for each pair-wise integration of objects to demonstrate the requisite functionalities. In essence, the integration planning with the aid of an integration model defines the components and interfaces of each function, the tests that need to be performed to demonstrate the requisite functionalities, and the level(s) of acceptability for the performances (and losses) of those functions.

Events

Planning for integration is foundational for knowing what tasks need to be done, who is responsible for the tasks, and what activities need to be accomplished to complete the tasks. Planning focuses on the events—those occurrences that result in progress toward satisfying the set of requirements for the product or service and that have measurable outcomes. Scheduling is the association of temporal knowledge with an event.

Planning and scheduling present different views of events. Planning shows sequencing, prescience, and concurrencies, that is, causal relation(s) between events. Scheduling reveals the duration of the tasks and activities leading up an event, the duration of an event, and the uncertainties in these durations.

The concept of "project" includes stakeholders, tasks, and processes. The result of a task is an event. For example, the project may be to hire people (Appendix 1: Outline 1.2.8.6.1). The stakeholders include the hiring staff, the applicants, the organizations who manage the advertising venues, the workers with whom the new hire will be working, the people who are in contact with the new hire (e.g., customers and people who associate the new hire with the hiring organization), and the family members of the new hire. The task for the hiring organization is to fill positions; the task for the applicants is to be hired. The processes of "hire" include "advertise," "review application," "interview," "check references," and associated activities (from the perspective of the hiring organization). The processes "to be hired" for the applicant include "learn of company intent to hire," "fill out application," "send application," "go to interview," and associated activities. The two tasks will result in the same event for the successful applicant and the hiring company: countersigned employment contract. Each of the processes involve stakeholders who receive inputs of EMMI, and transform those inputs into various outputs by following procedures, common practices, and various associated activities. Similarly, the function of 'to hire' involves the interaction between the applicant and the hiring individual. For example, during the interview, both the applicant and the hiring individual exchange EMMI (e.g., acoustic waves, termed talking). The interaction between the applicant (object A) and the hiring individual (object B) fulfills the function of 'to interview.' The function of 'to interview' is a joint function involving both the participants, the outcome of which is the event of INTERVIEW.* One of the processes that results from this interaction is 'to send' the offer letter. The event of SEND LETTER is the result of the function of 'to send' which summarizes many actions that result in placing the letter in a postal drop box. The interaction between the mailer and the postal drop box is the function of 'to send.' The outcome of each of the interactions that has occurred between the applicant and other objects (e.g., application and the computer; applicant and the transportation; applicant and the hiring manager; applicant and the reading of the company's offer letter) are all events. Each of these events occurred in some order at certain times, lasting for certain durations. In this manner, an event is exemplified by the properties or traits of the interacting objects at a certain time, and at a certain place (Kim 1966; Goldman 1970; Borghini and Varzi 2006). The precise location of an event can be considered to be the mereological sum of the locations of the objects (Borghini and Varzi 2006). Events are characterized as having causal effects on other events.

An event occurs because an object transforms EMMI into an output. Specifically, an event is the result of an interaction (driven by the mechanisms of the two objects). An event is any detectable output from an object. Events are antecedent or subsequent. The history and future of the project are summarized by events. The relation(s) between events as objects with antecedent, subsequent, or concurrent juxtapositions simultaneously in event-space and the temporal domain define the plan and schedule for the project. Therefore, events always have an orderliness that is discernable as a pattern of behavior. Events that do not occur at the time indicated by the plan are either "early" or "late," as measured by the temporal reference to the planned starting time of the event. Events that take shorter or longer than indicated by the plan are either "ahead of schedule" or "behind schedule," as measured by the temporal reference to the planned length of time. Events that are planned in a certain sequence are either "in order" or "out of order." Events that are planned as concurrent are either "in parallel" or "out of phase." Figure 6.6 illustrates planning and scheduling for objects as related by EMMI exchanges through subfunctions.

---

* Capital letters signify EVENTS.

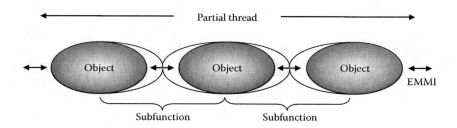

FIGURE 6.6
Objects exchanging EMMI through subfunctions.

Integration Planning and Scheduling Steps

The purpose of planning and scheduling for integration is to lay out the work tasks so that the appropriate and sufficient resources, skills, and facilities are brought together to complete the assigned work in pattern scheduling (lowest impact on the budget), network scheduling (determinable, but with a higher variance than pattern scheduling, impact on the budget), and ad hoc (undeterminable impact on the budget). The principle of planning (Principle 6) indicates that the most effective means of integrating objects requires an effective means of planning and scheduling. To achieve full pattern schedule with high certainty of the pattern of object and events, the starting times, and the durations requires knowing all the tasks, all the objects, and all the events that must occur to complete the project. For routine projects that have the benefit of historical precedence and relevant project team experience, the task of integration planning and scheduling is tractable (although it should never be considered routine). An integration plan is always needed to consider the possible scenarios that reflect project-specific details, such as the environment, suppliers, and time and budget constraints. When the project is confronted with a very complex system or system of systems, integration planning must be founded on scenarios of hope, false hope, and despair (usually three to four scenarios is sufficient to capture the degrees of uncertainty).

How well the project proceeds and is evaluated on the basis of the plan depend on knowing these factors and the confounding factors during the planning and scheduling stage. The integration plan is often limited by cost or schedule considerations. Both these limitations result in constraints that are also dependent on the skill set of the developers, integrators, and testers, the availability of resources and facilities, the plans, the expectations, and the technologies. It suffices to indicate that all that one needs and should be known is probably not known for any such planning and scheduling. Consequently, it is advisable to develop several plans and schedules based on "what if" scenarios. Scenarios describe events or activities (enactments of functions or processes) that synopsize a sequence of intentions or events to assist in accounting for a plausible future (Peterson et al. 2003). Scenarios can be used to explore the uncertainties of integration based on patterns of behavior (order and schedules of objects). The Bayesian interpretation of scenario development leads to the defining of object order, start time, and duration as probabilistic measures of the stakeholder beliefs that the pattern will occur, given all that is relevant, known, and included by the stakeholders. In this interpretation, the probability of an event occurring in the sequence specified, at the starting time indicated, and having duration stipulated is dependent on both the event and the stakeholder's knowledge of the event (Poh 1993). By this means, various scenarios can be scripted and evaluated, using the variables of object pattern, start time, and duration to represent the stakeholder knowledge.

There are four techniques for determining the pattern and behavior: expert opinion (including engineers, financial analysts, systems engineers, and managers); estimates of the work duration by similar types of tasks (e.g., by index lines of code in the case of software, standard part or off-the-shelf part delivery or manufacturing times); estimates of work by functions (e.g., using function points for software estimation or applying functional analysis to compare like functions with previous projects); and algorithmic comparisons with like-kind work packages or work sequences. While today these techniques are heralded as the best, they do not separate planning and scheduling using reverse view perspectives. Scheduling moves temporally into the future, whereas planning moves from the most future event (delivery of the product or service to the events of present day—the reverse view perspective). A promising approach is the use of augmented Lagrangian optimization (Guignard 1995; Li and Ierapetritou 2010). The last 20 years have shown promising results and this relatively unexplored approach deserves attention.

Integration Plan

An integration plan should include the strategy for bringing objects together to demonstrate functionality. The integration plan is based on the system design and the system architecture. Commensurate with the strategy and sequencing for joining objects is the preparation of integration resources and make-up of the integration teams. The integration plan also covers approach, testing, and verification of the integrated subsystems, as well as laying out the validation for the integrated system. The integration plan is the management guide for integration, followed by both the systems engineers and engineers. Integration is so important with large, complex, and expensive systems that in 2005, the U.S. Government Accountability Office that reports to the Subcommittee on Oversight of Government Management, The Federal Workforce, and the District of Columbia, Committee on Homeland Security and Governmental Affairs, United States Senate, recommended that the Senior Executive Service within the Department of Homeland Security have their strategy and execution of project integration be made part of their annual performance appraisals. The Under Secretary for Management in the Department of Homeland Security stated that the integration plan will be tied to such performance for the fiscal year 2010 performance cycle (GAO 2009). Integration is a top priority for acquiring systems.

Systems Integration Model

As the process for integration is summarized briefly as bringing the right objects together at the right time, there is a strong basis for developing a model for integration that is represented in an integration plan: Integration plans are often (inadvertently) said to characterize integration as an iterative process. Integration is anything but iterative. And sometimes it is stated that a complex project *may* (added for emphasis) need a written plan for integration. Systems integration (i.e., integration that will lead to a system) requires an integration plan. Integration planning would then focus on the objects that when integrated resulted in some bit of functionality that is integral to the final system. Further, it is often indicated that planning includes the sequence in which components are to be integrated. It is desirable to demonstrate links of functionality (i.e., small sequential subfunctions) that when brought together in an end-to-end fashion, reveal a system function. In carrying out this notion of links of functionalities, integration has a sequence where the sequence is focused on the end-to-end result, not necessarily any particular link comprised of subfunctions. There is an important distinction between providing links of functionality and attempting to demonstrate a particular set of subfunction in a particular order. If the system design is modularized by functions, then should a subfunction fail to be demonstrable, a "substitute" replicate function could be provided as a backup (referred to as plan B). Providing requisite functionality in such a modular form would most likely result in lesser performance than indicated at the onset of the work, but the system could be delivered with proven functionality, with an upgraded capability provided at a later time. The progression of demonstrating functionality is tracked typically by schedule and amount expended to achieve that functionality. Each demonstration of functionality is a verification of the specifications, the system design, the architecture, and the requirements of the stakeholders. Specifics of the objects that when combined result in the observed functionality is verification of the system architecture.

During the aftermath of integrating components, lessons learned are often discussed. For integration planning, the integration plan is more often than not a composite of sections from the systems engineering management plan, the development plan, the systems engineering plan, and the testing plan (other plans may also be included). The integration plan usually includes the development aspects (schedule constraints, resources, facilities, labor); the results of the development aspects (system design, preliminary design, detailed design, architecture, descriptions of physical objects, interfaces between physical objects, descriptions of product or service functionalities), and testing plans (early-to-late stages). The iterative nature of systems engineering transforms a set of initial requirements into design, architecture, and then development. Throughout these steps, the requirements continue to be refined, the design becomes more detailed, and the architecture matures. By the time the development effort is readying for integration, the system functionalities, the functional decomposition, and the allocation to physical objects should be substantially completed. By the end of the detailed design phase, the integration plan should be thought of as the baseline from which refinements will be added during development.

Integration strategies come under various names from do it all at once "big bang" or dividing the integration process into stages (Tahan and Ben-Asher 2005 citing Sommerville 2001) for incremental integration; or do it when you can, or do it top down and bottoms up, or do it according to some other rationale. Perhaps the reason why there are so many choices is that no one strategy seems to have proven very effective given the factors that confound integration (e.g., unknown impacts of boundaries) (i.e., complexity). The notion of iterative integration often means that one or both the components undergoing integration will need to be changed. A failed integration activity means just that, one or both components need to be changed. Whether the changes are to be localized or are pervasive throughout an object, the iterative notion of integration means failure. This is in sharp contrast to systems engineering that relies heavily for its success in dealing with unknowns through iteration.

Building an integration plan requires a strategy and a model for integration. The strategy is at once the path we undertake to develop and build pairs of objects that will provide the requisite functionalities. When these objects are linked through interfaces with proper inputs and outputs of EMMI, the resultant functions form the end-to-end chains (or threads) that demonstrate the system-level functions. A systems integration strategy usually involves planning for codevelopment of objects, orchestrated first at the component level, and then at the subsystem level. The strategy aims to match the needs of user's requirement for functionality with the priorities the user ascribes based on the minimally acceptable set of functions that demonstrate the basic elements of the product or service. To be clear, this is not the "wish" list of the user, nor is it what the user needs. The first objects to be integrated are only those that are necessary and sufficient to show end-to-end viability of a major system-level function. Integration activities focus on this one thread, not all threads in parallel (Figure 6.7).

Once the main system thread is completed and demonstrated, parallel threads, interacting threads, and subsidiary threads that add additional functionality are then worked on in a similar manner. Again, the integration

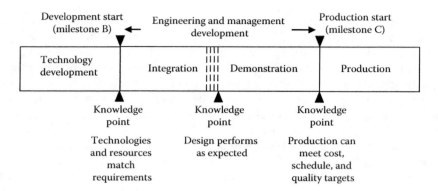

FIGURE 6.7
United States Department of Defense acquisition cycle and decision knowledge points.

task is to link end-to-end functionalities, but this time through the first demonstrated system thread (first demonstrated). Should there be completely parallel threads without any interactions, the system design and architecture is that of a system of systems and a different integration plan is necessary.

During the systems integration process, beginning with the most important system-level thread, the critical objects that must provide linkages between the subsystem functions are built and tested. Should an object that is required to demonstrate a critical function not be available, the adjacent objects that contribute to either of the two adjacent functions should be demonstrated. For every object there are always two adjacent objects from which to integrate and demonstrate functionality.

A systems integration model is a means of structuring the integration work to better measure the progress of integration. It is quite likely that the integration model that works for one set of objects and constraints will need to be tailored for a different set of objects and constraints. It may be that one model does not deal with all sensitivities that characterize a systems integration effort. This is not unlike the differences in systems engineering process models. The systems engineering process models strive to first elicit a complete, deliverable set of requirements that can be accomplished within the constraints of time, budget, and skills; second, provide the structure to manage development work so that it conforms and is totally responsive to this "legitimate" set of requirements; and third, deliver a product or service that respects the integrity of the budget, schedule, and needed performances. Within the management paradigm the work is presumed to be both tractable and reliably similar in nature throughout the development work. However, systems engineering is predominantly iterative for the bulk of its use, while systems integration is predominantly recursive. Yet, the dominant view of integration today retains its iterative actions—that of fix what needs to be fixed, then work with the next object to make both of them work (fixing what needs to be fixed), and so on through the components, subassemblies, assemblies, and subsystems. The recurring pattern is one of looking to the objects at hand to assure they work together, knowing that if they do not work together, then nothing will work with them later. In a recursive environment, the dominant theme would be to determine a pattern that looks to the future objects and how those objects will work together given the lessons learned from the existing objects. Specifically, the functional analysis in the systems engineering process model needs to be redone to accommodate the lessons learned from the integration of the first few objects. For example, if two objects are integrated successfully, they represent a function or set of functions that were called out in the functional analysis step performed as part of the systems engineering process model. Were it to be the case that all such functions were found to result from the integration work, there would be no problem or concern. But that is rarely, if ever, the case for systems of a complex nature. Instead, the functions that were agreed to during the functional analysis step were a "best match" with the physical entity, but not an entirely best match. The iterative nature of systems engineering may not have surfaced all of the functions and made the appropriate physical allocations. The means of finding these problems is usually left to modeling, simulation, and testing (first at the unit level, and then at the component level). The iterative nature of systems engineering results in an integrated system some percentage of the time. The recursive nature of systems integration should result in an integrated system a higher percentage of the time based on (1) having refined the set of functions from the functional analysis stage during the integration work (thereby better preparing and configuring the object for integration); (2) extrapolating the patterns discernable in the early systems integration work and apply those lessons learned to subsequent objects (thereby applying the lessons learned to subsequent integration activities); and (3) focusing the integration effort on the end-to-end system functionalities that are demonstrable by the concatenation of subfunctions. The extensive use for recursive systems integration is for complex systems and system of systems integration. The failures of iterative systems integration are revealed most notably with the most complex of systems or system of systems. The systems integration process model fits into the systems engineering process models in the same fashion as integration is currently conceived and enacted. For example, the number of U.S. Department of Defense programs that are major and complex (GAO 2011) total 98 with a total planned investment of $1.68 trillion.* One of those programs, the U.S. Army Future Combat System, was terminated. As with many efforts, the major difficulties seem to be recorded during integration, but in this case, the difficulties were determinable in the system architecture, the integration of new systems with legacy systems, and the degree of interoperability that

---

* The GAO reports $174 billion for 13 programs were removed from the portfolio of major defense acquisitions while 15 programs (est. $77 billion) were added (GAO 2011).

was required. This program is a prime example of the failure of iterative integration. The U.S. Army indicated that the Future Combat Systems was the "greatest technology and integration challenge it has ever undertaken" (GAO 2007). The management challenge was stated as "... reducing integration risk and demonstrating product design prior to the design readiness review ...." One of the particularly difficult issues for integration was the inclusion of "... adapting the [new] technologies to space, weight, and power demands of their intended environment" (GAO 2007). The term "adapting" is not an integration task, it is a development task. The consequence of including development activities in integration work simply compounds an already difficult job. The Future Combat System program was canceled in 2009.

Patterns in Systems Engineering and Patterns in Systems Integration

A clear demarcation should be drawn between development (systems engineering) and integration work (systems integration). The clear distinction is determinable through the patterns of activity and how those patterns are used. A pattern conveys the key aspects of design (Hennig and Cloutier 2011). Patterns can be used to facilitate the reuse of proven design knowledge (Gordijn 2002). A pattern is a test to determine if an object meets certain criteria, with the results of a test having meaning within the context of the test and of the object; as such, the existence of a pattern indicates either potential behavior (an increase in certainty) or to discounting previously hypothesized behavior (which is an increase in uncertainty) (Hollywood et al. 2004). Patterns are not just there, but rather can be created by (1) establishing a perspective from which to observe, (2) setting a set of contexts or circumstances that prescribe the boundaries and boundary conditions, (3) stipulating a set of independent and dependent variables that can be traced through causal factors, (4) specifying the constraints under which the event is to occur, and (5) advising as to how the measurement(s) should be accomplished. These five factors are called the "duties" of a pattern. Patterns are held in high regard as something that portends future event(s) or means something else that is either relative or deterministic. Patterns may have meaning that only applies to that which is at hand as these are restricted by their duties. It is the nature of establishing the duties, as it is the nature of looking for patterns that a single important factor in recognizing patterns is the experiences of the observers—the greater the number of experiences, the more accurate and intuitive the detection and interpretation of patterns. The intuitive quality arises when patterns are improved by testing them against our experiences (Alexander 1979). When engineers (and systems engineers) review tests and test results for patterns during development (as opposed to integration), the intent is to pass the test so the focus is on "fixing the problem" with the object. The expectation during systems engineering development is to uncover, face, and fix problems. Consequently, the thinking during systems engineering development is to loop back and fix problems. Should an object pass its tests during development the task is to then either move on to another object (repeating what was done on yet another object) or carry forward with the same object and bring another object together with it for another test. This bringing together of objects is (by definition) integration. But bringing together objects that result in a failed test again focuses attention on the problem that must be solved to pass the test. Iteratively, the test provides insight into the problem. Systems engineering is focused on iterative activities to pass tests.

The iterative nature of systems engineering is built into the structure of analyzing the results of the testing of an object. The purpose of analyzing patterns to perform retrospective analysis is focused on improving the chances of the object to successfully complete its prescribed tests. The view that passing the tests is needed to give confidence to the design and implementation of the object is predicted on the belief that tests accurately demonstrate some measure of meaningful progress toward project goals. The difficulty arises when those beliefs translate into some measure that is quantifiable in terms of schedule or rate of expenditure (i.e., earned value). If only one object is tested, then no function is enabled (as it takes two objects to comprise a function).

For acquisition, decision makers rely on events to help determine the status of a project (or program). The acquisition cycle and knowledge points (specifically for the U.S. Department of Defense (GAO 2011)) to aid with decisions to continue with development work are indicated as technology development, integration, demonstration, and production. Figure 6.8 illustrates these knowledge points in terms of a sequence of phases.

These are top-level categories that subsume a myriad of decision points. Various versions of acquisition cycles and decision points are routine for government and industry, some formalized while others are ad hoc.* Regardless of the manner or formality, management review of development work for new products and services is a key aspect of most projects. In the case of the U.S. Department of Defense, technology development means achieving a sufficient level of technology maturity by the start of system development, coupled with the project's resources matching its expected needs. These two factors have been shown by the U.S. Government Accountability Office to be key determinants of a project's success. The first decision point involves work under the auspices of the system engineering

---

* The systems engineering process models have many decision points, separated both by stage of development and by milestones that signify that a step is completed and the next step can begin.

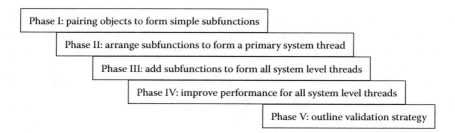

FIGURE 6.8
Technology development, integration, demonstration, and production sequence.

process model to carry the project to the start of development (sometimes referred to as milestone B). The second critical issue for decision makers is immediately prior to the stage of work called integration (which is often termed as development). Integration occurs when the system design is determined to satisfy the customer requirements.

According to the U.S. Government Accountability Office, the first knowledge point (knowledge point 1) encompasses the start of milestone B which occurs at the culmination of technology development and at the start of integration. This is the time at which engineering development phase has begun (sometimes referred to as milestone B). At this time, the requirements and resources should be matched, the requisite technologies should be demonstrated in their intended environments, and the preliminary design should be completed. The systems engineering process models bring the system design to this first major knowledge point with completion of the systems design, architecture, concept of operations, and an integration plan (to mention a few of the activities and a sampling of the plethora of documents and reports).

The second knowledge point (knowledge point 2) is determined when the system design of the product or service satisfies the customer requirements. Knowledge point 2 occurs at the critical design review between integration and demonstration. The design is proven stable through prototyping and no substantial changes are *permitted*. The manufacturing drawings are releasable to manufacturing. Requirements are met. Reliability rates are demonstrated. This knowledge point assumes that the system design and architecture are sufficiently stable to begin demonstrating various key subsystems of the system. When the manufacturing processes are mature, the product or service is shown to meet cost, schedule, performance, and quality targets in a manufacturing environment (knowledge point 3).

There is no one point of particular interest for physical integration in this description as it is masked in vague descriptions between knowledge point 1 and knowledge point 3. The systems engineering process models (not covered in this book) reveal more, but still present an inadequate case for what integration is and how to accomplish it. The grossness of the descriptions of these knowledge points and the details of what and when is expected as described in the systems engineering process models serve to reinforce the result in the inevitable leanings toward iterative thinking significantly past when recursive. For the most part, the systems engineering process models associate design verification with development and system verification with integration. There is much overlap in how the various stages are defined in the numerous systems engineering process models, so a definitive statement about when certain activities should occur is both fruitless and perhaps misleading. The consequence is that many decisions are left to the users of the systems engineering process models which allows for the necessary tailoring to be accomplished based on the project constraints, management preferences, and customer requests. The U.S. Government Accountability Office and the systems engineering process models reflect that integration is a work activity that fits within a domain called development: as integration is related to bringing together objects as part of the process of building a product or service. The integration framework (Chapter 2) captures that thinking using process management and product and service deliverables.

The first phase of the Systems Process Model for Integration (SysPMI) is the identification of objects (by pairs, where pairing of a single object with another object always involves two other objects at a minimum) that are required to demonstrate the top-level function(s), the subfunctions that support the top-level (end-to-end) functions, and the associated functional flow block diagrams (at the pertinent level). The identification of the objects and the match with functions is referred to as the mapping of functions to physical entities (or physical entities to functions). Additionally, the behaviors of the users are associated with the functions and the physical objects. From the description of the physical entities, the functions, and the user behaviors, the boundaries and the boundary conditions can be included in the descriptions of the system functions.

The second phase of the SysPMI is the recognition and characterization of the top-level system functions in the simplest form and structure of subfunctions that is necessary to demonstrate the end-to-end system-level thread. There can be multiple end-to-end system-level threads depending on the input EMMI and the current actions of the system. The simplest case, the simplest dataset, the simplest implementation, the simplest of physical structure, computer hardware, computer software, and infrastructure and support are the focus for demonstrating an end-to-end functionality. The few associated performance(s) are to be measured and set down as the benchmark performance(s) that demonstrate feasibility. Various types of models may be useful to establish a model of what an end-to-end system function might entail. The SysPMI accommodates both the modeling aspects and the benchmark execution as part of Phase II. Phase II spans the test planning, test execution, and benchmarking of the single-most important system-level functional thread. The culmination of Phase I is the demonstration of the subfunction levels that concatenate to the highest-priority system-level function (in its simplest form).

The third phase of the SysPMI covers the expansion of integration to the prioritized top-level system functions and the critical independencies. Phase III spans the test planning, test execution, and benchmarking of the prioritized system-level functional threads. The culmination of the third phase of the SysPMI is the demonstration of system-level functions in their simplest forms based on their set(s) of subfunctions. To complete the third phase, some object demonstrated in Phase I will need to have additional subfunctions built into their physical entities. As additional subfunctions are completed they will be tested to demonstrate additional functionality as well as to show improvement(s) in the end-to-end system-level functions. Should a system-level function be degraded from its benchmark performance by the addition of a new functionality, the objects that were modified will be analyzed to improve their handling of EMMI through adjustments of their mechanisms.

The fourth phase of the SysPMI is the overall improvement of performance(s) for all system-level threads by improving subfunctions. Mechanisms are tuned, enhanced, improved, or replaced with improved components or objects. Should new technologies or modifications to existing objects be necessary, they are performed in the fourth phase (after the system has been shown to provide the system-level functions and the optimized level of performance(s) associated with those performances). The culmination of the fourth phase of SysPMI is system-level testing.

The fifth phase of the SysPMI is preparation of the final version of the validation strategy, upgrade strategy, and maintenance strategy. Furthermore, the validation strategy is carried out in the fifth phase. Note that verification occurs during all phases of the systems engineering work, from conceptualization through development. Verification of the system design, architecture, concept of operations, and specifications must be completed before integration of physical objects to assure that only those objects and functions that satisfy the specifications and meet the requirements end up having time spent on them by the integration teams.

Three Tests for Iterative Thinking versus Recursive Thinking

As successive phases of the SysPMI are implemented and progress toward their objectives, the type of work transitions from iterative-centric thinking to recursive-centric thinking. To discern how much thinking needs to be involved with work in a particular stage, there are three questions that should be answered. The first question deals with the scalability of the objects to deal with inputs and outputs (including losses to achieve those outputs) through their mechanisms in the same fashion that is determinable from a sampling of data points (taken by modeling, simulation, or testing). The relation between successive domains of data, taken in various regions within the domain of validity for the variables on which the viability of the object is predicted, is termed as the scalability. Scalability is not simply doing more as the domain of interest increases. Scalability is not simply maintaining the same (or similar) relation between variables as the quantification of these variables increase or decrease. And, scalability is not simply maintaining some semblance of fidelity as the variables change in either a decreasing or an increasing manner. Scalability is the object's adaptability as the increases or decreases in the transformation of EMMI by its mechanism and control in a manner that retains the same ratio of input to output regardless of the amount of increase or decrease in EMMI. Scalability is the first question that should be asked. The second question is to determine which part of an end-to-end sequence (i.e., system-level function) is the object a part of. Since two objects comprise a subfunction, two objects (at a minimum) will be working together to form a subfunction. Testing both functions over the domains of a few variables within their specification limits does not expose the pertinent patterns that necessarily reveal inconsistencies in scalability. A convenient means of discovering patterns of inconsistency is to construct an end-to-end threat. Why just one? Because one key thread of system-level functionality establishes a baseline of operations that can be improved through better performances, added to additional subfunctions that build toward additional system-level functionalities, and carry with them the patterns that expose the constraints that impose limitations on scalability. The third question that should be asked deals with the architecture that supports all of the system-level threads. As designed, the system architecture provides various system resources to enact system functionalities. An equally convenient way of thinking about the contributions of architecture is to envision the allocations of resources or the partitioning of resources. While equivalent at the top level of thinking about the patterns to watch for, the skills of the systems engineer vary considerably. Each systems engineering speciality will view the patterns differently, depending on the observer's perspective, their skill set, and their experience. Asking the three questions sensitizes the engineering staff to the issues of integration and not just passing the tests laid out for objects. Merely focusing on the iterations to improve an object may have no impact on the issues of an object's integrability with other objects. One cannot presume that because an object passes one of its tests, it can be integrated with another object. In fact, the discernable patterns might show that scalability, the baseline for a basic thread of end-to-end system functionality, and a piece-wise contiguous means of bringing objects together is far more efficient than any other means of integration. But the ultimate test of thinking that must be done to complete systems engineering and to complete systems integration is clearly different—as distinguished by iterative versus recursive methods. If the answers to the three questions are a resounding "no time to deal with future implications of what is being done today to fix today's problems," the thinking is iterative—priorities placed on changing whatever needs to be changed to pass a test. However, if the answers to the three questions (scalability, one key thread, and multiple key threads) are suggestive of patterns that expose weakness in the system design, architecture, or implementation of specifications, then the objects may be difficult to integrate.

Besides the differences in activities between iterative thinking and recursive thinking there is a difference in the type of people that are performing the work. For building the objects, the focus is on the engineering skills and abilities necessary to be successful at building and testing. Iteration is key. Detecting patterns may require different skills and capabilities that are neither normally invoked nor discussed during critical times of "making the object work to pass a test." Granted, these particular events are stressful due to the strong dependencies between the constraints of schedule, budget, and performance. However, it is exactly at these times that the patterns are at their earliest stage of detection. After each failed or successful test, a review with a senior systems engineer (who has domain-specific knowledge and work experience) should be involved in the discussions (however brief). The more senior the systems engineer, the more likely the team is at detecting patterns that could help with integration. The percentage contribution to a project from a systems engineer (based on the number of years of experience) is related exponentially for requirements analysis, project management, and cost management (as examples of activities performed by systems engineers). All three of these activities depend on detecting and interpreting patterns and therefore reflect recursive thinking more than iterative thinking. de Souza (2008) showed that there was an exponential relation between the contribution to the project of a systems engineer with respect to the number of years of experience. For requirements analysis and management skills, a systems engineer with nearly 9 years of experience contributes four times as much as a systems engineer with 4 years of experience. However, cost management skills at 13 years of experience contribute twice as much to the project as that of a systems engineer with nearly 9 years of experience. The level of contribution for requirements analysis and project management skills are seemingly acquired faster than cost management skills, but rising to a high level of proficiency after 14 years of experience. The acquisition of systems engineering skills as gleaned from a triangulation methodology compared data from recruitment advertising, review of systems engineering-juried publications, and surveys of practicing systems engineers indicated that systems engineering is a skill that is learned through project work. Nearly all systems engineers have strong domain expertise in an engineering specialty.

The influence on integration work as derived from the sampling of systems engineering skills indicates the need for a structure of leadership (Honour 2004). Leadership in making the determination as to when thinking changes should be made to encourage a greater measure of recursive thinking to iterative thinking.

The recursive thinking associated with this part of development and development in general is most pronounced toward the end of integration (when objects have passed their individual tests but defy attempts to integrate smoothly). Iterative thinking at the end of development takes a lesser role in the discussion (although never disappears completely). At the beginning of the development and build process, iterative thinking dominates with recursive thinking left for water cooler talk. There is a gradual transition from one type of thinking to another, both important, both representative of the problem solving that must take place, and both determined by the task at hand. For systems integration of some complexity, the issue is to begin recursive thinking earlier in the integration process to prepare more effectively for what lies ahead. The challenge of management is to recognize that recursive thinking may not yet dominate, but one should always be mindful that they are near the near of iterative thinking.

The SysPMI ties into the systems engineering process models as products and services are prepared in their normal fashion and become available for integration planning. A common place for integration to begin is after subsystem verification (referring to the Vee process model (Forsberg and Mooz 1996)). Verification of the system design and implementations at the unit and component levels also occur before major integration activities.

References

Alexander, C. 1979. *A Timeless Way of Building*. New York: Oxford University Press.

Blanchard, B. S. 1998. *Systems Engineering Management*. New York: John Wiley & Sons, Inc.

Borghini, A. and Varzi, A. C. 2006. Event location and vagueness. *Philosophical Studies* **128**(2): 313–336.

Bornman, L. G. 1993. *Command and Control Measures of Effectiveness Handbook (C2MOE Handbook)*. Study and Analysis Center Study Directorate. For Leavenworth, United States Department of Army TRADOC Analysis Command 42.

Burkatzky, F. H.-H. 2007. *Development of Measurement Scales for Project Complexity and Systems Integration Performance*. PhD thesis, School of Management, Walden University, 161pp.

Choi, D. O. and Langford, G. O. 2008. *A General Quality Loss Function Development and Its Application to the Acquisition Phases of the Weapon Systems*. Monterey: United States Naval Postgraduate School, 87pp.

Chung, H. C. and Chao, Y. C. 2005. Determining a one-sided optimum specification limit under the linear quality loss function. *Quality and Quantity* **39**: 109–117.

de Souza, R. A. 2008. *Maturity Curve of Systems Engineering*. MS thesis, Systems Engineering. Monterey: The Naval Postgraduate School, 114pp.

Drucker, P. F. 1959. Long-range planning. *Management Science* **5**(April): 238–239.

Drucker, P. F. 1964. *Managing for Results*. New York: Harper & Row.

Forsberg, K. and Mooz, H. 1996. *Visualizing Project Management*. New York: John Wiley & Sons.

Fowler, J. H., Heaney, M. T., Nickerson, D. W., Padgett, J. F., and Sinclair, B. 2009. *Causality in Political Networks*. Carbondale: Southern Illinois University.

GAO. 2007. *Defense Acquisitions: Assessments of Selected Weapon Programs*. Washington, DC: United States Government Accountability Office, 178pp.

GAO. 2009. *Department of Homeland Security: Actions Taken Toward Management Integration, but a Comprehensive Strategy Is Still Needed*. Washington, DC: United States Government Accountability Office, 52pp.

GAO. 2011. *Defense Acquisitions: Assessments of Selected Weapon Programs*. Washington, DC: United States Government Accountability Office, 195pp.

Goldman, A. 1970. *A Theory of Human Action*. New York: Prentice-Hall.

Gordijn, J. 2002. *Value-Based Requirements Engineering: Exploring Innovative e-Commerce Ideas*. PhD thesis, Dutch Graduate School for Information and Knowledge Systems, 311pp.

Gosling, W. 1962. *The Design of Engineering Systems*. New York: John Wiley & Sons, Inc.

Guignard, M. 1995. Lagrangean relaxation: A short course. *Belgian Journal of Operation Research* **35**(Special Issue Francoro 3–4): 5–21.

Heckman, J. J. and Vytlacil, E. J. 2006. *Econometric Evaluation of Social Programs*. Amsterdam: Elsevier.

Hennig, C. and Kutlukaya, M. 2007. Some thoughts about the design of loss functions. *Statistical Journal* **5**(1): 19–39.

Hennig, J. T. and Cloutier, R. 2011. *System of Systems Architecture Patterns Which Enable Agility*. CSER 2011, Redondo Beach, University of Southern California.

Hitchins, D. K. 2007. *Systems Engineering: A 21st Century Systems Methodology*. West Sussex: John Wiley & Sons, Ltd.

Hollywood, J., Synder, D., McKay, K., and Boon, J. 2004. *Out of the Ordinary: Finding Hidden Threats by Analyzing Unusual Behavior*. Santa Monica: RAND Corporation, 187pp.

Honour, E. C. 2004. Understanding the value of systems engineering. *Proceedings of the 14th International INCOSE Symposium*. Toulouse: France.

Jansen, W. A. 1998. *Inheritance Properties of Role Hierarchies*. Gaithersburg: National Institute of Standards and Technology, 10pp.

Kaindl, H. and Dvetinovic, D. 2008. *Distinction between Requirements and Their Representations*. CSER 2008, Redondo Beach: University of Southern California.

Kaplan, R. S. and Norton, D. P. 1992. The balanced scorecard—Measures that drive performance. *Harvard Business Review* January–February: 75–85.

Kaplan, R. S. and Norton, D. P. 1993. Putting the balanced scorecard to work. *Harvard Business Review* September–October: 134–142.

Kaplan, R. S. and Norton, D. P. 1996. Using the balanced scorecard as a strategic management system. *Harvard Business Review* January–February: 75–85.

Kapur, K. C. and Wang, C. J. 1996. Economic design of specifications region for multiple quality characteristics. *IIE Transactions* **28**: 237–248.

Kerzner, H. 2009. *Project Management: A Systems Approach to Planning, Scheduling, and Controlling*. Hoboken, NJ: John Wiley & Sons.

Kim, J. 1966. On the psycho-physical identity theory. *American Philosophical Quarterly* **3**: 277–285.

Kumpe, T. and Bolwijn, P. T. 1994. Towards the innovative firm challenge for R&D management. *Research Technology Management* January–February: 38–44.

Langford, G. O. 2009. Product upgrades based on minimum expected quality loss. *Proceedings of the 19th Annual International INCOSE Symposium.* Singapore: International Council on Systems Engineering.

Lewis, J. W. 2006. Systems Engineering Model for Integrability (SEMI). *INCOSE 2006—16th Annual International Symposium Proceedings: Shining Light on the Tough Issues.* Toulouse: International Council on Systems Engineering.

Li, Z. and Ierapetritou, M. G. 2010. Production planning and scheduling integration through augmented Lagrangian optimization. *Computers and Chemical Engineering* **34**: 996–1006.

Machamer, P., Darden, L., and Carver, C. F. 2000. Thinking about mechanisms. *Philosophy of Science* **67**(1): 1–25.

MacKenzie, R. A. (1969, 1988). Software engineering project management. *IEEE Computer Science Press* (first appeared in *Harvard Business Review* November/December 1969:11–14).

Mason, R. L., Gunst, R. F., and Hess, J. L., Eds. 2003. *Statistical Design and Analysis of Experiments.* New York: Wiley-Interscience.

Park, D., Saavedra, R. H., and Moon, S., 1997. Adaptive granularity: Transparent integration of fine- and course-grain communication. *International Journal of Parallel Programming* **25**(5): 419–446.

Peterson, G. D., Cumming, G., and Carpenter, S. R. 2003. Scenario planning: A tool for conservation in an uncertain world. *Conservation Biology* **17**(2): 358–366.

Pikula, M. and Siemion, A. 2007. Design patterns in application integration based on messages. *Software Engineering.* Warsaw: Polish-Japanese Institute of Information Technology. Master, 101pp.

Poh, K. L. 1993. *Utility-Based Categorization Engineering-Economic Systems.* PhD thesis, Stanford: Stanford University, 238pp.

Reeves, C. A. and Bednar, D. A. 1994. Defining quality: Alternatives and implications. *Academy of Management Review* **19**(3): 419.

Sage, A. P. and Armstrong, J. E. 2000. *Introduction to Systems Engineering.* New York: John Wiley & Sons, Inc.

Shannon, C. 1948. A mathematical theory of communications. *The Bell System Technical Journal* **27**: 379–423.

Smartt, C. and Ferreira, S. 2011. *The Archetypes of Systems Engineering Strategy.* CSER 2011, Los Angeles: University of Southern California.

Sommerville, I. 2001. *Software Engineering.* New York: McGraw-Hill.

Sopha, B. M., Fet, A. M., Keitsch, M. M., and Haskins, C. 2010. Using systems engineering to create a framework for evaluating industrial symbiosis options. *Systems Engineering* **13**(2): 149–160.

Steinbit, J. P. 2002. *NATO: Code of Best Practice for C^2 Assessment.* Washington, DC: United States Department of Defense.

Steiner, G. A. 1969. *Top Management Planning.* Toronto: Collier-Macmillan Canada, Ltd.

Stevens, W. P., Myers, G. J., and Constantine, L. L. 1974. Structured design. *IBM Systems Journal* **13**: 115–139.

Taguchi, G. 1986. *Introduction to Quality Engineering: Designing Quality into Products and Processes.* Tokyo, Japan: Asian Productivity Organization.

Taguchi, G. 1990. *Introduction to Quality Engineering.* Tokyo, Japan: Asian Productivity Organization.

Taguchi, G., Chowdhury, S., and Wu, Y. 2005. *Taguchi's Quality Engineering Handbook.* Hoboken, NJ: John Wiley & Sons, Inc.

Taguchi, G., Elsayed, E. A., and Hsiang, T. C. 1989. *Quality Engineering in Production Systems.* New York: McGraw-Hill.

Tahan, M. and Ben-Asher, J. Z. 2005. Modeling and analysis of integration processes for engineering systems. *Systems Engineering* 8(1): 62–77.

Yahya, F. and Chanwut, P. 2007. A quartic quality loss function and its properties. *Journal of Industrial and System Engineering* 1(1): 8–22.

Yao, L., Kiran, K., Janet, K. A., and Farrokh, M. 1999. Robust design: Goal formulations and a comparison of metamodeling methods. *ASME Design Engineering Technical Conferences.* Las Vegas, NV.

Zhang, J. and Norman, D. A. 1994. Representations in distributed cognitive tasks. *Cognitive Science* **18**: 87–122.

Appendix 1: "To Manage" Decomposition

1.0 Enterprise
1.1 External
1.2 Internal
   1.2.1 Finance
   1.2.2 Operate
   1.2.3 Sell
   1.2.4 Legal
   1.2.5 Build
   1.2.6 Support
   1.2.7 Lead
   1.2.8 Manage
      1.2.8.1 Plan
         1.2.8.1.1 Determine the problem or "opportunity"
            1.2.8.1.1.1 Research current conditions
            1.2.8.1.1.2 Conditions of own organization
            1.2.8.1.1.3 Conditions of competitors
            1.2.8.1.1.4 Conditions of market
         1.2.8.1.2 Identify all problems or potential opportunities
            1.2.8.1.2.1 List
            1.2.8.1.2.2 Prioritize
            1.2.8.1.2.3 Select which one(s) to address
         1.2.8.1.3 Specify objectives
            1.2.8.1.3.1 Determine the beginning state
            1.2.8.1.3.2 Identify the expected end state
               1.2.8.1.3.2.1 State the success condition
                  1.2.8.1.3.2.1.1 Time phase the conditions
                  1.2.8.1.3.2.1.2 Event phase the conditions
               1.2.8.1.3.2.2 State the failure condition
                  1.2.8.1.3.2.2.1 Time phase the conditions
                  1.2.8.1.3.2.2.2 Event phase the conditions
            1.2.8.1.3.3 State reason for project
               1.2.8.1.3.3.1 Background
            1.2.8.1.3.4 Define the concept of operations, the operational model, the business model
               1.2.8.1.3.4.1 Develop the project statement
                  1.2.8.1.3.4.1.1 Design
                  1.2.8.1.3.4.1.2 Technology
         1.2.8.1.4 Determine requirements
            1.2.8.1.4.1 Determine course of action
               1.2.8.1.4.1.1 Identify work packages

1.2.8.1.4.1.2 Determine schedule
1.2.8.1.4.1.3 Determine budget
1.2.8.1.4.2 Estimate needed resources
    1.2.8.1.4.2.1 Personnel
    1.2.8.1.4.2.2 Funds
    1.2.8.1.4.2.3 Equipment
    1.2.8.1.4.2.4 Facilities
    1.2.8.1.4.2.5 Information
    1.2.8.1.4.2.6 Technology
    1.2.8.1.4.2.7 Time
    1.2.8.1.4.2.8 External support
    1.2.8.1.4.2.9 Infrastructure
1.2.8.1.4.3 Define stakeholders
1.2.8.1.4.4 Define boundaries
    1.2.8.1.4.4.1 Identify limitations
        1.2.8.1.4.4.1.1 Consult project audience and stakeholders
        1.2.8.1.4.4.1.2 Review relevant written materials
        1.2.8.1.4.4.1.3 Note source of identified limitations
    1.2.8.1.4.4.2 Identify environmental barriers
    1.2.8.1.4.4.3 Identify economic barriers
    1.2.8.1.4.4.4 Identify legal boundaries
    1.2.8.1.4.4.5 Identify social and cultural boundaries
    1.2.8.1.4.4.6 Identify ethical boundaries
    1.2.8.1.4.4.7 Identify regulatory boundaries
1.2.8.1.4.5 Determine measures of effectiveness and measures of performance
1.2.8.1.4.6 Create strategy
    1.2.8.1.4.6.1 Consider organization's usual approaches
    1.2.8.1.4.6.2 Conduct risk–benefit analysis
    1.2.8.1.4.6.3 Select strategy
    1.2.8.1.4.6.4 Develop backup strategies
1.2.8.1.4.7 Write plan
1.2.8.1.4.8 Promulgate plan
1.2.8.1.5 Identify risks
1.2.8.1.5.1 Define "risk"
    1.2.8.1.5.1.1 Technical solution
        1.2.8.1.5.1.1.1 Failure to provide required functionality
        1.2.8.1.5.1.1.2 Failure to provide required performance
    1.2.8.1.5.1.2 Schedule

1.2.8.1.5.1.2.1 Failure to assign required resources

1.2.8.1.5.1.2.2 Lack of available resources

1.2.8.1.5.1.3 Cost

1.2.8.1.5.1.3.1 Total cost exceeds budgeted cost

1.2.8.1.5.1.3.2 Rate of expenditures exceeds cash flow constraints

1.2.8.1.5.2 Implement practices to identify risks

1.2.8.1.5.2.1 Establish risk management board

1.2.8.1.5.2.2 Checklists for project risk areas

1.2.8.1.5.2.3 Lessons learned from previous projects

1.2.8.1.5.2.4 Resource availability lists

1.2.8.1.5.2.5 Resource training records for applicable skills

1.2.8.1.5.2.6 Peer review of project plans

1.2.8.1.5.2.7 Senior management review of project plans

1.2.8.1.5.3 Document all risks captured during definition and risk identification

1.2.8.1.6 Assess

1.2.8.1.6.1 Estimate probability of occurrence of each risk event identified

1.2.8.1.6.2 Estimate the consequences of the occurrence of each risk event

1.2.8.1.6.2.1 Estimate the impact on technical solution (measured in terms of impact on schedule and/or cost)

1.2.8.1.6.2.2 Estimate the impact on schedule (measured in terms of extra time required to complete the project)

1.2.8.1.6.2.3 Estimate the impact on cost (measured in terms of dollars)

1.2.8.1.6.3 Visualize risk assessments for technical solution, schedule, and cost using two-dimensional plots

1.2.8.1.6.3.1 Plot risk probabilities along the y-axis

1.2.8.1.6.3.2 Plot risk consequences along the x-axis

1.2.8.1.6.4 Categorize assessments based on data plots

1.2.8.1.6.4.1 "High" risks have the greatest consequences

1.2.8.1.6.4.2 "Medium" risks have less impact on the project than high risks

1.2.8.1.6.4.3 "Low" risks have the least impact

1.2.8.1.7 Mitigate risks*
  1.2.8.1.7.1 Develop contingency plans for risks
    1.2.8.1.7.1.1 High risks are first priority for contingency planning
    1.2.8.1.7.1.2 Medium risks are second priority for contingency planning
    1.2.8.1.7.1.3 Low risks have last priority for contingency planning
  1.2.8.1.7.2 Present contingency plans to senior management for review and approval
  1.2.8.1.7.3 Risk management board monitors risks as the project progresses and advises the project manager
  1.2.8.1.7.4 Implement contingency plans as directed
  1.2.8.1.7.5 Monitor success of contingency plans
1.2.8.1.8 Validation of the plan
  1.2.8.1.8.1 The achievability of schedule
  1.2.8.1.8.2 Planned expenditures are compatible with budget
  1.2.8.1.8.3 Workforce is necessary and sufficient
  1.2.8.1.8.4 Facilities, assets, and equipment are necessary and sufficient (appropriate)
1.2.8.1.9 Verification of the plan
  1.2.8.1.9.1 Schedule meets the objectives
  1.2.8.1.9.2 Planned expenditures satisfy the budget and rate
  1.2.8.1.9.3 Workforce is adequately trained and competent
  1.2.8.1.9.4 Facilities, assets, and equipment are adequate
1.2.8.1.10 Adjustments
  1.2.8.1.10.1 Performance
  1.2.8.1.10.2 Organization
  1.2.8.1.10.3 Communications
  1.2.8.1.10.4 Team building
  1.2.8.1.10.5 Direct
  1.2.8.1.10.6 Control
1.2.8.1.11 Phases of work stipulated for milestones and reviews (selected process model)
1.2.8.2 Communications—the activity of conveying information
  1.2.8.2.1 Knowledge management
    1.2.8.2.1.1 Convert
    1.2.8.2.1.2 Identify
    1.2.8.2.1.3 Access
    1.2.8.2.1.4 Create/capture
    1.2.8.2.1.5 Represent
    1.2.8.2.1.6 Leverage

---

* *Project Manager's Portable Handbook*, David I. Cleland and Lewis R. Ireland, pp. 339–344.

1.2.8.2.2 Specify the problem that needs to have a communication
1.2.8.2.3 Identify the need for the message
1.2.8.2.4 Determine the recipients
1.2.8.2.5 Conceptualize the form of the message content
1.2.8.2.6 Conceptualize the message content
1.2.8.2.7 Draft the message content
1.2.8.2.8 Type of content
    1.2.8.2.8.1 Physical
1.2.8.2.9 Trust
    1.2.8.2.9.1 Content
    1.2.8.2.9.2 Address
    1.2.8.2.9.3 Channel
    1.2.8.2.9.4 Delivery
        1.2.8.2.9.4.1 Act of
            1.2.8.2.9.4.1.1 Building, inputting, and out-
                    putting message
            1.2.8.2.9.4.1.2 Encrypting
            1.2.8.2.9.4.1.3 Decrypting
1.2.8.2.10 Finalize the message content
    1.2.8.2.10.1 Identify recipient(s)
    1.2.8.2.10.2 Specify address of recipient(s)
    1.2.8.2.10.3 Confirm address of recipient(s)
    1.2.8.2.10.4 Determine the available channels for conveying
        message to recipient(s)
        1.2.8.2.10.4.1 Media
            1.2.8.2.10.4.1.1 TV
            1.2.8.2.10.4.1.2 Visual (broadcast/cable)
            1.2.8.2.10.4.1.3 Digital storage media
            1.2.8.2.10.4.1.4 Radio
            1.2.8.2.10.4.1.5 Print advertising
            1.2.8.2.10.4.1.6 Word of mouth
    1.2.8.2.10.5 Determine the temporal requirements for
        delivery
        1.2.8.2.10.5.1 Start time
        1.2.8.2.10.5.2 End time
        1.2.8.2.10.5.3 Elapsed time
    1.2.8.2.10.6 Determine the needs of the channel to accommo-
        date the message
    1.2.8.2.10.7 Select the channel(s) for the message to recipient(s)
    1.2.8.2.10.8 Encrypt
    1.2.8.2.10.9 Transmit
    1.2.8.2.10.10 Receive
    1.2.8.2.10.11 Acknowledge receipt
    1.2.8.2.10.12 Decrypt
    1.2.8.2.10.13 Acknowledge content

1.2.8.3 Organize
    1.2.8.3.1 Create organizational structure
        1.2.8.3.1.1 Hierarchy
        1.2.8.3.1.2 Flat
    1.2.8.3.2 Create team
        1.2.8.3.2.1 Define roles
            1.2.8.3.2.1.1 Assign personnel to tasks
                1.2.8.3.2.1.1.1 Negotiate agreements to assure availability
            1.2.8.3.2.1.2 Define responsibilities
            1.2.8.3.2.1.3 Give and explain tasks
            1.2.8.3.2.1.4 Integrate team
            1.2.8.3.2.1.5 Create operating practices and procedures
                1.2.8.3.2.1.5.1 Define processes
                1.2.8.3.2.1.5.2 Define interactions between processes
            1.2.8.3.2.1.6 Set up tracking systems
                1.2.8.3.2.1.6.1 Financial
                1.2.8.3.2.1.6.2 Personnel
                1.2.8.3.2.1.6.3 Sales
                1.2.8.3.2.1.6.4 Production
                1.2.8.3.2.1.6.5 Support
1.2.8.4 Control
    1.2.8.4.1 Determine performance standards/requirements
    1.2.8.4.2 Monitor actions and results
        1.2.8.4.2.1 Stay within time constraints
        1.2.8.4.2.2 Stay within budget constraints
        1.2.8.4.2.3 Personnel
        1.2.8.4.2.4 Facilities
        1.2.8.4.2.5 Resource expenditures
    1.2.8.4.3 Address problems
        1.2.8.4.3.1 Time problems
        1.2.8.4.3.2 Budgetary problems
        1.2.8.4.3.3 Personnel problems
            1.2.8.4.3.3.1 Hire employees (redundant)
            1.2.8.4.3.3.2 Fire employees
            1.2.8.4.3.3.3 Hire contractors
        1.2.8.4.3.4 Share information (reporting)
            1.2.8.4.3.4.1 Inform superiors
                1.2.8.4.3.4.1.1 Progress
                1.2.8.4.3.4.1.2 Problems
            1.2.8.4.3.4.2 Keep project team members informed
            1.2.8.4.3.4.3 Keep clients informed
        1.2.8.4.3.5 Manage resources
            1.2.8.4.3.5.1 Time

1.2.8.4.3.5.2 Money
1.2.8.4.3.6 Conduct audits
1.2.8.4.3.6.1 Financial
1.2.8.4.3.6.2 Production
1.2.8.4.3.6.3 Operations
1.2.8.5 Direct
1.2.8.5.1 Obtain assets
1.2.8.5.2 Project start-up activities
1.2.8.5.2.1.1 Baseline
1.2.8.5.2.1.1.1 Design
1.2.8.5.2.1.1.2 Technology
1.2.8.5.3 Assign tasks
1.2.8.5.4 Close-out
1.2.8.5.4.1 Develop action plan to resolve problems
1.2.8.5.4.2 Obtain approvals for tested and final deliverables
1.2.8.5.4.3 Communicate lessons learned to staff
1.2.8.5.4.4 Assist in reassigning staff
1.2.8.5.4.5 Celebrate project success
1.2.8.6 Team build
1.2.8.6.1 Hire
1.2.8.6.1.1 Prepare
1.2.8.6.1.1.1 Planning for human resources
1.2.8.6.1.1.1.1 Appropriate staff available
1.2.8.6.1.1.1.1.1 Number
1.2.8.6.1.1.1.1.2 Type
1.2.8.6.1.1.1.2 Adequate level of training and skills
1.2.8.6.1.1.1.2.1 Legal
1.2.8.6.1.1.1.2.2 Technical
1.2.8.6.1.1.1.2.3 Administrative
1.2.8.6.1.1.1.3 Appropriate facilities
1.2.8.6.1.1.1.4 Required infrastructure and support
1.2.8.6.1.1.2 Planning for hiring
1.2.8.6.1.1.2.1 Determination of needs
1.2.8.6.1.1.2.1.1 Short-term
1.2.8.6.1.1.2.1.2 Mid-term
1.2.8.6.1.1.2.1.3 Long-term
1.2.8.6.1.1.3 Resources available for hiring
1.2.8.6.1.1.3.1 Budget
1.2.8.6.1.1.3.2 Facilities
1.2.8.6.1.1.3.3 Support
1.2.8.6.1.1.3.4 Equipment
1.2.8.6.1.1.3.5 Tools
1.2.8.6.1.1.4 Start date determined

1.2.8.6.1.1.4.1 Minimum range
1.2.8.6.1.1.4.2 Maximum range
1.2.8.6.1.1.5 Status of new hire
1.2.8.6.1.1.5.1 Permanent
1.2.8.6.1.1.5.2 Temporary
1.2.8.6.1.1.5.3 Consultant
1.2.8.6.1.1.6 Contract guidance (if applicable)
1.2.8.6.1.1.7 Identify reporting manager
1.2.8.6.1.1.8 Roles and responsibilities
1.2.8.6.1.1.9 Job title and position
1.2.8.6.1.1.10 Compensation
1.2.8.6.1.1.10.1 Money
1.2.8.6.1.1.10.2 Equity
1.2.8.6.1.1.10.3 Perquisites
1.2.8.6.1.1.11 Organizational identity
1.2.8.6.1.1.12 Legal and company requirements
1.2.8.6.1.1.12.1 EEO
1.2.8.6.1.1.12.2 Policy
1.2.8.6.1.1.12.3 Legal
1.2.8.6.1.1.12.4 Positions restrictions
1.2.8.6.1.1.13 Security
1.2.8.6.1.2 Advertise/search
1.2.8.6.1.2.1 Direct placement
1.2.8.6.1.2.1.1 Duration (specific days)
1.2.8.6.1.2.1.1.1 Start date
1.2.8.6.1.2.1.1.2 End date
1.2.8.6.1.2.1.2 Content
1.2.8.6.1.2.1.3 Image
1.2.8.6.1.2.1.3.1 Size
1.2.8.6.1.2.1.3.2 Location
1.2.8.6.1.2.2 Contract
1.2.8.6.1.2.2.1 Services provided
1.2.8.6.1.2.2.2 Start date
1.2.8.6.1.2.2.3 Duration
1.2.8.6.1.2.2.4 Termination
1.2.8.6.1.2.2.5 Payment
1.2.8.6.1.2.2.6 Intellectual property rights
1.2.8.6.1.3 Screen and select
1.2.8.6.1.3.1 Minimum requirements
1.2.8.6.1.3.2 Desired requirements
1.2.8.6.1.3.3 Excessive requirements
1.2.8.6.1.3.4 Letters of reference
1.2.8.6.1.3.5 CV or resume
1.2.8.6.1.3.6 Cover letter
1.2.8.6.1.3.7 Portfolio of work

1.2.8.6.1.4 Interview
    1.2.8.6.1.4.1 Follow process
        1.2.8.6.1.4.1.1 Legal
        1.2.8.6.1.4.1.2 Company
1.2.8.6.1.5 Checking
    1.2.8.6.1.5.1 Technical
    1.2.8.6.1.5.2 References
    1.2.8.6.1.5.3 Citizenship
    1.2.8.6.1.5.4 Summation of opinions
1.2.8.6.1.6 Select
    1.2.8.6.1.6.1 Comparison and threshold
        1.2.8.6.1.6.1.1 Internal
        1.2.8.6.1.6.1.2 External
1.2.8.6.1.7 Approval from hiring authority
    1.2.8.6.1.7.1 Updated plan
        1.2.8.6.1.7.1.1 Budget check
        1.2.8.6.1.7.1.2 Skills inventory
    1.2.8.6.1.7.2 Updated organization
1.2.8.6.1.8 Offer
    1.2.8.6.1.8.1 Confirmation
        1.2.8.6.1.8.1.1 Salary
        1.2.8.6.1.8.1.2 Conditions
        1.2.8.6.1.8.1.3 Location
        1.2.8.6.1.8.1.4 Position
        1.2.8.6.1.8.1.5 Title
        1.2.8.6.1.8.1.6 Perquisites
1.2.8.6.1.9 Orientation
    1.2.8.6.1.9.1 Confidentiality agreement
    1.2.8.6.1.9.2 Employment agreement
    1.2.8.6.1.9.3 Rights agreement
    1.2.8.6.1.9.4 Assignment agreement
    1.2.8.6.1.9.5 Company policy concurrence
    1.2.8.6.1.9.6 Set-up account for work
    1.2.8.6.1.9.7 Benefits
    1.2.8.6.1.9.8 Ownership
    1.2.8.6.1.9.9 Orientation
    1.2.8.6.1.9.10 Security
1.2.8.6.2 Sustain
    1.2.8.6.2.1 Train
    1.2.8.6.2.2 Education
    1.2.8.6.2.3 Promotion and advancement
    1.2.8.6.2.4 Assignments
    1.2.8.6.2.5 Surveying
    1.2.8.6.2.6 Security
1.2.8.6.3 Exit

1.2.8.6.3.1 Notice of intentions
    1.2.8.6.3.1.1 Schedule
    1.2.8.6.3.1.2 Use of benefits
1.2.8.6.3.2 Status change
    1.2.8.6.3.2.1 Account termination
    1.2.8.6.3.2.2 Account suspension
    1.2.8.6.3.2.3 Account accrual
1.2.8.6.3.3 Notifications
    1.2.8.6.3.3.1 Planning
    1.2.8.6.3.3.2 Budgeting
    1.2.8.6.3.3.3 Organization
    1.2.8.6.3.3.4 Legal
    1.2.8.6.3.3.5 Property accounting
    1.2.8.6.3.3.6 Security
1.2.8.6.3.4 Exit interview
1.2.8.6.4 Maintain focus
1.2.8.6.5 Maintain motivated team
    1.2.8.6.5.1 Recognize performance
        1.2.8.6.5.1.1 Salary increases
        1.2.8.6.5.1.2 Promotions
        1.2.8.6.5.1.3 Job assignments
    1.2.8.6.5.2 Oversee/approve requests for annual/administrative leave
1.2.8.6.6 Mentor team members
    1.2.8.6.6.1 Complete regular performance appraisals
1.2.8.6.7 Train team members
1.2.8.6.8 Provide technical expertise and guidance
1.2.8.6.9 Establish values

Appendix 2: Product Upgrades Based on Minimum Expected Quality Loss

Introduction (Langford 2009)

Maintenance and sustainment costs are typically one-third of the development costs, as was briefed to the Government Accountability Office (Chaplain 2008) for the Space Shuttle, to 70% of the lifecycle costs for the general category of software (Boehm and Basili 2001). Often, managers responsible for maintenance and sustainment target costs reductions of the order of 15–20% to improve product profitability. Perhaps such actions assume that customers are pleased by both the gesture to reduce costs and the company's interests in supporting fielded products. However, for customers, perhaps the most meaningful consideration of continued product support is lower cost of ownership.

The authors posit that the upgrade cycle for fielded products could be based on the expected quality loss that results from the period of the upgrade. The consequence would be a Pareto-efficient determination of the upgrade period. To achieve Pareto-efficiency (based on the principle that one-sided benefit to a party to a negotiation results in an inequitable distribution of losses), losses for all stakeholders must be considered and incorporated into a cooperative exchange of benefits and losses. A common distinction between the interests of stakeholders can be depicted graphically as leaning toward either smaller or larger than some position that will eventually be the negotiated settlement. That is, the agreement between two stakeholders is defined as the position whereby neither side to a negotiation has an unfair or disproportionate advantage. For the purpose of this paper, the mathematics simplifies by assuming an idealized negotiation (Figure A2.1) where two parties incur equal losses about a center point target value, m.

The minimum loss depicted as the quality loss function in Figure A2.1 defines the *target value* of the critical performance characteristic, m, as a negotiation between two strategies with opposite demands on quality for a given investment. One party to the negotiation determines that more performance is better (considered as larger-the-better (LTB) strategy) while the other party considers that smaller-the-better (STB) demands on performance is required.

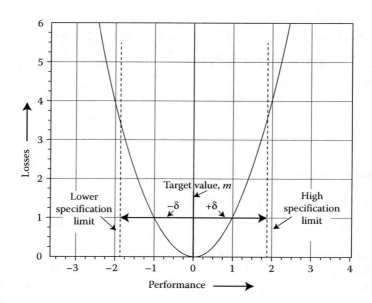

FIGURE A2.1
Pareto-efficient negotiation.

The LTB (buyer's strategy) benefits from larger values of performance, m, coupled with lower loss. Alternatively, the STB (seller's strategy) faces higher losses from a higher-performance requirement, m. For example, a seller might want to deliver more product performance but is unwilling to accept increased costs which may lead to reduced marketshare, while the buyer might expect more product performance for lower costs. Figure A2.2 illustrates LTB and STB strategies plotted with performance indicated on the x-axis and the loss on the y-axis.

Simple addition of the two curves, x and $1/x$, results in a pictorial representation of negotiation, based on both parties achieving the minimum loss. Figure A2.3 shows the resultant quality loss function. The competition between one party espousing STB and another party posturing LTB is in essence a negotiation between two parties that results in defining a working regime that reflects their mutual interests, solution, and requirements. The property of Pareto-efficiency (that one-sided benefit to a party to a negotiation results in an inequitable distribution of losses) should guide the selection and agreement of m. The result of a Pareto-efficient determination of m is a minimum loss for that negotiation (Figure A2.3). Such a negotiation is representative of the desire by the buyer and seller to have a product upgrade at exactly the Pareto-efficient point, m.

From Figure A2.3, the resultant quality loss distribution has a minimum at $m = 1$, representing the minimum loss that can be caused after the upgraded

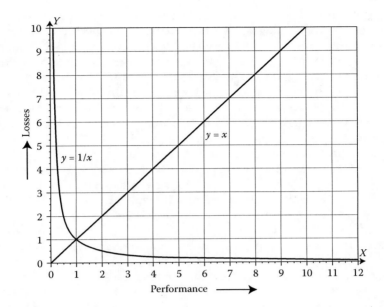

FIGURE A2.2
Smaller-the-better (STB: $y = x$, seller) and larger-the-better (LTB: $y = 1/x$, buyer).

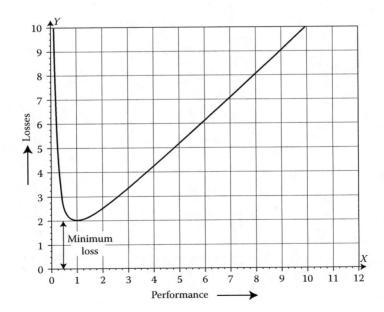

FIGURE A2.3
Combining two loss distributions that compete for a definitive product upgrade period, m.

product is shipped. This minimum loss represents the most effective periodicity for the product upgrade within the product's specified environment, given the conflicting constraints of STB and LTB. The goal of the stakeholder negotiations is to minimize the total system losses due to and during the course of upgrading and releasing upgraded products. There are an infinite number of quality loss functions, each with a minimum loss determined by cooperative negotiation between buyer and seller. But there is only one Pareto-efficient quality loss function that reflects the optimized minimum loss during a product upgrade cycle (Figure A2.4).

Losses to the buyers can result from an early release of an upgraded product that may not take full advantage of better technology. These losses may manifest through lower performances of product functions, relative to the lost opportunity available from more current and relevant technologies. Later release of an upgrade product may deprive customers of productivity that could have been achieved given an earlier release of the upgrade. Premature release of an upgraded product may require fixes and patches to achieve an acceptable operational effectiveness, while perfectly functioning upgrades may be function rich, but performance poor.

The result of an early or late release is in effect to slide the performance target parameter m horizontally as the quality loss functions show an

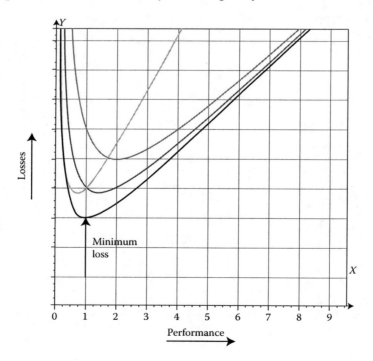

FIGURE A2.4
Pareto-efficient quality loss function optimized for minimum loss.

increase in their losses about the point m (Figure A2.4). Therefore, the resultant quality loss function only has a minimum value when Pareto-efficiency is achieved. The range of possible curves (Figure A2.4) can be determined by applying Bayes' framework to predict the Pareto-efficient solution either if the true distribution of past observations is known or if one computes a universal distribution, defined as the weighted sum of distributions based on the complexity (Hutter 2001). Hutter goes on to prove that by using a universal probability distribution where lower weights are assigned to more complex distributions, the universal distribution is nearly as good as using the unknown true distribution. By applying Hutter's approach, the Pareto-efficient solution can be defined.

Outline of the General Quality Loss Function (Choi and Langford 2008)

Quality Characteristics

To achieve the desired level of quality and to determine the period for upgrading a product, stakeholders pose the following question—how much loss can I incur for various upgrade periods? This question can be answered by considering the results of an analysis based on a general quality loss function. We introduce a shape parameter that governs the amount of losses as a function of the periodicity, m, for the product upgrade. Since the product upgrade has competing interests between the user and the developer whose product is to be upgraded, we present a function which covers nominal-the-better (NTB).

Traditionally, quality is viewed as a step function that signifies a good product from a bad product. A good product is distinguishable by achieving its performance with fewer losses than that of a bad product. This view assumes that product quality is uniformly good between the lower specification and the upper specification. Sometimes traditional decision makers and those using Taguchi's loss function will make the same judgments if both the positions of the average and the variance as well as the averages are equal and/or the variances are equal. Both the average performance and variation from a target value are measures of quality (Taguchi et al. 1989).

The principle of Taguchi quality is based on the observation that customers become increasingly dissatisfied as performance falls further away from a specified target value. His work with industry over the last 35 years suggests that a quadratic curve best represents this customer's dissatisfaction with a product's performance. When the target value is set to zero, the first derivative of a Taylor series expansion taken about the target value is of quadratic form. The best achievable performance at the curve's minimum is centered on the target value. However, identifying the appropriate performance measures and selecting the best target value is not an easy task. And further while it has been productive to improve product quality by reducing the performance variability, defining and implementing appropriate testing to achieve only a handful of nonconforming products are confounded by misclassifying items as rejected/accepted or rejecting conforming items (Arts 1998). Testing is an early feedback that presupposes and validates specifications that are indeed sometimes the designer's best guess at the customer's interests. In actuality, the quadratic form was chosen by Taguchi because it was both simple, and as it turned out, useful. Further, after the Taylor expansion, higher powers in the series change the loss at the target value by a very small margin, and for practical purposes can be ignored within experimental error. We constructed a general quality loss function that would satisfy Taguchi conditions when the quadratic order was satisfied.

The quality loss function developed by Taguchi (1990) is used to describe quality in terms of smaller-the-better, larger-the-better, and nominal-the-best characteristics. A smaller-the-better output response results when it is desirable to minimize the performance, with the ideal target for performance being zero. Examples of smaller-the-better output responses are the wear on a component, the amount of engine audible noise, the amount of air pollution, and the amount of heat loss. The larger-the-better output response reflects cases when it is desirable to maximize the result, the ideal target being infinity. Examples of larger-the-better output responses are strength of materiel or fuel efficiency. The nominal-the-best characteristic results when there is a finite target point (or domain of cooperative agreement) to achieve, often associated through a negotiated outcome. In this case there are typically upper and lower specification limits on both sides of the performance target, representing the maximum or minimum acceptable bounds for the parties of the negotiation. Examples of nominal-the-best characteristics are the plating thickness of a component, the length of a part, and the output current of a resistor at a given input voltage.

The loss function is a means to quantify the benefits achieved for the customer by reducing variability around the target. It can help justify a decision to invest, determine how much process improvement is needed when the product is already capable of satisfying specifications, or determine the appropriate period for a product upgrade. Accomplishing the goal of improving quality by minimizing the effects of variations in performance did not necessarily need to come at the expense of eliminating the causes of that variation. The aim was to immunize the product design to variations that imparted customer value without an associated loss (Yao et al. 1999).

Types of Quality Loss Functions

Taguchi considers three cases of quality loss functions, including nominal-the-best, smaller-the-better, and larger-the-better. The methodology used to deal with the larger-the-better case is slightly different from that for the smaller-the-better and nominal-the-better cases. However, for each quality characteristic there exists some function that uniquely defines the relationship between economic loss and the deviation of the quality characteristic from its target value. Taguchi found the quadratic representation of the quality loss function to be an efficient and effective way to assess the loss due to deviation of a quality characteristic from its target value. For a product with a target value m, from a customers' perspective, $m \pm \Delta_0$ represents the deviation at which functional failure of the product or component occurs. When a product is manufactured with its quality characteristic at the extremes, $m + \Delta_0$ or $m - \Delta_0$, some measure to counter the loss must be undertaken by the customer. The loss function L (average loss) with characteristic of nominal-the-best (NTB) is described in Equation A2.1.

Nominal-the-best

$$L = k(y - m)^2 k = \frac{A_0}{\Delta_0^2} \qquad \text{(A2.1)}$$

where k is a proportionality constant and could be the cost of each unit (returned, modified, reworked) divided by the range limits of process variability divided by 2, y is the measure of performance (e.g., output) for a given function, m is the target value of y, and A_0 is the cost of the countermeasure. The loss function can also be determined for cases when the output response is a smaller-the-better response. The formula is a little different, but the procedure is much the same as for the case of nominal-the-best. For the case of smaller-the-better (STB), where the target is zero, the loss function is described as the following:

Smaller-the-better

$$L = ky^2 k = \frac{A_0}{y_0^2} \qquad \text{(A2.2)}$$

where A_0 is the consumer loss and y_0 is the consumer tolerance.

For a larger-the-better (LTB) output response where the target is infinity, the loss function can be written as the following:

Larger-the-better

$$L = k \frac{1}{y^2} k = A_0 y_0^2 \qquad \text{(A2.3)}$$

Assumptions

The following seven assumptions are made to develop a general quality loss function:

A1: The total quality loss ($L_n(x)$) consists of the stakeholders' loss plus unknown losses.

A2: If the level of quality equals the target value of the quality (i.e., m), the total quality loss is to be zero (or the minimum loss that is inherent in the system).

A3: If the acquisition phase is production and deployment, the value of shape parameter n is equal to 2.

A4: The minimum value of a shape parameter is close to zero and the value of the shape parameter in the concept refinement phase of the acquisition phases varies from 0 to 1.

A5: When the acquisition phases are the technology development or system development and demonstration phase, the range value of shape parameter varies from greater than one to less than two.

A6: After the production and deployment phase, the value of the shape parameter is greater than two.

A7: The probability distribution of the quality response remains the same regardless of the acquisition phases.

Notation

C_b: Baseline cost with a constant value.

C_s: If the type of quality characteristic is smaller-the-better, this means a proportionality constant of stakeholder's loss per response of quality. Additionally, if the type of quality characteristic is larger-the-better, it means a proportionality constant of developer's or manufacturer's loss per response of quality.

C_l: If the type of quality characteristic is larger-the-better, this means a proportionality constant of developer's or manufacturer's loss per response of quality. Additionally, if the type of quality characteristic is smaller-the-better, it means proportionality constant of the stakeholder's loss per response of quality.

n: Shape parameter for representing an acquisition phase of a weapon system ($n > 0$).

x: Response of quality.

$L_n(x)$: Total quality loss per piece in the case of shape parameter n and quality response x.

L_n: Expected quality loss per piece in the case of shape parameter n and quality response x.

According to the assumption A1 and Equations A2.1, A2.2, and A2.3, a general quality loss function can be described as the following Equation A2.4. Equation A2.4 covers all quality characteristics such as nominal-the-best, smaller-the-better, and larger-the-better.

General form

$$L_n(x) = C_b + C_s x^n + C_1 x^{-n} \tag{A2.4}$$

After applying the assumption A2 into Equation A2.4, we can get Equations A2.5 and A2.6 as follows. If the response of quality equals to the target value (i.e., m), the total quality loss is to be zero (Equation A2.5) and the result of differentiation for the response of quality having the target value (i.e., m) is also to be zero as Equation A2.6.

$$L_n(m) = C_b + C_s m^n + C_1 m^{-n} = 0 \tag{A2.5}$$

$$L'_n(m) = nC_s x^{n-1} - nC_1 x^{-n-1} = 0 \tag{A2.6}$$

If we incorporate the specific value of n into Equations A2.5 and A2.6, we obtain the general loss function as follows. If the value of n equals to 1, we obtain the following results:

$$L_1(m) = C_b + C_s m^1 + C_1 m^{-1} = 0 \tag{A2.7}$$

$$L'_n(m) = C_s m^0 - C_1 m^{-2} = 0 \tag{A2.8}$$

After solving Equations A2.7 and A2.8, we obtain the following results:

$$C_1 = C_s m^2, \quad C_b = -2C_s m$$

If n equals to 2, we obtain the following results:

$$L_2(m) = C_b + C_s m^2 + C_1 m^{-2} = 0 \tag{A2.9}$$

$$L'_2(m) = 2C_s m - 2C_1 m^{-3} = 0 \tag{A2.10}$$

TABLE A2.1

Results of Iterative Process for Generating a Quality Loss Function

n	C_1	C_b	$L_n(x)$
1	$C_1 = C_s m^2$	$C_b = -2C_s m^1$	$L_1(x) = -2C_s m^1 + C_s x^1 + C_s m^{2 \times 1} x^{-1}$
2	$C_1 = C_s m^4$	$C_b = -2C_s m^2$	$L_2(x) = -2C_s m^2 + C_s x^2 + C_s m^{2 \times 2} x^{-2}$
3	$C_1 = C_s m^6$	$C_b = -2C_s m^3$	$L_3(x) = -2C_s m^3 + C_s x^3 + C_s m^{2 \times 3} x^{-3}$
4	$C_1 = C_s m^8$	$C_b = -2C_s m^4$	$L_4(x) = -2C_s m^4 + C_s x^4 + C_s m^{2 \times 4} x^{-4}$
n	$C_1 = C_s m^{2n}$	$C_b = -2C_s m^n$	$L_n(x) = -2C_s m^{2n} + C_s x^n + C_s m^{2n} x^{-n}$

After solving Equations A2.9 and A2.10, we obtain the following results:

$$C_1 = C_s m^4, \; C_b = -2C_s m^2$$

After iterating in the above manner, we generate a quality loss function as shown in Table A2.1.

As shown in the last row of Table A2.1, we present the general quality loss function, detailed as follows:

$$L_n(x) = -2C_s m^n + C_s x^n + C_s m^{2n} x^{(-n)} \tag{A2.11}$$

$$= -2C_s m^n + C_s x^n (1 + m^{2n} x^{(-2n)})$$

Shapes of Quality Loss Function

As the value of n (shown in Equation A2.11 for the general quality loss function) changes, the shapes of quality loss function also change. To illustrate these changes, we plot the value of quality loss versus the response of quality with $C_s = 2$ and $m = 3$, in Figure A2.5, according to the change in the value of n.

As shown in Figure A2.5, the quality loss function with the red line is for $n = 1$, the blue line is for $n = 2$, the green line is for $n = 3$. By plotting the loss functions with different values of n, we observe that the width of the quality loss function depends on the value of n. In order words, the larger the value of n, the narrower the performance width of the quality loss functions.

In order to clearly see the proportionality between the quality losses as the value of n changes, we calculate all the related values in Table A2.2.

Expected Quality Loss

Now, suppose that the probability density function of X is normal with mean μ and variance σ^2. The probability density function of X will be of the following form:

$$f(x) = \frac{1}{\sigma\sqrt{2\pi}} \exp\left(-\frac{(x-\mu)^2}{2\sigma^2}\right), \quad -\infty \le x \le \infty \tag{A2.12}$$

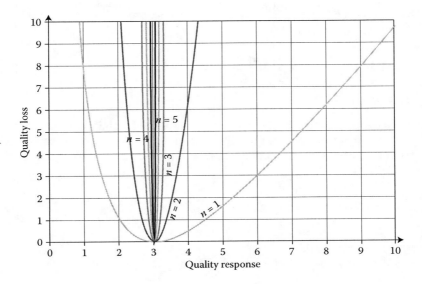

FIGURE A2.5
Shapes of the general quality loss function.

TABLE A2.2

Values of Quality Loss According to the Value of n

Response of Quality (x)	$n = 1$	$n = 2$	$n = 3$	$n = 4$	$n = 5$
1.00	8.00	128.00	1352.00	12,800.00	117,128.00
1.50	3.00	40.50	330.75	2278.13	14,595.20
2.00	1.00	12.50	90.25	528.13	2782.56
2.50	0.20	2.42	16.56	90.05	432.64
3.00	**0.00**	**0.00**	**0.00**	**0.00**	**0.00**
3.50	0.14	1.72	11.76	63.57	303.29
4.00	0.50	6.13	42.78	239.26	1191.33
4.50	1.00	12.50	90.25	528.13	2782.56
5.00	1.60	20.48	153.66	947.00	5315.79
5.50	2.27	29.86	233.51	1520.47	9117.15
6.00	3.00	40.50	330.75	2278.13	14,595.20
Quality loss function	$L_n(x) = -2 \times 2 \times 3^n + 2 \times x^n + 2 \times 3^{2n} \times x^{-n}$				
Baseline cost					
$-2 \times C_s \times m^n$	-12	-36	-108	-324	-972
$= -2 \times 2 \times 3^n$					

The expected loss per item is calculated according to Equation A2.13.

$$E[L_x(n)] \equiv L_n = \int_{-\infty}^{\infty} L_x(n)f(x)dx \qquad (A2.13)$$

where $f(x)$ is the probability density function of the normal random variable.

By substituting the general quality loss function and probability density function into Equation A2.13, the equation can be rewritten as follows:

$$L_n = \int_{-\infty}^{\infty} L_x(n)f(x)dx = \int_{-\infty}^{\infty} C_s(-2m^n + x^n + m^{2n}x^{(-n)})\frac{1}{\sigma\sqrt{2\pi}}\exp\left(-\frac{(x-\mu)^2}{2\sigma^2}\right)dx$$

$$= \int_{-\infty}^{\infty} C_s(-2m^n)\frac{1}{\sigma\sqrt{2\pi}}\exp\left(-\frac{(x-\mu)^2}{2\sigma^2}\right)dx + \int_{-\infty}^{\infty} C_s x^n\frac{1}{\sigma\sqrt{2\pi}}\exp\left(-\frac{(x-\mu)^2}{2\sigma^2}\right)dx$$

$$+ \int_{-\infty}^{\infty} C_s m^{2n}x^{(-n)}\frac{1}{\sigma\sqrt{2\pi}}\exp\left(-\frac{(x-\mu)^2}{2\sigma^2}\right)dx$$

$$= L_{n1} + L_{n2} + L_{n3} \qquad (A2.14)$$

where

$$L_{n1} = \int_{-\infty}^{\infty} C_s(-2m^n)\frac{1}{\sigma\sqrt{2\pi}}\exp\left(-\frac{(x-\mu)^2}{2\sigma^2}\right)dx = -2C_s m^n \qquad (A2.15)$$

$$L_{n2} = \int_{-\infty}^{\infty} C_s x^n\frac{1}{\sigma\sqrt{2\pi}}\exp\left(-\frac{(x-\mu)^2}{2\sigma^2}\right)dx \qquad (A2.16)$$

$$L_{n3} = \int_{-\infty}^{\infty} C_s m^{2n}x^{(-n)}\frac{1}{\sigma\sqrt{2\pi}}\exp\left(-\frac{(x-\mu)^2}{2\sigma^2}\right)dx \qquad (A2.17)$$

Because it is difficult to integrate Equations A2.16 and A2.17 in a closed-form solution, we adopt Taylor series expansion as the following. Taylor series for x^n and x^{-n} at target value of x (i.e., m) is as follows:

$$x^n = \sum_{k=0}^{\infty} \frac{f^{(k)}(m)}{\Pi k}(x-m)^k = \sum_{k=0}^{n} \frac{f^{(k)}(m)}{\Pi k}(x-m)^k + R_n$$

$$= \sum_{k=0}^{n} \frac{1}{\Pi k} \frac{\Pi(n)}{\Pi(n-k)}(m)^{n-k}(x-m)^k + R_n$$

$$x^{-n} = \sum_{k=0}^{\infty} \frac{f^{(k)}(m)}{\Pi k}(x-m)^k = \sum_{k=0}^{n} \frac{f^{(k)}(m)}{\Pi k}(x-m)^k + R_n$$

$$= \sum_{k=0}^{n} \frac{1}{\Pi k}(-1)^k \frac{\Pi(n-1+k)}{\Pi(n-1)}(m)^{-n-k}(x-m)^k + R_n$$

R_n: Error after n terms

By ignoring terms higher than the fourth order, we get the following forms:

$$x^n \approx \sum_{k=0}^{4} \frac{1}{\Pi k} \frac{\Pi(n)}{\Pi(n-k)}(m)^{n-k}(x-m)^k \tag{A2.18}$$

$$x^{-n} \approx \sum_{k=0}^{4} \frac{1}{\Pi k}(-1)^k \frac{\Pi(n-1+k)}{\Pi(n-1)}(m)^{-n-k}(x-m)^k \tag{A2.19}$$

After substituting Equations A2.18 and A2.19 into Equation A2.14, we obtain the following results:

$$L_{n1} = -2C_s m^n$$

$$L_{n2} = \int_{-\infty}^{\infty} C_s x^n \frac{1}{\sigma\sqrt{2\pi}} \exp\left(-\frac{(x-\mu)^2}{2\sigma^2}\right) dx$$

$$\approx C_s \int_{-\infty}^{\infty} \sum_{k=0}^{4} \frac{1}{\Pi k} \frac{\Pi(n)}{\Pi(n-k)}(m)^{n-k}(x-m)^k \frac{1}{\sigma\sqrt{2\pi}} \exp\left(-\frac{(x-\mu)^2}{2\sigma^2}\right) dx$$

$$= C_s \left\{ \begin{array}{l} m^n + nm^{n-1}(E(X) - m) + \dfrac{n(n-1)}{2}m^{n-2}(E(X^2) - 2mE(X) + m^2) \\[2ex] + \dfrac{n(n-1)(n-2)}{6}m^{n-3}(E(X^3) - 3mE(X^2) + 3m^2E(X) - m^3) \\[2ex] + \dfrac{n(n-1)(n-2)(n-3)}{24}m^{n-4}(E(X^4) - 4mE(X^3) \\[2ex] + 6m^2E(X^2) - 4m^3E(X) + m^4) \end{array} \right\}$$

$$L_{n3} = \int_{-\infty}^{\infty} C_s m^{2n} x^{(-n)} \frac{1}{\sigma\sqrt{2\pi}} \exp\left(-\frac{(x-\mu)^2}{2\sigma^2}\right) dx$$

$$\approx C_s m^{2n} \int_{-\infty}^{\infty} \sum_{k=0}^{4} \frac{1}{\Pi k} (-1)^k \frac{\Pi(n-1+k)}{\Pi(n-1)} (m)^{-n-k} (x-m)^k \frac{1}{\sigma\sqrt{2\pi}} \exp\left(-\frac{(x-\mu)^2}{2\sigma^2}\right) dx$$

$$= C_s m^n \left\{ \begin{array}{l} 1 - nm^{-1}(E(X)-m) + \dfrac{n(n+1)}{2} m^{-2}(E(X^2) - 2mE(X) + m^2) \\[2mm] -\dfrac{n(n+1)(n+2)}{6} m^{-3}(E(X^3) - 3mE(X^2) + 3m^2 E(X) - m^3) \\[2mm] +\dfrac{n(n+1)(n+2)(n+3)}{24} m^{-4}(E(X^4) - 4mE(X^3) \\[2mm] + 6m^2 E(X^2) - 4m^3 E(X) + m^4) \end{array} \right\}$$

Therefore, the expected quality loss in the case of the normal distribution of quality characteristic is of the following form:

$$L_n = C_s \left\{ \begin{array}{l} n^2 m^{n-2}(E(X^2) - 2mE(X) + m^2) \\[2mm] -n^2 m^{n-3}(E(X^3) - 3mE(X^2) + 3m^2 E(X) - m^3) \\[2mm] +\dfrac{(n^4 + 11n^2)}{12} m^{n-4}(E(X^4) - 4mE(X^3) + 6m^2 E(X^2) - 4m^3 E(X) + m^4) \end{array} \right\}$$

(A2.20)

where

$$E(X^n) = \int_{-\infty}^{\infty} x^n \frac{1}{\sigma\sqrt{2\pi}} \exp\left(-\frac{(x-\mu)^2}{2\sigma^2}\right) dx$$

$$E(X^0) = 1$$

$$E(X^1) = \mu$$

$$E(X^2) = \mu^2 + \sigma^2$$

$$E(X^3) = \mu^3 + 3\mu\sigma^2$$

$$E(X^4) = \mu^4 + 6\mu^2\sigma^2 + 3\sigma^4$$

In order to show the trend of the expected quality loss according to the position of the target value, we consider three cases of the mean of quality output, by using a numerical example.

Case 1: the target value of the quality characteristic is equal to the mean of quality.

Case 2: the target value of the quality characteristic is greater than the mean of quality.

Case 3: the target value is less than the mean of quality.

Before suggesting the results of the application, we should assume the inputs for demonstrating the trend of the expected quality loss as indicated in Table A2.3.

First, consider Case 1 and observe the trend of expected quality loss as the value of n varies, through a numerical example. After substituting the data from Table A2.3 into Equation A2.20, we obtain the expected quality loss, as shown in Equation A2.21.

$$L_n = C_s(m^{n-2}n^2\sigma^2 + (n^4 + 11n^2)\sigma^4/4)$$
$$= 0.3(0.25 \times 10^{n-2}n^2 + 0.25^2 \times (n^4 + 11n^2)/4) \tag{A2.21}$$

For Cases 2 and 3, after applying the same method used in Equation A2.21, we obtain the expected values, respectively. In order to compare the expected

TABLE A2.3

Data for Application with Normal Distribution

Given data for the general quality loss function	
• Baseline cost (C_b)	$-2C_s m^n$
• Cost incurred in the case of smaller-the-better (C_s)	0.3
• Cost incurred in the case of larger-the-better (C_l)	$C_s m^{2n}$
Given data for the normal distribution	
• Mean of quality (μ)	10
• Variance of quality (σ^2)	0.25
• nth moment of the probability distribution is given by the Riemann–Stieltjes integral	1st: 10
	2nd: 100.25
	3rd: 1007.5
	4th: 10150.1875
Three cases	
• Case 1: $m = \mu$	10
• Case 2: $m > \mu$	11
• Case 3: $m < \mu$	9

quality loss among three cases, we need to display the expected quality loss to the value of *n* as shown in Figure A2.6.

The expected quality loss function for Case 1 is shown by the black line, Case 2 by the dark gray line, and Case 3 by the light gray line. By plotting the expected quality loss functions having different values of *n*, we show that the amount of the expected quality loss depends on the value of *n*, regardless of the position of the target value. In order words, if the value of *n* is increasing, then the slope of the function is increasing.

The amount of the expected quality loss change is proportional to the value of *n*. We show all the related values in Table A2.4.

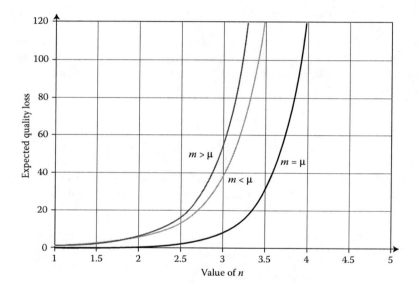

FIGURE A2.6
Expected quality loss with normal distribution.

TABLE A2.4

Expected Value of Quality Loss with Normal Distribution

n	Case 1 ($m = \mu$)	Case 2 ($m = \mu$)	Case 3 ($m = \mu$)
1	0.06375	0.84468	0.841435
2	0.58125	5.72216	5.29792
3	7.59375	53.9438	37.7437
4	122.025	847.425	439.425
5	1879.22	14,126.7	5831.72

Conclusion

A general quality loss function having a shape parameter is developed which is applicable to evaluate the expected quality loss for quality characteristics such as nominal-the-best, smaller-the-better, and larger-the-better. Additionally, we present an appropriate range of shape parameter values in the proposed general quality loss function to accommodate the impacts of upgrades to fielded systems.

By plotting the loss functions with different values of shape parameter n, we show that the width of the quality loss function depends upon the value of n. In order words, if the value of n is increasing, then the slope of the expected quality loss function is increasing. When we calculate the expected quality loss, we consider the normal probability distribution. Similar results are obtained with the exponential distribution, truncated exponential distribution, and truncated normal distribution. To show applicability of the proposed general quality loss function to the periodicity of upgrading fielded systems, we present the quality loss function and demonstrate a process for determining acceptance level of periodicity through numerical examples.

Therefore, the proposed general quality loss function can be used to justify a decision to release a product upgrade, and determine a specification limit on the release dates that minimizes the expected quality loss.

The limitations of this study are as follows: (1) We adopt Taylor series expansion for the general quality loss function. Due to this, the expected quality loss using the proposed function has nominal errors. (2) Since we have difficulty in obtaining actual data for quality loss associated with upgrading products and periodicity, we cannot present a validation of shape parameter value, n.

References

Arts, G. R. J. 1998. *Test Limits in Quality Control Using Correlated Products Characteristics*, ISBN 90-365-1129-1, The Netherlands: University of Twente.

Boehm, B. and Basili, V. R. 2001. Software defect reduction top 10 list. *Software Management Journal* January: 135–137.

Chaplain, C. T. 2008. GAO -08-581T 24, U.S. Government Accountability Office, April.

Choi, D. O. and Langford, G. 2008. A General Quality Loss Function Development and Its Application to the Acquisition Phases of the Weapon Systems, NPS-SE-08-007, Technical Report, Naval Postgraduate School, November.

Hutter, M. 2001. General Loss Bounds for Universal Sequence Prediction, Technical Report IDSIA-03-01, Instituto Dalle di Studi sull'Intelligenza Artificiale, Manno-Lugano, Switzerland, April.

Taguchi, G., Elsayed, E. A., and Hsiang, T. C. 1989. *Quality Engineering in Production Systems*, Tokyo: McGraw-Hill.

Taguchi, G. 1990. *Introduction to Quality Engineering*. Tokoyo: Asian Productivity Organization.

Yao, L., Kiran, K., Janet, K. A., and Farrokh, M. 1999. Robust design: Goal formulations and a comparison of metamodeling methods. *ASME Design Engineering Technical Conferences*, September 12–15, Las Vegas, NV.

Glossary of Terms

Abstraction: Abstraction is an insufficiency of details to describe completely all that is needed so that the EMMI that is necessary to enact a mechanism is available as needed.

Abstraction classification of integration: We refer to the class of integration as an abstraction classification of integration (or "cognitive integration")—that is two things connected conceptually and cognitively considered as entities in object or process thinking.

Act, activity: An act is the change in an object that is distinguishable as one increment in a sequence of acts, which combine to make an activity.

Action: An action is the release or receipt of something due to the enactment of a mechanism.

Aggregation: An assemblage of discrete objects or processes that have relations of convenience, but not in any particular way or pattern. Aggregations do not have relations due to interaction, but show no evidence of integration.

Alignment: Objects or processes (and their logical derivatives, e.g., functions or procedures, respectively) having cooperative association and affiliation.

Archetype: The detailed extractions from an original pattern or roothold that is related through levels of abstractions.*

Architecture: Conceptual and logical structures of objects and processes (and their logical derivatives, e.g., functions or procedures, respectively).

Artifact: An object or process made by humans.*

Attribute: Attribute is a measure and measurement, configuration and structure, and constraint (e.g., time, cost, and scope), performance, and loss due to achieving the performances of a function.

Axiom: Assumed to be self-evident.*

Behavior: A behavior is describable in terms of observed reactions to influences of energy, matter, material wealth, or information. A behavior is the movement of objects by processes; processes that result in objects; objects interacting with other objects. Behaviors are defined by the use of an operational definition (Kerlinger and Lee 2000) that particularizes objects and processes in ways that are measurable.

---

* The definition is adapted from or modeled after the dictionary resources provided through www.wordweb.co.uk, WordWeb Pro thesaurus/dictionary, Version 5, 2007, Anthony Lewis; these sources include the *Oxford Dictionary of English*, Oxford University Press, 2005, and the *Chambers Dictionary* (11th Edition), Chambers Harrap Publishers Ltd., 2008, and WordNet database, Princeton University, 2006.

Behavioral boundary: Behaviors that are defined as having no more influence as a result of the object(s) through their functions and interfaces.

Black box testing: Black box method of testing is the analysis of input transformation to output based on not knowing the internal workings, logic, or configuration of the object.

Body of knowledge: A body of knowledge is the collection of data and information that has been formed into broadly applicable and widely accepted cognitive structures that are derived from theory, approach, method, procedures, tools, and techniques. These cognitive structures may be based on causal beliefs, convictions that suggest validity, broadly considered methodology which inculcates the study of research and interpretive practices applied to data and information, and the shared notions about the appropriateness of mereological and ontological thinking.

Boundary: Boundaries mark the end of one factor, but not necessarily the beginning of something else. Boundaries are predicated on a perspective.

Boundary condition: Boundary conditions can be defined as mediation of capabilities that enact across boundaries. Boundary conditions are a way of limiting how EMMI affects a bounded object. Boundary conditions mediate the flow of EMMI across interfaces at boundaries.

Business model: A business model is descriptive of the management of value for an entity, for example, that of a product or service, or that which pertains at the business or enterprise level. The essential characteristics of a business model require the enterprise be describable in terms of its key traits, for example, managing the enterprise; delineating the needs of the enterprise; prioritizing the relative importance of these needs; evaluating the scalability and externalities of the internal operations and external processes; identifying the efficacies of the products and services in the user environments; determining the causal boundaries and boundary conditions; and identifying all interactions both internal and external to the enterprise with internal-to-external and external-to-internal delineations. (Enterprise or business are used interchangeably, whereas, project can also be applicable and serve equally well given the context of the discussion.) The business model encompasses the concepts of enterprise, business, and project. The business model is concerned with the interactions, events, objects, and organization of survival. The business model must be true and faithful to all types and significances of interactions, capturing all interactions to expose the structural inherences and opportunities for risk.

Case study: A behavioral model might be constructed that represents a more detailed examination of a portion of the lessons, grounded in a set of perspectives, measurement theory, and the objective actions. The set of perspectives is referred to as a case study.

Causality: The sequence of events we term as causality—event by event. Causality requires that the relation between two objects be modeled as the change in the sending object, the change in the receiving object, and the context of both the sending and receiving objects. Causal events have both provenance and pertinent specificity. Causality is formed from the modal threads of events leading to the proximate events (nearby in space and time) from which the conditions are stipulated to select the necessary and sufficient events.

Change: Change means that an object or process is different one instance from the previous or next instance.

Characteristic: The qualities that are descriptive of something.*

Classifications of integration: The combination of the abstraction, social, and model classifications of integration constitutes the whole of the integration description. These classifications (or classes) link the common set of limitations, the constraints allocated within the project and temporal constraints that synchronize the interpersonal relations, intellectual discussions, and the various corporeal representations for the product or service.

Cognition: The acts and activities of knowing.*

Cognitive domain: The cognitive domain involves the abstractions and reasoning that takes place when thinking about a particular subject.

Cohesion: The manner in and degree to which the objects or processes relate to each other.

Combination: An arrangement or assemblage of discrete objects or processes that are in totality thought of in a particular way construct, but where the objects or processes are interacting, and show evidence of some integration.

Complexity: Complexity results from emergent properties of integrated objects, number and types of processes, and the number, types, and frequency of interactions between and within processes. Complexity can be seen as relative to the level of abstraction in which one views two objects.

Concept: A thing that is thought of and generally described.*

Condition: A condition is the circumstance that encompasses an object; the factors that affect the manner and ways in which the object interacts; the situation in which the object operates; or the terms under which an object behaves.

Conditional causality: The most apparent direct cause. Related modal causality to proximate causality. Constrains the context and circumstances surrounding the sequences and trails of events.

Connectivity: The joining of objects or processes by interaction or to facilitate interaction.

Constraint: Constraints are a structural property of the solution. Constraints are the results of boundary conditions. Constraints are conditions of allocations that once established are changeable, however, vicissitudinous. Constraints are flexible within the overall limitations set.

Context: Context is the situation or framework (Aerts et al. 2003) in which the interaction between two objects takes place.

Correlation: Correlated events have nexus, without satisfying the three types of causal events required for strict demands of recognizing cause(s) and effect(s). That correlation depends on developing metrics for both subjective measures and for objective measures that are related to the same concomitant object and process (i.e., procedure, activity, or act (in decreasing level of sophistication and complexity)).

Cost: What was spent in money or material wealth to accomplish or produce an object or process. Cost is a consummate, independent measure of a cycle.

Coupling: The degree of dependency between objects or between processes.

Customer: The person or organizational entity (object or process) that buys products or services (objects or processes).

Cycle: An event or series of events that are regularly repeated in the same order.*

Deduction: An inference based on a principle, rule, or process.*

De facto: The accidental embodiment of systems that are caused by human or natural workings (that which is unintended). De facto systems can be thought of as emergent systems that have developed as a result of circumstances.

Definition: Descriptive, explanatory, ostensive, stipulative, explicative, recipe, or examples of a word (sometimes referred to as a term) or a concept (Swartz 1997).

Definitional: Defined as limiting the extent of a meaning of a noun.*

De jure: The rightful (that which is intended) embodiment of human work in the form of products and services.

Deonitic: Study of duty, ethics, and piety.*

Dependent: Enabled or sustained by interaction with an object or process.

Design: Design is the conceptualization of the needs of the stakeholders.

Domain: Domain is the sphere of activity that includes the physical entities, functions, processes as related through their relations and context.

Element: An element is an object and process, combined or integrated.

Embodiment: The tangible or material form.*

Emergence: Any effect that produces a change in intrinsic properties, traits, or attributes that results by combining objects through the interactions of objects with EMMI. Emergence is due to the traits of an object or objects, process or processes.

EMMI: Energy, matter, material wealth, and information. EMMI expresses the interactions between objects.

Energy: Energy is the capacity to do work, that is, to make change.

Engineering: A workable definition of engineering is found on the Accreditation Board for Engineering and Technology (ABET) site www.abet.org and interpreted functionally by Andy Sage (Sage and Armstrong 2000, p. 30). "Engineering is that profession in which knowledge of the mathematical and natural sciences gained by study, experience, and practice is applied with judgment to develop ways to utilize, economically, the materials and forces of nature for the benefit of mankind."

Enterprise: The enterprise aims to lay out the nature, vision, and boundaries of work. The enterprise has two requirements for business that a project must satisfy: (1) providing revenue and profits consistent with the enterprise policies, and (2) the operating within the limited and constrained environment imposed by the realities of the project.

> ... enterprise is a mental image of that organization's current and future reason for existing (Morris and Pinto 2004)

To enterprise means to undertake the actions and activities necessary to carry out the business of an individual, group, or legal entity. To enterprise means to attain effectiveness.

Epistemology: The theory of knowledge that provides for measures and measurement of properties, traits, and attributes of objects. Epistemology aims to quantify the relation between an object and an observer's cognitive structures (Ferris 1997).

Event: An event is the enactment of a mechanism by input EMMI which is then transformed into output EMMI (i.e., performance). Events transpire as a result of every enactment of a mechanism. The result of a set of processes (referred to as the result of a mechanism or procedure) is called an event. One process could be referred to as an event, a set of processes can be referred to as an event (or set of events), or an abstraction of processes (or a single process) can be referred to as an event. An event implies the activation of a mechanism that is embodied within an object.

Force: Force is defined as the influences of EMMI on objects. Were there no influence on an object, there would be no force. That the influence should be measurable or detectable is of no consequence to this definition, as influence is relative. The test for influence is determined by the net of power (i.e., work done) on an object as observed by the outputs of that object's mechanism; changes in the object's properties, traits, or attributes; or other such changes in boundary, boundary conditions, physical issues, and functional or behavioral issues (Kocsis 2008).

Frame: A set of concepts that together represent a unique, singularly characterized perspective that is different than other perspectives.

Framework: A framework is the logics and consistency of method for a group of frames. The structure of concepts and narrative is termed as a framework. A framework is characterized by its (1) consistency of logic, (2) continuity of method, (3) applicability across disciplines and fields involved in the frames, (4) scalable from the interdomain's micro to macro instances and events, and (5) showing the requisite capaciousness to convey the needs of the scope of intentions that are inherent in its needs. Most importantly, the framework maintains focus on the eventual goal of describing a definitive theory of systems engineering.

Function: A function is defined as an action that is realized when objects interact. Specifically, the exchange of EMMI between two objects and satisfaction of the interface boundary conditions create a function that did not exist before the connection. A function is described relative to a particular stakeholder's perspective. A function is the essence of interaction between two objects; and for integration, a function is a structural property of the relations between objects. A function is the result of the interaction or integration of two objects. A function manifests itself as a trait of interaction. A function provides for a use.

Functional boundary: Determined by two objects and their interfaces with cognitive structures and thinking. Functional boundaries are formed at the interface of objects. No interface, no function. Interface, then function.

Fundament: The foundation, essential nature, or basis of something.*

General systems thinking: Thinking in terms of systems to bring partial patterns into full view by changing perspective, granularity, and the abstraction of cognitive structures to a generality that is applicable across all observations, fields, disciplines, and frameworks.

Granularity: Granularity is the result of partitioning objects or processes into some sort of heterarchical or hierarchical conceptualization of the relation between granules.

Granule: That which is within its limits or constraints is termed as a granule.

Group: A combination of objects or processes that interact.

Heterarchy: A classification schema which depicts relations between objects or between processes or between objects and processes as self-referenced, self-adjudicated as appropriate to the context in which the relation exists. Unlike hierarchical arrangements, heterarchical structures are unilevel, planar representations of the various aggregations and combinations of objects and processes.

Hierarchy: A classification scheme which depicts relations between objects or processes as successively subordinate (e.g., by detail or trait) in conjunction with like-kind classes of relations between objects or processes.

Holism: Holism is defined as the fundamental principle of a whole made up of parts, interconnected parts that cannot exist independently without the whole.

Illative: Making or stating an inference from premise(s).*

Independent: Unencumbered by influences outside of an object's boundary or process limitations.

Induction: The logic of deriving a general law from an observation, perspective, measurement theory, and causalities framework. Induction assumes that (1) knowledge can be represented by rules that govern conditions, (2) rules are based on current and future states (suggesting causality is pluralistic), (3) rules are defined and enacted similarly in hierarchical or heterarchical structures, (4) subsets and supersets of objects or processes identify with the higher levels (in hierarchical thinking) and with primary reference objects and processes (in heterarchical thinking), (5) synchronic and diachronic rules promote superordinate relations, (6) interactions between objects that are of inconsistent frequency or are barely over thresholds necessary to activate mechanisms result in measurable outputs, (7) two classes of mechanism are possible—those that revise parameters and those that generate plausibility and rules, (8) mechanisms require interactions that lie within the range of the parameters that control the mechanisms to result in measurable output, and (9) a depth and breadth of knowledge associated with actions and events, processes, and activities are structured within frames that are consistent with the causalities framework (or integration framework). This definition is extracted from and inspired by Holland et al. (1986).

Influence: The effect(s) on an object or process.

Information: Data with context.

Instability: Instability results in loss of functionality or performance. The consequences instability is generally correlated with loss of value. Instability is not the opposite of stability. Instability is the tendency to change in a manner that does not maintain either the systemic properties and traits of a system (objects and processes) or the individual properties and traits of an object or process.

Instance: A single occurrence of an enactment of a mechanism.

Integration: Integration is the unification of the objects through their interactions of energy, matter, material wealth, and information to provide system-level functionalities and performances. Integration is a coalesce of objects, interacting in perhaps unpredictable ways. Integration is the combining of a systematic series of actions that take place in a definite manner, directed to bring about a particular interaction between objects and sets of objects. Integration is the method of setting up or by chance satisfying the conditions that lead to a set of objects we refer to as a system. Integration is a collaborative, value-enhancing approach to demonstrating functionalities and performances of products and services. Integration is a method that facilitates outcomes that are beyond what an individual object can do either individually or by a number of objects acting independently, that is, make things happen that would otherwise not happen.

Integration requires the structures of knowledge, the benefit of information, and meaningful data to determine the alternative ways in which to integrate a product or service.

Integration is defined variously as a unifying process (Kirk et al. 2009), the progressive linking of system components to merge functional characteristics into an interoperable system (Haskins 2007), the whole is greater than the sum of its parts.*

Integration frame: The integration frame is one dimension of the integration framework. In the case of the causalities framework, one of the integration frames is the set of processes that portray the mechanisms of integration. The other integration frame is the objects and their associated representations as functions and behaviors that represent the product or service. An integration frame is a constituent part of the framework that represents the structural member that defines the framework.

Integration framework: An integration framework provides the basis for identifying principles that have substantial theoretical foundation(s). Integration framework that reflects theory and best practices in systems fields—engineering, sociology, psychology, biology, cybernetics, computer science, economics, management, and the like. The integration framework provides the venue and rules for combining the items within the integration frames (termed frames).

Integrative mechanism: The mechanisms of integration construct, bind, and instigate (or allow for) change in the natural and social world. The mechanisms of integration are a universal "adhesive."

Intellectual objects: Intellectual objects are entities by reason or principle. All that is not physical is intellectual. A person is a physical object, but the thoughts of a person are intellectual objects.

Interaction: Interaction is defined as the transfer of EMMI. Interaction is characterized by the transfer of something from one object (sender) to another object (receiver).

Interface: An interface is within the boundary that separate two objects or two processes. Physical connection of two objects results in an interface. Functions occur at the interface between two objects.

Iterative: To do again or to do something similar to that which was done before with the aim of improving on what was done before. Iterative processes are typified as refinements on existing objects or existing processes.

Juxtapose: Being placed or placed close together physically, functionally, or behaviorally.

Key stakeholder: Key stakeholders are those who represent the totality of the people who have various needs associated with the product or service that is to be built, used, and disposed. Those needs could be embodied in some form of substantial risk, consequential opportunity, significant influence, or essential support.

Knowledge: Knowledge is an object. A mechanism of knowledge is the mental procedures to build cognitive frames and the enactment of procedures to carry out placing the knowledge in tangible form. The corporeal representation(s) of both the knowledge and the mental procedures and physical activities is shown as the piece of paper on which the knowledge is represented.

> Knowledge is not a series of self-consistent theories that converges towards an ideal view; it is not a gradual approach to the truth. It is rather an ever increasing ocean of mutually incompatible (and perhaps even incommensurable) alternatives, each single theory, each fairy tale, each myth that is part of the collection forcing the others into greater articulation, and all of them contributing, via this process of competition, to the development of our consciousness. Nothing is ever settled ... (Feyerabend 1993)

Law: A law is a recurring rule or a collection of recurring rules, proven empirically.

Lifecycle: Lifecycle is associated with the product or service as that which occurs from conception to disposal, from the earliest moment of thought to the settlement of the last lawsuit. It is more appropriate from a product's or service's perspective to think of lifecycle as various stages of interaction and integration. The lifecycle paradigm for a product or service is from the perspective of the customer, developer, and user. Lifecycle is the result (or a symptom) of integration.

Lifecycle costs: Lifecycle costs consist of the costs associated with the processes (and it can be said, with the results of the processes) or an object (and the consequences of the object) within a cycle or the product's lifecycle.

Lifecycle success: Lifecycle success means not overrunning cost budgets ascribed to delivering the expected product or service.

Lifetime: The duration in which the product or service, object, or process exhibits properties and traits that are indicative of the characteristic nature of the product or service, object, or process.

Limitation: The limitations are given by the domain of the problem. Limitations are conditions of boundaries, and once imposed they are immutable.

Logic: Logic for integration is (1) defined as sufficient to support detailed analysis and interpretation of within a framework of relevant variables; (2) based on a consistent set of assumptions; (3) stipulated as the ontology of formalisms that translate into each other; (4) residing within the traditions of epistemology; (5) agreeing on a narrative that elicits particular interpretations of phenomenology; (6) being a consistent set of metaphysical facts that relate "phenomena as a whole to other genera of existence" (Lewes 1875); (7) supporting a set of value structures that are at least partially, piece-wise predictable;

and (8) applying methodology to define and transform relations into knowledge (Lazarsfeld 1993).

Loss: Loss is the relative, quantifiable difference in EMMI between the performance of a function at its target value and that measurement at any other value of performance. Loss can be thought of in terms of a generalized loss function that attributes EMMI losses to deviations from a target performance value, and as a result of not having a target performance value (meaning that a function was not provided or available for use and therefore had no performance value).

Management: Management is the set of processes used to deal with people to enterprise with products and services within the context-appropriate limitations and constraints. The five processes of management are "to plan," "to communicate," "to direct (or command)," "to control," "to organize," and "to team-build." In total, these five processes form the overarching management process "to manage."

Mapping: To place in one object or process within the context of another object or process, where the correspondence between sets of objects or processes is repeatable, persistent, and valid by comparison with standards, metrics, or shared representations.

Material wealth: Material wealth can be thought of as cash, investments (e.g., stocks, bonds, and notes), and other equivalents (credit and debit cards). Material wealth includes all that has the capacity to be converted into cash or cash equivalents. Therefore, one's time could be considered material wealth. Exchanging time, exchanging information, exchanging matter, and exchanging energy for remuneration in money are examples of the fungibility of material wealth. Broadly speaking, material wealth creates the financial capacity to do work through money, similarly to that of energy representing the capacity for work through mass. Material wealth has its place in both human endeavors as well as for natural processes. Material wealth is all that is referenced by abundance or plenitude.

Matter: Matter may not be the only substance that is comprised of energy or mass (or its derivatives such as force and momentum), but for convenience we loosely interpret all such "things" as matter.

Measurement: Measurement is the process of quantifying properties, traits, or attributes of a domain. A measurement model must be defined to reach a consensus about the particular meaning of the measurement and the relation of the measurement to other meanings. A measurement model reflects a certain perspective of the measurement (which encompasses its meaning and relations). One must be concerned about the (1) relationships between the measures (empirical properties) and the properties, traits, and attributes; (2) mapping of properties, traits, and attributes into a numerical relationship (the measure) to codify the empirical relationship; and (3) meaningfulness of statements and measurements made about the properties, traits, and attributes. The validity of a measurement is concerned with the relations between the measurement and the theories and concepts behind the application of the theory in which the measurement is interpreted and made predictive. Specifically, validation of a measure of stability is the process of ensuring the measure is a proper characterization (e.g., numerical) of the claim (i.e., trait) of stability. Measurement is generation of conclusions about the observed (Ferris 1997).

Measures: Measures are properties, traits, and attributes that are qualitatively and quantitatively determinable.

Mechanism: Mechanisms are the means by which objects and processes change. The effect of a mechanism is to transform an input EMMI into an output EMMI. A mechanism is that which operates in the context of forces.

Mereology: The formulation and study of relations for parts and their whole.

Metastability: Metastable combinations of objects and processes are stable for most practical purposes due to their long-lived properties and traits. Yet the system theoretically is unstable as large disturbances due to EMMI result in perturbations that show the more common (nonemergent) characteristics of the objects and processes.

Method: A generalized set of specifications for accomplishing objectives, from the perspective of the person designing or carrying out the method. Method is the overall plan by which work progresses from what one thinks they know to what others do not know.

Metrics: A metric is a convenient grouping of variables in which each variable is a partial determinate of a property, trait, or attribute. As such, metrics are representative of perspectives that are used to appreciate the difference between one instance or event and another instance or event. Metrics represent the shared value of what the common goal needs to be. Metrics determine how well work is proceeding according to a set of measures (or measures in absentia, i.e., not present by direct observation, but acknowledged as significant), standards, and measurement theory. Metrics are not about trade-offs between what best to do versus what is expedient. Metrics are used to represent that state of being, the determinant of "how is it going?" Metrics are not measurements, and measures are not metrics.

Milestone: An event that is recognized as significant.

Modal causality: Modal causality is the root cause of all events. Modal causality limits what is causally possible. Modal causality is the basic source of events (the historical provenance) that provide the foundational causes from which local circumstances (proximate causality) and the apparent most direct event (conditional causality) arise.

Model: A model is a relation or set of relations between variables that are representative of an object or process (termed the objective part of the model). A model is based on a value or a set of values and a principle or set of principles that form the (subjective) basis for the relations making up the objective part of the model.

Model classification of integration: Model or representation classification of integration (referred to as model or representation of integration) deals with the intended functionality of combining things into a whole.

Nature: The nature of an object is that set of characteristics that distinguish one object from another object. It is an object's character and behavior as determined by its intrinsic properties. By inheritance, we differentiate from that which is a trait or an attribute, both of which are a consequence of experiencing interactions with other objects.

Need: Need is something you must believe will solve the problem, is possible, is affordable, can be provided when desired, and does not cause another problem of such significance that offsets the benefit of solving the original problem. A need is absolute and unconditional.

Nexus: A linkage or series of connections linking two or more objects or processes.*

NotaSystem: Objects that are grouped and actively sending EMMI as an aggregation, in other words—not a system. There are no characteristics of NotaSystem that differ from two objects interacting occasionally. Both objects retain their individual characteristics and exhibit only perhaps a temporary change in attributes. A single (by definition) noninteracting object is NotaSystem (by definition). NotaSystems are stable.

Object: We commonly think of an object as a fundamental element, entity, or representation. It may be atomic or an aggregation of entities. Objects are or represent material structures, material wealth, and information. From these physical entities comes energy or matter. Objects can be physical or abstract (e.g., intellectual). Objects may be conceptual, phenomenological, or ideological. Objects may be comprised of other objects, each of which is related by interactions. Objects can be ordinary or elemental. Objects have boundaries.

Object frame: The frame whose domain is objects and their relations and behaviors.

Object type: Objects are differentiable by their input and output characteristics (indicative of the type of object). Regardless of the type of object that is interacting, the mechanical processes that carry out the actions of the "send" and "receive" functions are limited by the object's capacity to initiate a "send" or respond to a "receive." Further, the mechanics of interaction also preserve the constraints of the objects.

Objective causalities: Combining modal with proximate causalities form the *sine qua non* of causes. Objective causalities are posited to be both necessary and sufficient to render a complete explanation of an event—substantiating the causal connection. Objective causalities are posited to be both necessary and sufficient to render a complete explanation of an event—substantiating the causal connection between objects and processes.

Objective frame: The objective frame follows the objective user behaviors, the product's functions, and the physical entities. The objective frame of the framework of objective causalities is the final result of the work efforts managed under processes—the physical objects, the product or service functions, and the objective behaviors that were determined by the development team to result from the use of the product or service or in anticipation of the product or service.

Objective measure: An objective measure is that which is quantifiable in terms of performance. Objective measures include any item or combination of items that are categorized as EMMI. Sometimes, the terms "objective measures of performance" or "performance measures" are used (United States Department of Defense 2010) instead of objective measures. Every object has at least one objective measure (and most often several), since there is something physical that is usually measurable (in the classic engineering and physics sense). Objective measures are used in testing to determine how well the object performs to a target value within the bounds of a specified variance that is deemed to satisfy an objective for stability (or quality).

Objective ontology: The objective ontology defines the semantics for objective frame. Specifically, the nature of being and reality are embodied in the conceptualization of an object.

Objective value: Objective value is often characterized by measures of amount (by numerical counting).

Performance: Performance is the action associated with a function. Performance is measured by the extent to which various standards are met.

Performance-based value: Performance-based value is the recognition of utility by the objective measure of some aspect of a function, or the subjective measure of some aspect of a process or procedure. There are different types of value spanning use, esteem, cost, exchange, scrap, and various performances as compared with standard references. Value can be thought of both for objects and processes.

Performance measures: Performance measures are observed and measured according to a reference scale or standard of measurement.

Perspective: The way of determining, identifying, considering, and using facts and their relative importance.*

Physical boundary: Determined by the limits of matter of one object.

Power: At the fundamental level of interaction, power is the limit imposed by one object on its EMMI. The result of power is change or status quo. Said another way, power is both an object's EMMI and the object's constraints that limits another object's access to EMMI.

For humans, power is EMMI and access to EMMI. Objects value EMMI as their means to make things happen. The rate of doing work is another form of power, as mediated by an object's output EMMI. Fundamentally, every object has a mechanism to transform EMMI into an output; therefore every object has some measure of power.

Prediction: A prediction is a probabilistic statement that something will happen in the future based on what is known today. A prediction generally assumes that future changes in related conditions will not have a significant influence. In this sense, a prediction is most influenced by the "initial conditions"—the current situation from which we predict a change.

Price: The amount of EMMI that is required to acquire an object or process.

Principle: Principles are general and fundamental statements that are comprehensive in their applicability and in general agreement with what people observe.

Problem: In general, a problem is perceived to be a source of difficulty, harm, unwelcomeness, or perplexity that someone or something needs to remedy. The determination of a need by an affected stakeholder to identify the set(s) of problems is an essential ingredient to identifying possible alternative solutions that are deemed adequate and acceptable. Disagreement over what is the problem will carry over into which solutions are adequate. Disagreement as to the solutions will indicate which problems are acceptable to solve. If there is agreement over what the problem is, yet the solutions are found to be unacceptable by various stakeholders, then the problem is either misstated or symptomatic of another more fundamental problem.

Procedure: Procedure is a step-by-step outline of what must be done.

Process: A process can be articulated as a systematic pattern, a coordinated set of procedures, tasks, activities, or acts that result from the conversion of inputs into outputs. Process is the amalgamation of activities and tools that combine ideas. A process requires all things that are both necessary and sufficient to accomplish or achieve an intended output. Processes are comparable to other processes, subjectively. From an integration perspective, processes guide the work.

Generalizing from definitions of software processes by Humphrey (1989) and Lonchamp (1993), a process is a partially organized set of activities, tools, and practices carried out by humans who are constrained by, for example, resources, budgets, schedules, scope, and policies.

Process frame: The frame whose domain is processes and their relations and behaviors.

Process model: The systems engineering process model describes the stages in which the project team focuses on various milestones and deliveries. The process model signifies what stage is next and what events constitute that stage.

Property: A property is embodied in an object that is physical or represents something that is physical. A property can be real (physical or material) or intellectual (conceptual, nonphysical, or intangible). A physical property of matter is mass. Intellectual property is a representation of real, physical property, such as software (which represents a process that is enacted through physical objects).

ProtaSystem: Prototypical systems exhibit some changes in properties, traits, and attributes due to interactions between objects. Before integration occurs, parts of a system may exist in various forms (termed a ProtaSystem). ProtaSystems are unstable.

Proximate causality: Proximate causality focuses through localization in time and space, that is, further limits the likelihood of an event.

Quality: Quality is the achievement of a level of acceptable variability of each measure of performance, and the variability in performance is representative of the user's perception of quality.

Recursive: Recursive implies that the answer is not known, but rather is discoverable in the process of objectifying the requirements and procedures into objects that satisfy the needs of key stakeholders. Recursiveness is the property of a process that is identifiable by an event (i.e., interaction between objects), deemed causal for that event, and that has predictive qualities for the next event or chain of events. Recursive processes are not iterative. Recursive processes are typified as refinements on future objects or future processes.

Risk: Risk is a structural property of the interactions between objects, whereas specifically, risk is inherent in the interactions involving the enterprise, business, and project. As stated by Kuwabara (2011) in discussing social exchanges, referencing Molm et al. (2000), "Risk is a structural property of exchange …".

Scalability: Scalability is all about doing what you do with either more people doing the same thing, or being able to do more with one person. Scalability in the first instance (more people doing the same thing with the same product) implies each person requires a product, that is, scalability by single-user products. Scalability in the second instance (being able to do more with one person) is through efficiency by using a service. Scalability in this second instance implies perhaps a single product that (through services) provides a similar functionality as with multiple products. So, by either increasing the number or speeding up a service, scalability is achieved.

Scenario: A scenario is a descriptive narrative of possible futures. A scenario represents a history of the future. The set of boundary conditions that is used in conjunction with making a projection is often called a scenario, and each scenario is based on assumptions about how the future will develop. A projection is a probabilistic statement that it is possible that something will happen in the future if certain conditions develop. In contrast to a prediction, a projection specifically allows for significant changes in the set of "boundary conditions" that might influence the prediction, creating "if this, then that" types of statements.

Scenarios describe events or activities (enactments of functions or processes) that synopsize a sequence of intentions or events to assist in accounting for a plausible future (Peterson et al. 2003). Scenarios can be used to explore the uncertainties of integration based on patterns of behavior (order and schedules of objects). The Bayesian interpretation of scenario development leads the defining of object order, start time, and duration as probabilistic measures of the stakeholder beliefs that the pattern will occur, given all that is relevant, known, and included by the stakeholders. In this interpretation, the probability of an event occurring in the sequence specified, at the starting time indicated, and having duration stipulated is dependent on both the event and the stakeholder's knowledge of the event (Poh 1993).

Scope: The extent of one's purpose or aim within a boundary. Scope is determined by work that will satisfy the stakeholders. Scope of a product or service is defined spatially by its physical, functional, and behavioral boundaries.

Social classification of integration: The classification of integration focused on the procedures (or social mechanisms (Moody and White 2003; Reed 2008)) of carrying out the cognitive issues of process and the physical realization of the objects and procedures as a means of documenting the ideas. Social classification of integration (referred to as social integration) captures the dynamics of how the products are used by people and how the context and content of communication is intonated through vocalization and gestures for others to hear and see.

Specifications: Based on a requirement that stipulates what is needed, specifications are defined as the detailed description of how to make something according to a set of procedures, by employing certain resources, by applying various standards and practices, and staying within a set of guidelines, policies, and rules.

Stability: Stability is defined as the ability to apply restoring forces to mitigate events that trigger changes in the status quo. Stability is achievable through interactions in certain circumstances, that is, when properties and traits are only sustainable.

Stakeholder: More general than key stakeholders, stakeholders are all those people and their agency agreements or obligations to represent others and affect a performance related in any way to the lifecycle of a product or service object or process.

Strategy: Strategy is a process that provides for achieving an intended result.

Structure: The arrangement of parts; the relations between parts; the manner of grouping, aggregating, or combining parts; and the classification and categorizing of parts.

Subjective frame: The subjective frame represents the cognitive part (however, that is conceived or formulated), the procedures to carry out the cognition, and the thinking about the models or representations of the cognitions and procedures. The set of factors that make up this social thinking process (subjective frame) are consistent with scientific sociological investigations that divide the factors into planning, procedure (mechanisms), and models of the plans and procedures (mechanisms) for clarity and to avoid confusion with dual use of the term "mechanism"; "mechanism" is used in the engineering and science sense while "procedures" are used instead of social mechanisms. Subjective implies things influenced by personal feelings, biases, or intuitive thoughts.

Subjective measure: Subjective measures are measures that are based on personal beliefs or reflect biases (as determined in an objective manner). The belief that an object will have various functions with performances (with some level of quality) is a subjective measure. User's anticipations of an object are determinable and identifiable as subjective measures. These subjective measures include behaviors. Subjective measures are used to determine how a constraint or condition impacts development work or integration.

Subjective ontology: Ontological structures based on method or processes.

Subjective value: Subjective value is often characterized by esteem, opportunity, or some form of intangibles.

Synthesis: Synthesis joins and merges the results of interactions between system elements to sustain the emergent properties that distinguish ProtaSystems de jure or de facto. Synthesis is that intermediate step which encompasses emergence. However, emergence by itself does not result in a system, but rather a ProtaSystem. Synthesis is founded on the notion of action at a distance—the impact of forces acting on things possibly displaced in time and place from the original action.

System: A system is a bounded, stable group of objects exhibiting intrinsic emergent properties that through the interactions of energy, matter, material wealth, and information provide functions different from their archetypes. Said more abstractly and succinctly (but with loss of precision), a system is a bounded, stable group of objects exhibiting intrinsic emergent behaviors based on interactions of energy, matter, material wealth, and information. And finally, paired down to its barest abstraction (with loss of precision and accuracy), a system is a group of stable objects showing intrinsic emergence based on interactions.

A system is comprised of objects that are interconnect and exchange EMMI. A system is comprised of pair-wise objects that have continuity of functionality. No system is possible without interaction. The system is the result of integration, or conversely, integration is the result of achieving system behaviors.

Following the definition of a system according to Palmer (2009), a system can be conceptualized in terms of the behavior of its objects (descriptive of the essence of their system); the context of the minimum energy structures (reflective of the design and architecture); the perspective of the definer (providing a referenced view); and the methods that epitomize its functioning (the socioeconomic realities of projects). "In the behavioral approach to system theory a system is regarded as a subset of a function space, the behavior, containing the input/state/output-trajectories ..." (Trumpf 2002).

Systemic behaviors: Systemic behaviors may include (1) equifinality (various sorts of degeneracy that are inherent in the existing structure of objects); insufficient density of objects of the kind and location needed to sustain interactions; inadequate EMMI needed for sustainment, including squelching of the mechanism due to saturation, below threshold inputs, and insufficient suffusion; and incompatibility of EMMI and mechanisms (Troncale 2011); (2) isomorphicity—the similarity of mixes and kinds of objects observed in systems (von Bertalanffy 1968); and (3) inadequate or inappropriate emergence(s) and unsustainable losses of EMMI from any cause. Integration is the process of setting up or by chance satisfying the conditions that lead to an integrated set of objects (i.e., system).

System of systems: A system of systems is a set of systems that are both integrated and interoperable to achieve a set of metasystem functions in which all the component systems participate (to varying degrees).

Systemness: Systemness is a sufficiency of sustained interactions that arise from an adequate density of the appropriate types of objects with the appropriate types of mechanisms, fed by the appropriate types of EMMI. The factors that are significant to systemness are object, boundary, function, property, trait, attribute, output, self-reliance, control, and performance. The duration of lifetime and stability of systemness are determined by the boundary conditions and variances about the performances of the functions. Within these categories of factors we find emergence (trait) and trust (self-reliance).

Systems engineering: The charter of systems engineering is to create and express ideas and integrate components into systems that are referred to as products or services. The essence of systems engineering is to unbound the seemingly bounded, broaden the concepts to beyond recognition, open the solution domain to include the ridiculous, and consider the issues and problems in an abstract space rather than as they are posed or presumed to be real. No other discipline or field carries with it that worldview.

Systems integration: Systems integration is the unification of the objects and their interactions of energy, matter, material wealth, and information to provide system-level functionalities and performances.

Target value: The designated performance requirement.

Technology: Technology is the scientific, mechanical, electronic, or chemical means of improving people's performances or by providing or enhancing their indigenous functions. These improvements provide for (1) making better decisions, (2) doing more work faster, and (3) doing work that could not be accomplished before by any one individual.

Testing: Testing is a process to determine the difference(s) between an object's properties, traits, and attributes under certain conditions in a given set of circumstances with that of a representation (or test model) of what is desired. The representation includes the test setup; the test procedures; the test plan; the test personnel; the test objectives; the data analysis (and tools); the theory in which the measurements are planned, executed, and interpreted; and the biases (of all parties). It is tacitly assumed that all factors not included in the representation (or test model) are factually extraneous (and therefore not significant either to a specific test or a concatenation or totality of tests). Testing highlights the test object's performance at a particular instance under certain conditions. Testing needs to be done as an audit to determine what functions are demonstrated to some level of proficiency. Testing indicates whether functions are operative and to what degree (i.e., performance). Testing is a means of comparison: comparing an object to the test model. Tests do not prove anything; they only show a correspondence to an expected result (Aerts 1983).

Theory: A theory, broadly defined and widely recognized, is expressive through its structure and narrative, its cognitive substance. The role of theory is to organize, explain, and predict actions and events. While theory inspires various frameworks from which to view and interpret empirical data and qualitative aspects that imbue various values (e.g., cultural, societal, and individual), it is the practices of systems engineering, systems engineering integration, and systems integration that indurate the applicability of theory's use.

Time: Time is a linear sequence of real numbers based on a quantity measured by the angle through which the earth turns on its axis. Time is a universal standard by which observations of various measures are compared (Feigenbaum 2008).

Trait: A trait is the nexus of the property along with its conditions that distinguishes it from other traits.

Type 0 interactions: Type 0 interactions are the results of stored energy used to drive an internal mechanism. Type 0 interactions are typical of "sources" of energy. Type 0 interactions are one type of energy source to enable or sustain the interactions necessary for a system. The outputs of EMMI can be received by other objects and interact with those objects.

Type 1 interactions: Type 1 interactions result from the complete absorption of EMMI without external provocation, resulting only from the nature and enactment of its internal mechanism. A type 1 interaction is potentially receivable by objects, but is either not received, is received and not recognizable as an accurate representation of the sending object, or is received and the receiving object does not respond to the sending object. Type 1 interactions are inhibited or masked by physical, functional, or behavioral reasons (internal or external to the receiving object). A type 1 interaction is initiated from within. Type 1 interactions reflect the internal needs or intentions of an entity, for example, the self-initiated requirements for survival. Type 1 interactions are in response to internal processes, the mechanically induced self-regulation for fulfilling basic needs.

Type 2 interactions: Type 2 interactions are sent and received. A type 2 interaction eliminates (or discharges or "sends") EMMI due to some external receipt of EMMI. Type 2 interactions are the responses to external stimuli, the simultaneous or reflexive reactions based on are capabilities within the entities structure.

Types of requirements: Requirements are either stated or unstated. Unstated requirements are the result of not knowing (i.e., unanticipated or unforeseen); or knowing but not incorporating. Misstated requirements result from an inadequate appreciation for relations or context, and adequate appreciation, but with a different interpretation, or an adequate appreciation, but with a different priority. Pretermitting requirements are acknowledged, but unincorporated.

Use: Use is the putting of the product into service for a particular purpose. The use of an object is the result of the building process and the functions that are enabled by design, by application, and by accident.

Validation: Validation is an assessment of the operational system that exposes and quantifies the systems' limitations. The intent of validation is to determine if the user's needs are satisfied for different uses (often referred to as scenarios). When the functions are provided, the physical entities are adequate, and the user's behaviors are as needed, the product or service is deemed fit for the uses intended by the set of requirements. The concept of validation suggests that requirements can be mapped into physical, functional, and behavioral needs of key stakeholders. Validation is also the process of demonstrating the effectiveness of the new product or service. Validation is direct evidence that the new product or service meets the requirements through its design, architecture, and implementation. From the perspective of integration, validation is the confirmation that integration had satisfactory results.

Value: "… an economic good has value; and the height of this value is measured according to the—marginal utility which one can obtain with the unit of labor" (von Böhm-Bawerk 2005).

Variables: Properties, traits and attributes that change due to interactions, that is, with EMMI.

Verification: The process of confirming the truth or accuracy by describing the characteristics of interactions, the enactments of mechanisms or procedures, or the consequences of EMMI.

Vignette: Vignettes are more detailed sequences of events that highlight particulars about a scenario (a possible set of circumstances, conditions, and constraints, e.g., the environment of the future).

Want: A want is something that will solve the problem, but is not necessarily possible, affordable, deliverable, or acceptable. A want is a desire, as yet unfulfilled.

White box: White box method of testing focuses on the identification and evaluation of the internal logic and procedures that are based on knowledge of the workings and configuration of the internals of the object.

Whole: The whole is the totality of the group, the aggregation, the combination, or the integration. The whole is always greater than the sum of its parts. A whole can be represented as an object, either physically or intellectually, or as a process, either procedurally or as activities (or acts).

References

Aerts, D. 1983. *The Description of One and Many Physical Systems Brussels.* Vrijo Universiteit.

Aerts, D., Broekaert, J., and Gabora, L. (Ed) 2003. A case for applying an abstracted quantum formalism to cognition. *Mind in Interaction.* Amsterdam: John Benjamins.

Feigenbaum, M. J. 2008. *The Theory of Relativity—Galileo's Child.* arXiv:0806.1234v1 [physics.class-ph]. Ithaca, NY: Cornell University Library, 31pp.

Ferris, T. L. 1997. *Foundation for Medical Diagnosis and Measurement.* PhD thesis, School of Physics and Electronic Systems Engineering, University of South Australia, 350pp.

Feyerabend, P. 1993. *Against Method.* Third edition. New York: Verso.

Haskins, C. 2007. A systems engineering framework for eco-industrial park formation. *Systems Engineering* **10**(1): 83–97.

Holland, J. H., Holyoak, K. J., Nisbett, R. E., and Thagard, P. R. 1986. *Induction: Processes of Inference, Learning, and Discovery.* Cambridge, MA: The MIT Press.

Humphrey, W. S. 1989. *Managing the Software Process.* Reading, MA: Addison-Wesley.

Kerlinger, F. N. and Lee, H. B. 2000. *Foundations of Behavioral Research.* Fourth edition. Belmont, CA: Cengage Learning.

Kirk, G. S., Raven, J. E., and Schofield, M. 2009. *The Prescocratic Philosophers.* Cambridge: Cambridge University Press.

Kocsis, J. G. 2008. *Determining Success for the Naval Systems Engineering Resource Center.* MS thesis, Department of Systems Engineering. Monterey, CA: United States Naval Postgraduate School, 101pp.

Kuwabara, K. 2011. Cohesion, cooperation, and the value of doing things together: How economic exchange creates relational bonds. *American Sociological Review* **76**(4): 560–580.

Lazarsfeld, P. F. 1993. *On Social Research and Its Language.* Chicago: University of Chicago Press.

Lewes, G. H. 1875. *Problems of Life and Mind.* Boston: James R. Osgood and Company.

Lonchamp, J. 1993. A structured conceptual and terminological framework for software process engineering. *Proceedings of the 2nd International Conference on the Software Process (ICSP 2).* Berlin, Germany: IEEE Computer Society Press.

Molm, L. D., Takahashi, N., and Peterson, G. 2000. Risk and trust in social exchange: An experiment test of a classical proposition. *American Journal of Sociology* **105**: 1396–1427.

Moody, J. and White, D. R. 2003. Structural cohesion and embeddedness: A hierarchical concept of social groups. *American Sociological Review* **68**(1): 103–127.

Morris, P. W. G. and Pinto, J. 2004. *The Wiley Guide to Managing Projects.* Hoboken: John Wiley & Sons, Inc, p. 213.

Palmer, K. D. 2009. *Emergent Design: Explorations in Systems Phenomenology in Relation to Ontology, Hermeneutics and the Meta-dialectics of Design.* PhD thesis, Division of Information Technology, Engineering, and the Environment. Mawson Lakes: University of South Australia, 679pp.

Peterson, G. D., Cumming, G., and Carpenter, S. R. 2003. Scenario planning: A tool for conservation in an uncertain world. *Conservation Biology* **17**(2): 358–366.

Poh, K. L. 1993. *Utility-Based Categorization Engineering-Economic Systems.* PhD thesis, Stanford: Stanford University, 238pp.

Reed, I. 2008. Justifying sociological knowledge: From realism to interpretation. *Sociological Theory* **26**(2): 101–129.

Sage, A. P. and Armstrong, J. E. 2000. *Introduction to Systems Engineering.* New York: John Wiley & Sons, Inc.

Swartz, N. (1997, November 8, 2010). Definitions, Dictionaries, and Meanings. Retrieved July 19, 2011, from http://www.sfu.ca/~swartz/definition.htm.

Troncale, L. 2011. Would a rigorous knowledge base in systems pathology add significantly to the SE portfolio? *Conference on Systems Engineering Research (CSER).* Redondo Beach: University of Southern California.

Trumpf, J. 2002. *On the Geometry and Parametrization of Almost Invariant Subspaces and Observer Theory.* PhD thesis, Mathematics Department. Wurzburg: Universität Wurzburg, 206pp.

United States Department of Defense. 2010. *Defense Acquisition Guidebook, Version 5 May.* Department of Defense. Washington DC: Defense Acquisition University, p. 310.

von Bertalanffy, L. 1968. *General System Theory: Foundations, Development, Applications.* New York: George Braziller.

von Böhm-Bawerk, E. 2005. *Basic Principles of Economic Value.* Grove City, PA: Libertarian Press, Inc.

Index

Note: n = Footnote

A

Abstract concepts, 64
Abstraction, 135n, 136, 288, 353. *See also*
        Granularity; Integration;
        Procedure; Process model
    of categories, 166
    EMMI, 288
    hierarchical subprocesses, 289, 290
    integration classification,
        160–161, 353
    "to command" process
        decomposition, 289
    "to control" process
        decomposition, 289, 290
    "to manage" process, 289
    of processes, 85
    and reasoning, 89
Acceptance, 152, 278
    marketplace, 217, 223
    purchasing as user's, 238, 239
Acquisition cycle, 312
Acquisition process, 24–25
Action, 48, 134, 174, 353. *See also*
        Process
Active objects, 166
Activity model, 287
Adaptive optics, 5n
Agglomerate, 208, 209
    power structure, 210, 211
Aggregate, 208, 209
    power structure, 210
Aggregations, 112, 353
Agile-driven strategies, 303
Alignment of strategies, 11–12, 190
Archetype, 353
Architecture, 18, 274, 276, 277, 353
    building, 275
    changing business, 210–211
    constraints on, 66

expression of, 169–170
for integration, 15, 178
and interactions, 205
in lifecycle stage, 237
for power, 91
and system, 16, 277
Artifact, 353
Attribute, 34n, 55–56, 353
Automation, 118
Awareness, 30–31
Axiom of integration, 353
    action, 174
    degrees of freedom, 176
    inaction, 174
    inactivity, 175
    interaction, 174–175
    mechanisms, 174

B

Behavior, 353
    boundaries, 31, 35, 36, 354
    functions and, 46
    in integrative framework, 88, 89
    rules of, 15
Best practices, 104n, 177–178
Big-bang integration, 306
Black box approach, 146–147, 354
Bleed-throughs, 38
Body of knowledge, 224, 354
Bottom-up integration, 229, 306
Boundary, 14, 30, 32, 354
    analysis, 122
    awareness, 30–31
    behavioral boundaries, 31
    bleed-throughs, 38
    condition, 42–43, 354
    connectivity, 30
    declaration, 37
    determination of objects, 122